Genetically Engineered Viruses

Development and applications

Genetically Engineered Viruses

Development and applications

Edited by
Christopher J. A. Ring and Edward D. Blair

Glaxo Wellcome Research and Development, Stevenage, UK

CRC Press is an imprint of the
Taylor & Francis Group, an informa business

First published 2001 by BIOS Scientific Publishers Ltd

Published 2020 by CRC Press
Taylor & Francis Group
6000 Broken Sound Parkway NW, Suite 300
Boca Raton, FL 33487-2742

ISBN 13: 978-1-85996-103-2 (pbk)

Visit the Taylor & Francis Web site at
http://www.taylorandfrancis.com

and the CRC Press Web site at
http://www.crcpress.com

A CIP catalogue record for this book is available from the British Library.

Typeset by Mackreth Media Services, Hemel Hempstead, UK

Cover images kindly supplied by Kenneth Lundstrom, Markus U. Ehrengruber and J. Grayson Richards.

Contents

Contributors

Barclay, W.S. School of Animal and Microbial Sciences, University of Reading, PO Box 228, Whiteknights, Reading RG6 6AJ, UK

Bataille, D., Molecular Medicine and Gene Therapy Unit, Stopford Building, Oxford Road, Manchester M13 9PT, UK

Blair, E.D., Clinical Virology and Surrogates Unit, Glaxo Wellcome R&D, Gunnels Wood Road, Stevenage, Herts, SG1 2NY, UK

Carroll, M.W., Oxford BioMedica (UK) Ltd., Oxford Science Park, Oxford OX4 4GA, UK

Castro, M., Molecular Medicine and Gene Therapy Unit, Stopford Building, Oxford Road, Manchester M13 9PT, UK

David, A., Molecular Medicine and Gene Therapy Unit, Stopford Building, Oxford Road, Manchester M13 9PT, UK

Evans, D.J., Institute of Virology, Church Street, Glasgow G12 8QQ, UK

Jones, I.M., Institute of Virology, Oxford, UK

Jones, M.D., Department of Infectious Diseases and Microbiology, Imperial College School of Medicine, Du Cane Road, London W12 0NN, UK

Lacomme, C., Scottish Crop Research Institute, Invergowrie, Dundee, DD2 5DA, UK

Leib, D.A., Departments of Ophthalmology and Molecular Microbiology, Washington University School of Medicine, St. Louis, MO, USA

Lowenstein, P., Molecular Medicine and Gene Therapy Unit, Stopford Building, Oxford Road, Manchester M13 9PT, UK

Lundstrom, K., F. Hoffmann La Roche Research Laboratories, Basel, Switzerland

Melcher, A., Molecular Medicine, Mayo Clinic, Rochester, MN, USA

Pogue, G.P., Large Scale Biology Corporation, 333 Vaca Valley Parkway, Vacaville, CA, USA

Ring, C.J.A., Clinical Virology and Surrogates Unit, Glaxo Wellcome R&D, Gunnels Wood Road, Stevenage, Herts, SG1 2NY, UK

Santa Cruz, S., Scottish Crop Research Institute, Invergowrie, Dundee, DD2 5DA, UK

Stone, D., Molecular Medicine and Gene Therapy Unit, Stopford Building, Oxford Road, Manchester M13 9PT, UK

Thomas, C., Molecular Medicine and Gene Therapy Unit, Stopford Building, Oxford Road, Manchester M13 9PT, UK

Vile, R., Molecular Medicine, Mayo Clinic, Rochester, MN, USA

Warne, S.R., Health and Safety Executive, Magdalen House, Bootle, Merseyside L20 3QZ, UK

Wilkinson, G.W.G., Department of Medicine, University of Wales College of Medicine, Heath Park, Cardiff CF14 4XX, UK

Wilson, T.M.A., Horticulture Research International, Wellesbourne, Warwick CV35 9EF, UK

Abbreviations

AAHL	Australian Animal Health Laboratory
AAV	adeno-associated virus
AcMNPV	Autographa californica multiple nuclear polyhedrosis virus
ACMV	African cassava mosaic virus
ALMV	alfalfa mosaic virus
APC	antigen-presenting cell
ARD	acute respiratory disease
BAC	bacterial artificial chromosomes
BaMV	bamboo mosaic virus
BDV	Borna disease virus
BMNPV	Bombyx mori nuclear polyhedrosis virus
BMV	brome mosaic virus
BVDV	bovine viral diarrhoea virus
BYV	beet yellows colosterovirus
CAEV	caprine arthritic encephalitis virus
CaMV	cauliflower mosaic virus
CAT	chloramphenicol acetyl transferase
CEF	chick embryo fibroblast
CMV	cytomegalovirus
CP	coat protein
CPMV	cowpea mosaic virus
CPV	canarypox virus
CTL	cytotoxic T cell
CyRSV	cymbidium ringspot tombusvirus
DC	dendritic cells
ddNTP	dideoxynucleoside triphosphate
DHFR	dihydrofolate reductase
DI	defective interfering
ds	double strand(ed)
EBV	Epstein-Barr virus
EGF	epidermal growth factor
EGFR	epidermal growth factor receptor
EIAV	equine infectious anaemia virus
EPV	entomopoxvirus
ER	endoplasmic reticulum
FGF	fibroblast growth factor
FIV	feline immunodeficiency virus
FMDV	foot-and-mouth-disease virus
FnPB	fibronectin binding peptide
FPV	fowlpox virus

GDV	gene delivery vector
GFP	green fluorescent protein
GPCR	G protein-coupled receptor
GPI	glycosylphosphatidyl-inositol
GPUT	galactose-1 phosphate uridyl transferase
GTAC	Gene Therapy Advisory Committee
HDV	hepatitis delta virus
HIV-1	human immunodeficiency virus type 1
HMCV	human cytomegalovirus
HSE	Health and Safety Executive
HSV	herpes simplex virus
HVS	herpesvirus saimiri
IRES	internal ribosome entry site
kbp	kilobasepairs
KSHV	Kaposi's sarcoma herpesvirus
LAT	latency-associated transcript
LTR	long terminal repeat
MD	mucosal disease
MEV	mink enteritis virus
MHV	mouse hepatitis virus
MMCV	mouse cytomegalovirus
MoMLV	Moloney murine leukaemia virus
MP	movement protein
MSV	maize streak virus
NA	neuraminidase
NCR	non-coding region
NDV	Newcastle disease virus
NGF	nerve growth factor
NNS	non-segmented negative strand
orf	opening reading frames
ORSV	odonroglossum ringspot virus
PAL	phenylalanine ammonia-lyase
PCR	polymerase chain reaction
PEG	polyethylene glycol
PIB	polyhedron inclusion body
PNS	peripheral nervous system
PrV	pseudorabies virus
PTGS	post-transcriptional gene silencing
PVX	potato virus X
RCR	replication competent retrovirus
RF	replicative form
RNP	ribonucleoprotein
RSV	respiratory sclerosing panencephalitis
rvv	recombinant vaccinia virus
scTCR	single chain T-cell receptor

SIN	Sindbis virus
SIV	simian immunodeficiency virus
SMV	Semliki Forest virus
ss	single strand(ed)
SSPE	subacute sclerosing virus
TAA	tumour associated antigen
TBE	tick-borne encephalitis
TEV	tobacco etch virus
TGEV	transmissible gastroenteritis virus
TGMV	tomato golden mosaic virus
TM	transmembrane
TMV	tobacco mosaic virus
VEE	Venezuelan equine encephalitis virus
VSV	vesicular stomatitis virus
VV	vaccinia virus
VZV	varicella-zoster virus
WDV	wheat dwarf virus
YAC	yeast artificial chromosomes

Preface

The genesis of this book arose when we worked within the Virology Unit at Glaxo Wellcome. Our small group worked on viral vectors for cancer gene therapies whilst the vast majority of our colleagues were attempting to develop anti-viral therapies. So while this majority were trying to stop viruses from causing disease, we were encouraging infections! Our argument through these curious times was that "*viruses can be good as well as bad*". We are therefore indebted to the contributing authors of this book who have helped exemplify such a claim while also considerably enhancing our knowledge. It is now possible to construct viruses with desired characteristics and use them not only to study basic biology and pathogenesis, but also to prevent and treat disease. Finally, from a regulatory and ethical standpoint, we suggest that scientists conducting studies with genetically modified viruses are obliged to consider the many issues raised in the final chapter of this volume. Failure to do so – through ignorance, insularity or whatever – would be foolish and dangerous. We are only too aware that viruses can be bad as well as good!

Chris Ring and Eddie Blair
Stevenage, November 2000

Acknowledgements

We would like to thank all of the authors for their hard work. In addition our thanks go to Rachel Offord, Will Samson, Victoria Oddie and Andrea Bosher at BIOS for their patience and commitment to this project.

Chapter 1

Introduction: viruses as vehicles and expressors of genetic material

Christopher J.A. Ring and Edward D. Blair

Viruses are most widely known as being agents of disease, however of all the viruses studied to date, only a very small proportion are pathogenic. From a biological standpoint a more appropriate definition of viruses is that they are sub-microscopic intracellular parasites whose genetic material, consisting of either RNA or DNA, is packaged in a protective coat and transmitted from one host cell to another. Indeed, viruses can be considered merely as the 'software' containing instructions for their own replication. The life cycle of a typical virus can be summarized by the attachment to and entry of the virions (the extra-cellular form of the virus) into the host cell, uncoating of the genetic material, the expression of viral proteins, assembly of progeny virions and their release from the cells (see *Fig. 1.1*).

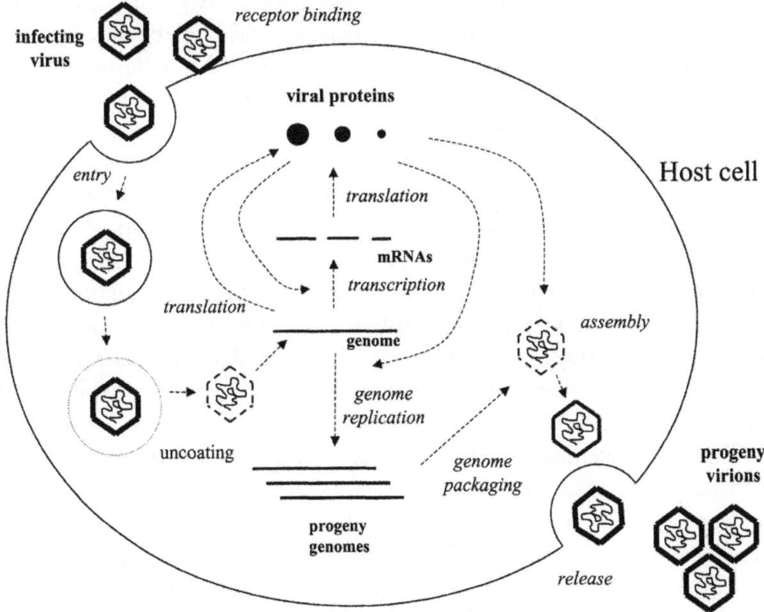

Figure 1.1. The life cycle of a typical virus.

Genetically Engineered Viruses: Development and Applications, edited by C.J.A. Ring and E.D. Blair

In order to survive viruses have had to develop highly efficient mechanisms to reproduce themselves. To achieve this viruses have to effectively introduce their genomes into cells and compete with the cell's biosynthetic machinery in order to express their proteins, produce progeny virions and exit the host cell (for a detailed description see Flint *et al.,* 2000). Virus replication often results in devastating effects on the host cell which can lead to disease in the host organism. The high efficiency in which viruses transmit and express their genetic material has led to viruses being adopted as gene transfer vectors. Viruses use a variety of cell surface molecules to attach to their host cells and either fuse with the cell membrane or enter *via* the endocytic pathway. Viruses have evolved a number of mechanisms to release virions or genomes from endosomes and in some cases import them into the cell nucleus. Viruses often possess powerful promoter and enhancers to direct high levels of transcription of their genes. Moreover, many viruses encode novel polymerases that selectively transcribe their genomes. Indeed viral proteins that serve to modify the promoter specifically of cellular polymerases, so that they preferentially recognize viral promoters, have also been reported. A number of mechanisms exist for inhibiting cellular gene expression. These include inhibition of transcription, polyadenylation and nuclear export of cellular mRNAs and their destabilisation (Flint *et al.,* 2000). Furthermore, many viruses exhibit mechanisms of inhibiting cap-dependent translation. The genomes of some positive-stranded RNA viruses contain structures called internal ribosomal entry sites (IRES) enabling them to be translated in a cap-independent manner. Thus, inhibition of cap-dependent translation by virally-encoded gene products results in selective translation of viral RNA over cellular RNA species. It should be noted however, that inhibition of cellular functions must be regulated so as not to deprive the virus of cellular proteins or pathways required for viral replication. Many enveloped viruses are released from cells by budding out of the plasma membrane, while others exit the cell by budding into internal compartments of the secretory pathway. Non-enveloped viruses and a few examples of enveloped ones generally rely upon cell lysis for release. This is often brought about by the inhibition of essential cellular functions but can also occur by the expression of specific lysis proteins. In order to facilitate replication by preventing the premature death or removal of infected cells, viruses have developed a number of mechanisms to block the apoptotic response (O'Brien, 1998) and counter the immune response of the host (Janeway *et al.,* 1999).

As viruses are relatively simple biological entities that are dependent upon the biochemical pathways of their host cells, the study of viral gene expression has resulted in fundamental breakthroughs in our understanding of the mechanisms of cellular gene expression. Important elements of the eukaryotic transcriptional machinery such as the basal and accessory transcription factors and enhancers have been elucidated using viruses. Indeed, much of our knowledge on the steps of mRNA processing, namely splicing, polyadenylation and 5′ capping and methylation, has been gained by studying viruses (reviewed by Levine, [1996]). In addition to broadening our knowledge of transcription and post-transcriptional modification, the study of viruses has also led to breakthroughs in our understanding of translation, interferon action, major histocompatibility complex (MHC) restriction and oncogenesis (reviewed by Levine [1996]). The discovery of reverse transcriptase not only helped to prove how retroviruses replicate, but also provided an essential tool to synthesize complementary DNA (cDNA). This, in turn, has allowed the detailed characterization of cellular and viral mRNA species. Besides reverse transcriptase, viruses have also proved to be useful sources of other reagents used in biotechnology. These include enzymes used for the manipulation of nucleic acids (kinases, ligases, exonucleases and RNA polymerases) and genetic elements used in the construction of expression vectors (promoters, enhancers, transcriptional terminators, polyadenylation signals, introns, replication origins and IRES elements). A review of ways in which viruses have broadened our understanding of cellular and molecular biology can be found in Levine (1996).

Rational approaches to the prevention and treatment of virus disease requires a thorough understanding of the replicative and pathogenic process. Recombinant viruses can be of use in this context as viruses bearing specific mutations can now be constructed so that the roles played by individual viral gene products in viral replication and pathology can be evaluated in the context of viral infection rather than when isolated genes are transfected into cells (see Chapters 6 and 7). The manner in which recombinant viruses can be exploited in the prevention and treatment of disease is dicussed in Chapters 4, 5 and 8. In this context, one may actually be able to exploit the pathogenic or cytolytic properties of viruses. Examples of this are the potential use of bacteriophages for the treatment of bacterial infections (Barrow and Soothill, 1997) (see Chapter 2), baculoviruses to control insect pests (see Chapter 3), herpes simplex viruses as oncolytic agents in the treatment of brain tumors (see Chapter 8) and recombinant vesicular stomatitis viruses to lyse HIV-infected cells (Schnell *et al.,* 1997). Viruses have been used extensively as prophylactic vaccines. Such viruses have usually been subjected to multiple rounds of passage in culture in order to attenuate them, however virulent revertants are known to arise. With our increasing understanding of the genetic determinants of virulence and the development of recombinant DNA and reverse genetics techniques, rational design of more stable, and hence safter, vaccine viruses may be possible.

Molecular techniques now make it possible to introduce near any type of mutation into many viruses. The first step is to clone the viral genomic region of interest as a recombinant DNA molecule. The cloned sequence is then mutated *in vitro*, re-introduced back into the viral genome and then the modified viruses is recovered. For those viruses that can be prepared in quality such as adenviruses, genomic manipulations can be conducted on purified virion DNA. The genomes of the larger DNA viruses such as herpesviruses are generally not cloned in a single piece but as overlapping subgenomic fragments. As RNA is not so amenable to manipulation as DNA, RNA virus genomes are cloned following transcription by reverse transcriptase into cDNA. In the case of retroviruses, however, the DNA replicative intermediates, either in the form of the unintegrated viral DNA or the integrated provirus, are cloned. DNA viruses, with the exception of poxviruses, can be recovered following transfection of a single molecular clone representing the complete genome, or a library of overlapping subclones that represent the genome, into permissive cells (see Chapters 5, 6 and 8). As poxvirus DNA is not infectious, recombinant viruses can only be recovered by transfection of a cloned fragment of vaccinia DNA into permissive cells and homologous recombination with a co-infecting wild-type vaccinia virus (see Chapter 5).

Even though poliovirus has been recovered following transfection of cloned cDNA (Racaneillo and Baltimore, 1981) recovery of positive-stranded RNA viruses usually requires transfection of RNA transcripts that have been generated *in vitro* from cloned genomes. Since the vRNA of the negative-stranded RNA viruses is by definition in antisense orientation and thus requires transcription into mRNA, the permissive cell has to be provided with polymerase components in order to recover infectious virus. This is done either by transfecting ribonucleoprotein or molecular clones encoding the polymerase proteins (see Chapter 7). Detailed descriptions of the methods used to construct recombinant viruses can be found in the subsequent chapters of this book.

As our knowledge of the mechanisms by which viruses replicate and cause disease increases, so does the potential to construct viruses with desirable characteristics. With this technology and the knowledge gained from studies of pathogenesis we have the potential to construct viruses that are safer and more effective prophylactic and therapeutic agents. We may be able to design better prophylactic vaccines by constructing viruses that are more stably attenuated and that present antigens in a more efficient and appropriate manner. Furthermore, we may be able to design safer and more effective therapeutic agents by constructing viruses whose ability to infect and express genes or replicate is restricted to the desired cell type and that expresses the therapeutic or pathogenic effects of such a virus would be targeted to the cells of interest, thus minimizing toxic effects. We may be able to design better biological control agents by modifying viruses so to maximize lethality to the target species and minimize the potential for resistance or the possibility of the virus crossing to other species. The following chapters describe how viruses can be genetically manipulated to generate reagents for studying molecular biology and pathogenesis, and for constructing prophylactic, therapeutic and biological control agents. The final chapter deals with the safety and regulatory issues pertinent to the genetic manipulation of viruses.

References

Barrow PA and Soothill JS. (1997) Bacteriophage therapy and prophylaxis: rediscovery and renewed assessment of the potential. *Trends Microbiol.* **5:** 268–271.

Coen DM and Ramig RF. (1996) *Viral genetics.* In: *Fields Virology* (eds Fields *et al.*) Lippincott-Raven Publishers, Philadelphia, pp. 113–151.

Flint SJ, Enquist LW, Krug RM, Racaniello VR and Skalka AM. (2000) *Principles of Virology: Molecular biology, Pathogenesis and Control.* ASM Press, Washington, D.C.

Janeway CA, Travers P, Walport M and Capra JD. (1999) Immunobiology: The immune system in health and disease. 4th edn. Current Biology, Garland.

Levine AJ. (1996) *The origins of virology.* In: *Fields Virology* (eds Fields *et al.*) Lippincott-Raven Publishers, Philadelphia. pp. 1–14.

Racaniello VR and Baltimore D. (1981) Cloned poliovirus complementary DNA is infectious in mammalian cells. *Science* **214**: 916–919.

O'Brien VO. (1998) Viruses and apoptosis. *J. Gen. Virol.* **79**: 1833–1845.

Schnell MJ, Johnson JE, Buonocore L and Rose JK. (1997) Construction of a novel viral that targets HIV-1-infected cells and controls HIV-1 infection. *Cell* **90**: 849–857.

Chapter 2

Prokaryotic viruses

Michael D. Jones

2.1 Introduction

Prokaryotic viruses or bacteriophages, are viruses that infect bacteria, and were independently discovered by Edward Twort (1915) and Felix d'Herelle (1917), who both reported the isolation of filterable entities that were capable of killing bacteria. These filtrates yielded small clear areas on lawns of bacterial cells, and were first called 'bacteriophages' by d'Herelle (d'Herelle *et al.*, 1922). The molecular biological study of phages has been pursued since the 1940s, and was helped immensely by the training courses initiated by Max Delbruck and Salvador Luria in 1945 at Cold Spring Harbor. The Nobel laureate James Watson was trained as a phage researcher, before journeying to Cambridge to discover the structure of DNA with Francis Crick (Watson, 1970).

This chapter concentrates upon prokaryotic viruses, and almost exclusively deals with bacteriophage λ and the filamentous phages that infect *Escherichia coli*. That is not to say other phages and their hosts are not important, but that most of the work on the development of bacteriophage vectors has been carried out employing these phages. It is to be hoped that the pioneering work on *E. coli* phages will inspire future workers to utilize other bacteriophages/bacterial host combinations for the development of a new generation of phage vectors. At this point I must make an apology. The incredible burst of research papers published continues to grow unabated, making it nearly impossible for any reviewer to read and cover everything. For example, searching the Medline database with the phrase 'bacteriophage lambda' for the years 1970–1999 yielded 6216 references, the phrase 'phage display' yielded 738. To do the field full justice would require several volumes in its own right, so I have been forced to be selective in the topics and papers chosen, and represent areas which I feel are exciting. My apology is to those researchers whose excellent work it has not been possible to include here.

This chapter covers two major areas, firstly the use of bacteriophages to clone recombinant DNA for analysis, and secondly the field of 'phage display'. The first topic, which has developed from the late 1960s up to the present time, can be considered a golden era of molecular and cellular biology. Genes could be cloned and analysed. The culmination of this era will be the determination of the complete human genome DNA sequence, approximately 3×10^9 base pairs, expected in 2001 (if not sooner). The second topic, that of phage display, is itself now over 10 years old. This technology is based upon the ability of bacteriophages to display foreign protein sequences on the surface of virion particles, allowing the foreign protein to be detected and to interact

Genetically Engineered Viruses: Development and Applications, edited by C.J.A. Ring and E.D. Blair

Table 2.1. Bacteriophage vectors

Bacteriophage	Vector size[a]	Insert size range[b]
Filamentous phage vectors	~7000	0–2000
Filamid vectors	~3000	0–5000
λ insertion vectors	~43 000	0–8000
λ replacement vectors[c]	~48 000	9000–15 000
Cosmid vectors	~5000	~30 000–45 000
BAC	~7000	100 000–200 000
YAC	~10 000	200 000–1 000 000

[a] Approximate size of the parental phage genome in base pairs (without a cloned insert).
[b] Approximate size range of DNA inserts, in base pairs, that can be cloned into the phage vector.
[c] In these vectors, the central stuffer fragment of the vector is replaced with foreign DNA.

with its biological partners. This technology has had a profound effect in the field of immunology, where it challenges traditional cell fusion methods for the production of monoclonal antibodies. *Table 2.1* shows the size of inserts that can be cloned into the various phage vectors. For comparison the typical range of inserts sizes for BACs and YACs is included.

2.2 Bacteriophage λ vectors

2.2.1 Bacteriophage λ

This bacteriophage has been a workhorse of molecular biology, and its main use has been, and continues to be, the construction of phage libraries containing DNA sequences from the target organism in discrete sized fragments. The libraries are either genomic, containing DNA fragments encompassing the complete genome of the organism, or cDNA, whereby all of the expressed genes are represented. These libraries are then searched for the 'gene of interest', which can then be analysed subsequently in more detail.

Bacteriophage λ was first discovered by Esther Lederberg when she found a K12 strain which had lost its prophage (Lederberg, 1951). λ can exist as a prophage whereby the viral DNA is integrated into the bacterial genome, the lysogenic state. However, under certain conditions, usually stressful for the bacteria, the phage is released as an infectious virus particle when it enters its lytic state. The control mechanisms which govern the life cycle of λ are a whole book in themselves and are not covered here (for those interested in the biology of λ see Hendrix *et al.*, 1983 and Gottesman, 1999).

The discovery of specialized transduction led the way to the development of λ as a cloning vector. Specialized transduction occurs when inaccurate excision of an integrated prophage leads to loss of phage DNA that is replaced by bacterial genes. The resultant phages are usually defective and require helper phage to provide replication functions in *trans*. Generalized transduction is the process whereby bacterial DNA is assembled into phage particles, allowing transfer of genes between bacteria. Specialized transduction meant that λ was capable of accommodating foreign DNA, and although formed naturally *in vivo*, these phages can be considered the first recombinant DNA clones, particular those phages which did not require helper phages for growth.

The double stranded (ds) DNA genome of bacteriophage λ*c*I*ind*1*ts*857*S*7 is 48 514 bp in length (Sanger *et al.*, 1982), encoding ~60 genes. The ends of the linear genome have 12 base long single strand (ss) DNA overhangs, part of the cos sequences, responsible for packaging viral DNA into preformed empty head assemblies. The sequences are complementary and *in vivo* and *in vitro* base pair together to circularize the genome. λ DNA replicates by a rolling circle mechanism generating long concatemers, which are the substrate for the packaging machinery. In essence, the *cos* site on concatemeric DNA is recognized by the phage packaging enzymes and is introduced into empty phage heads. Once this is full, the DNA is cleaved at the following *cos* site, thus producing a full phage head with a complete genome. There is a limit to the size of DNA that can be packaged into a phage head and thus produce viable phage particles, which is roughly 40–52 kbp. Packaging efficiently measures the distance between two successive *cos* sites as the DNA is taken up into the head assembly. If they are the correct distance apart, the DNA is cleaved at the following *cos* site and the head can subsequently form a viable phage particle. If the head is filled but no following *cos* site is found for cleavage then no virus particle is produced.

The λ genome can effectively be split into three regions, conventionally from the left is the head and tail region, encoding the genes involved phage assembly, then in the middle are genes involved in recombination and regulation, and finally at the right-hand side are the genes responsible for DNA replication and lysis. The middle section, ~18 kbp (~36% of the λ genome), can be replaced with foreign DNA and still yield lytic phage particles (*Figure 2.1*).

The ability to assemble viable bacteriophage particles *in vitro* utilizing protein extracts from λ-infected bacterial cells utilizing concatemeric λ DNA paved the way to the development of λ as a cloning vector (Becker and Gold, 1975; Hohn and Hohn, 1974).

2.2.2 Insertion vectors

Insertion vectors as their name implies are λ vectors that allow the insertion of foreign DNA into them, and still form viable phage particles. The parental vector has been manipulated such that it is viable despite deletion of phage DNA. Two classic insertion vectors are λgt10 and λgt11 developed by Huynh *et al.* (1985), which can accommodate inserts from 0 to 7 000 bp (*Figure 2.1*). For most higher eukaryotes, the majority of the expressed genes are less than 10 000 bases, and given that reverse transcription of mRNA generally produces cDNA which is 500–5000 bp in size, means that λgt10 and gt11 are ideally suited for the construction of cDNA libraries. The extremely efficient *in vitro* packaging of λ DNA into viable particles means that it is relatively easy to produce libraries of $>10^6$ individual recombinant clones. The number of clones required to generate a full representative cDNA library is debatable. This is due to the fact that usually the most interesting genes are expressed in low abundance, perhaps less than 10 mRNA molecules per cell. Thus, to ensure that your library contains all genes, cDNA libraries in excess of 10^7 are sometimes required. With efficient *in vitro* packaging this is not necessarily problematic.

(a) λ Vectors

wild type λ
0 10 20 30 40 48.5 kb
λ left arm (genes *nul* to *J*)
central region (genes *lom* to *cII*)
λ right arm (genes *O* to *bor*)

0 10 20 30 *imm* 434 40 43.3 kb
Eco RI
λ gt10

0 10 20 *lac* 5 30 40 43.7 kb
Eco RI
λ gt11

Bluescript
0 10 20 30 40.8 kb
polylinker
λ ZAP

central stuffer fragment
0 10 20 30 40 41.9 kb
Sal I
Bam HI
Eco RI
Sal I
Bam HI
Eco RI
λ EMBL 3

(b) *cos* packaging

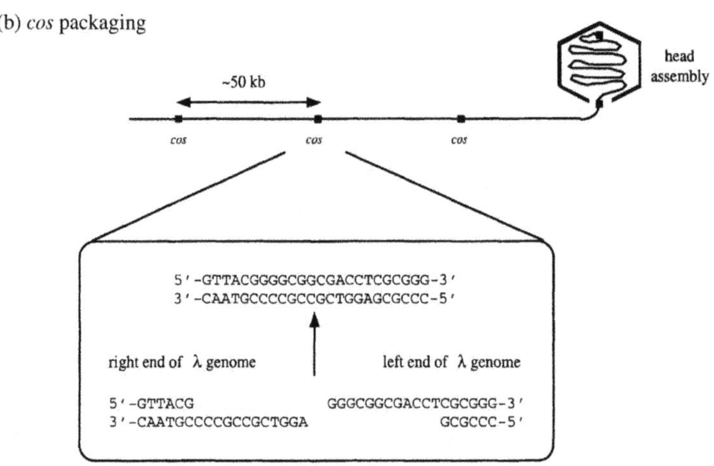

Figure 2.1. λ vectors. (a) Map of wild-type λ genome, together with that of the cloning vectors λgt10, λgt11, λZAP and λEMBL3. The genome of λDASH is similar to that of λEMBL3, except it has a more extensive polylinker flanking the stuffer fragment. (b) *cos* packaging, showing the sequence of the 12 base ss DNA overhang at the ends of the linear λ genome.

λgt10 has a unique *Eco*RI cloning site in the *imm*434 gene, and means that suitably *Eco*RI adapted cDNA can be cloned at this site. The *imm*434 gene, which encodes for the cI repressor protein, allows the use of λ biology to your advantage. Parental λgt10 phage (no cloned inserts) form cloudy or turbid plaques on lawns of *E. coli* C600 cells, because the phage can enter the lysogenic state. Cloning of foreign DNA into the *Eco*RI site effectively eliminates functional cI protein and the plaques formed on C600 cells are clear. This is because the phage replicates exclusively as a lytic virus. This difference in plaque morphology allows parental and recombinant phages to be easily distinguished. One problem with generating λ cDNA libraries is to ensure that every last bit of hard-earned cDNA is cloned, so an excess of vector DNA is generally used in the ligations. This results in the production of viable phage with no insert (parental vector); in some λgt10 libraries only 10% of phage particles contain cloned inserts, the remaining 90% is wild-type vector. This would mean constructing a library of 100×10^6 clones to ensure that you had 10^7 phage with inserts, and the subsequent screening of just 10^6 recombinant clones would require plating out a total of 10^7. Filter screening of 10^7 plaques is quite an undertaking, requiring 100 bacterial dishes which are 150 mm in diameter each containing 10^5 plaques (Allday and Jones, 1987; Benton and Davis, 1977).

Yet again λ biology comes to the rescue. *E. coli* strains carrying *hfl*A mutations (hfl stands for high frequency of lysogeny) when infected with λ expressing a functional *cI* gene leads to repression of lytic plaque formation. Recombinant λgt10 phage, with no functional cI protein, produce lytic plaques as normal. Plating your *in vitro* packaged λgt10 cDNA library on *hfl*A strains of *E. coli* efficiently eliminates the parental phage. Thus, for efficient construction of cDNA libraries which will be screened by filter lift hybridization with labeled DNA probes, then λgt10 is an ideal vector to use.

However, screening of cDNA libraries can take into account the fact that the cDNA represents the expressed genes of an organism or tissue, and libraries can be screened for expression of the encoded protein using antibody probes. The vector of choice would be a λ expression vector such as λgt11. This insertion vector contains the *lac*Z gene with a unique *Eco*RI site 53 bp upstream of the β-galactosidase termination codon. As with λgt10, this vector can accommodate inserts up to 7 kbp in size. Insertion of DNA into the *lac*Z gene inactivates the gene, and it is no longer able to hydrolyse the colorless indicator dye, X-gal (5-Bromo-4-chloro-3-indolyl-β-D-galactopyranoside) to release the intense blue indole moiety (see section 2.3.3). Thus, parental λgt11 forms blue plaques and recombinant λgt11 clones form clear/white plaques on lawns of cells containing X-gal.

Unfortunately, compared to λgt10, there is no biological way to eliminate the background of parental λgt11 phage. The construction of such expression cDNA phage libraries has to include *in vitro* manipulations to reduce the proportion of wild-type λgt11 present in the final library. These include phosphatase treatment of the *Eco*RI digested vector to prevent religation of vector arms. Also, when constructing an expression library in λgt11 you need ~6 times as many clones for expression screening compared to DNA hybridization screening. This is because reverse transcription of cDNA results in random synthesis of the cDNA and with cloning into a single restriction enzyme site there are 6 possible open reading frames (orf) to fuse with the β-galactosidase sequence. Only one of these will be the correct in-frame fusion of the actual encoded protein. Three possible fusions will be in the wrong orientation and from the correct orientation two fusions will be in the wrong reading frame. Thus, you need six times as many λgt11 clones as λgt10 clones for expression screening.

Recently improvements in the transfection of DNA into bacterial cells by electroporation have meant that many workers utilize plasmid vectors for constructing cDNA libraries. One can incorporate adapters onto the cDNA that allow directional cloning, and the cloning sites in the plasmid vectors are more easily manipulated, particularly for the construction of expression libraries.

One of the biggest problems with constructing cDNA libraries in λ vectors is the fact that the insert is small relative to the cloning vector arms. A typical clone may have a 3 kbp cDNA insert, with 43 kbp of λ vector. Manipulating small inserts in λ is not easy, and it is usual for the cDNA insert to be recloned into a plasmid vector for further characterization and manipulation.

2.2.3 Replacement vectors

Replacement λ vectors such as λEMBL3 (Frischauf *et al.*, 1983), λ2001 (Karn *et al.*, 1984), CHARON 4 (Blattner *et al.*, 1977) and commercial vectors such as λFIX and λDASH (Stratagene) have been used in the construction of genomic DNA libraries (*Figure 2.1*). In general these cloning vectors have been engineered to contain a stuffer fragment between left and right phage arms. DNA of interest replaces this central fragment with insert DNA in the range 10–20 kbp. Due to the size constraint imposed by the packaging enzymes, ligation of the left and right λ arms together, with no insert, yields a DNA molecule that is too small to be packaged. Thus, a central DNA fragment, either the original stuffer or foreign DNA, is required to position the *cos* sites the required distance apart for efficient *in vitro* packaging. The stuffer fragment in many vectors has been engineered to be selected against on appropriate *E. coli* strains. For example, wild type EMBL3 vector does not grow on P2 lysogens due to the *Spi* phenotype (Frischauf *et al.*, 1983). Phage vectors which are *gam*$^+$ do not form plaques on *E. coli* which are lysogenic for bacteriophage P2. Inclusion of a functional *gam* gene in the stuffer fragment biologically eliminates parental phage when the library is plated on P2 lysogens. This alleviates the need to physically remove the central stuffer fragment, from the left and right arms, prior to ligation with size-selected foreign genomic DNA. The vectors have simple polylinker sequences flanking the stuffer fragment to allow use of several restriction enzymes for cloning and subsequent analysis.

In general genomic DNA libraries rely upon partial digestion with a frequent cutting restriction enzyme such as *Sau*3A I, which recognizes the 4 base sequence GATC, to generate the DNA fragments for cloning. The 4 base overhang is compatible with the overhangs produced by the enzymes *Bam*HI and *Bgl*II. The target DNA is generally size-selected for fragments in the range 15–20 kbp. Due to the size constraint on λ phage DNA, cloning of more than one insert would produce nonviable particles as the *cos* sites would be greater than 50 kbp apart.

The number of recombinant λ clones required to give a 99% probability of containing any particular human genomic sequence is roughly 10^6 for an average insert size of

15 kbp. *In vitro* packaging easily allows the construction of genomic libraries of over 10^6 clones. However, there are two problems with cloned DNA in λ replacement vectors. One has already being mentioned above, that manipulation of λ DNA is difficult and the insert size is about 50% the size of the vector arms. The second problem is the fact that although most mammalian genes are up to 20 kbp in size, some are bigger and so the complete gene has to be cloned in several overlapping λ clones, particularly if the upstream promoter and downstream sequences are required. Thus, cloning 15–20 kbp fragments will most likely yield clones that do not contain the complete gene. It can be very difficult to analyse and manipulate a large gene if it is cloned as several overlapping λ clones. For these reasons λ replacement vectors have gone out of fashion to some extent.

Vectors capable of accepting larger genomic fragments have been developed, with the aim of containing complete genes. Cosmid vectors utilize the *cos* sequences in a plasmid vector for cloning and *in vitro* packaging to yield phage particles containing inserts of up to 40 kbp in a 5 kbp vector. In the last 10 years with the incentive to map and sequence the human genome, even larger fragments of DNA have been required, leading to the development of YAC (yeast artificial chromosomes; capable of cloning 500–1000 kbp inserts) and BAC (bacterial artificial chromosomes; able to clone inserts between 100–300 kbp) (see Coulson *et al.*, 1988; Shizuya *et al.*, 1992).

Bacteriophage λ has been a workhorse of molecular biology and although it has been surpassed by newer vectors capable of accepting larger inserts or easier manipulation of inserts, it has to be remembered that screening of libraries on filters is much easier with phage lifts rather than lifts from lysed colonies. However, λ is back in fashion in whole genome sequencing strategies (see section 2.3.3).

2.3 Filamentous bacteriophage vectors

2.3.1 Filamentous bacteriophages

Filamentous bacteriophages were discovered in the 1960s (Hofschneider, 1963; Loeb, 1960; Marvin and Hoffman-Berling, 1963; Zinder *et al.*, 1963). These are rod-shaped viruses which infect only F^+ *E. coli*. The three major filamentous phages are called f1, fd and M13, and all three have been sequenced (Beck *et al.*, 1978; Hill and Petersen, 1982; van Wezenbeck *et al.*, 1980). The genome of these viruses is composed of a circular single-stranded DNA molecule, approximately 6400 bases in size, and the genomes of M13, f1 and fd are ~99% identical. The genome encodes for 10 viral proteins with a very tight fit of genes (*Figure 2.2*). The only non-protein coding region is the intergenic region, ~500 bases, which contains the origins of viral replication. Many of the protein open reading frames overlap or the sequence between them is only a few bases. The filamentous tube of capsid proteins surrounding the ss DNA genome is composed of ~2700–3000 copies of the phage gene VIII protein (gVIIIp). The tips of the phage rod are composed of the minor proteins gIIIp, gVIp, gVIIp and gIXp. There are ~5 copies each of gIIIp and gVIp at the end of the rod involved in attachment and morphogenesis of the phage and ~5 copies of gVIIp and gIXp at the opposite end, which is involved in virion assembly. gVIIIp is synthesized with a 23 amino acid signal

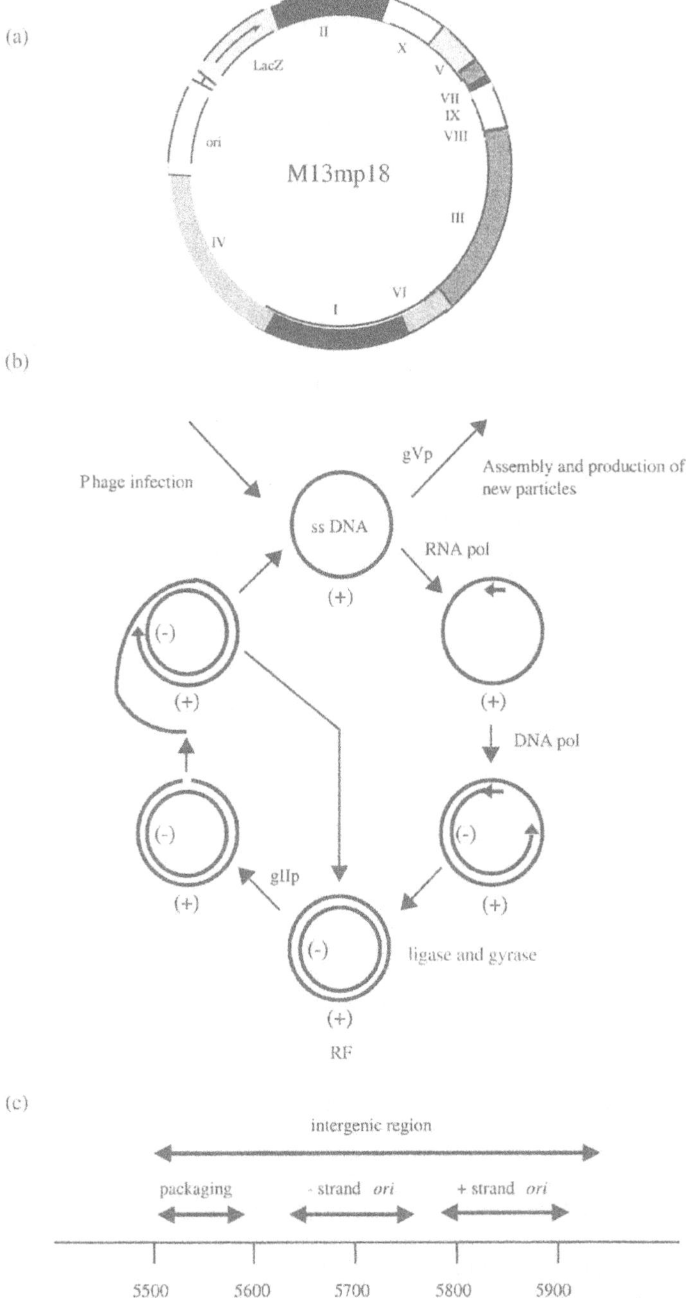

Figure 2.2. M13 bacteriophage. (a) Map of the phage vector M13mp18 showing the position of the 10 phage genes and the lacZ polylinker region inserted into the intergenic region. (b) Schematic outline of the replication of filamentous phage genomes (see text for details and Denhardt *et al.*, 1978). (c) Representation of the major features of the intergenic region.

sequence which is cleaved into the mature 50 residue capsid protein during its translocation into the outer membrane. Protein gIIIp at the tip of the phage particle is responsible for binding to the F factor on *E. coli* cells. Recent work on the structure of gIIIp have identified three domains in the mature protein (Chatellier *et al.*, 1999; Deng *et al.*, 1999; Holliger *et al.*, 1999). gIIIp is initially synthesized with an 18 amino acid leader sequence which is subsequently cleaved off, followed by domain D1 (residues 1–68) involved in binding to the *E. coli* TolA receptor, then after a glycine rich spacer is domain D2 (residues 87–218) which binds to the F pilus. This is followed by another glycine rich spacer then the final domain, D3 (residues 257–406) which is embedded in the phage particle.

Once the phage has injected its ss DNA genome into the bacterial cell, host enzymes convert the phage genome into a ds circular molecule. The intergenic region of ss viral DNA (the viral 'plus' strand) can form a hairpin structure which acts as a weak promoter. The RNA produced serves to prime synthesis of the 'minus' strand to yield a ds DNA molecule, which interacts with other host enzymes such as gyrase to become a supercoiled replicative form DNA (RF). Subsequent transcription and translation of phage genes results in synthesis of the gene gIIp protein, which recognizes the intergenic region of the genome and introduces a nick in the 'plus' strand. The nick provides the 3′- end required for DNA polymerase to copy the 'minus' strand. Synthesis displaces the original 'plus' strand and with the action of gIIp, which remains associated with the origin, releases the original 'plus' strand as a ss DNA circular molecule and generating a ds DNA RF molecule. The RF DNA molecule continues to produce ss DNA 'plus' templates, and during the initial stages of infection these ss DNA molecules are converted into ds DNA molecules. Thus, there is a production of about 100 RF molecules per cell. The next stage in phage production occurs upon accumulation of the gene gVp protein. This protein binds to ss DNA and thus sequesters the ss DNA and effectively prevents the produced ss 'plus' strands being converted into RF molecules. The gVp bound ss DNA is then converted into mature phage particles with release of gVp and assembly with the mature virion proteins gVIIIp, gIIIp, gVIp, gVIIp and gIXp (*Figure 2.2*). These mature particles are extruded from the bacterial cell without causing bacterial lysis. For more information on filamentous phage replication and structure see Denhardt *et al.* (1978). Filamentous phages in effect exist inside infected cells as a double stranded replicative form DNA circle, exactly like any ds DNA plasmid molecule. Outside, in the phage particle, the DNA is a ss DNA circle. Further basic work on the structure and replication of the filamentous phages is essential and will clearly assist in the future developments of these bacteriophages as phage vectors.

2.3.2 DNA sequence analysis

The development of filamentous phages as recombinant DNA cloning vectors goes along hand in hand with the development of the DNA sequencing technologies. The chain termination or dideoxy-method developed by Frederick Sanger relies upon a ss DNA molecule annealed with an oligonucleotide primer to act as a template for DNA polymerase copying, with incorporation of 2′,3′-dideoxynucleoside triphosphates (ddNTPs) to delineate the sequence of the target DNA (Sanger *et al.*, 1977). In the presence of all four normal 2′-deoxynucloside triphosphates and one ddNTP, say ddATP, the Klenow DNA polymerase copies the template incorporating nucleotides as dictated by Watson-Crick base pairing. Whenever the enzyme recognizes a T on the template it has, in effect, two choices. Firstly, it could incorporate a normal A nucleotide residue and then continue copying the template, or secondly incorporate a ddA nucleotide and terminate the growing chain. This results in the production of a series of synthesized DNA fragments, all commencing with the primer sequence at the 5′-end and at the 3′-end a ddA nucleotide. Separation on a urea denaturing polyacrylamide gel yields a ladder of fragments indicating the position of the A bases in the target sequence. Repeating this reaction with the other ddNTPs gives four reactions, which when separated on the gel, yields the base sequence of the DNA.

Construction of recombinant DNA molecules relies upon ds DNA cloning vectors, and restriction enzymes and ligases to cut and join together DNA fragments. It is not easy to convert plasmid ds DNA into ss DNA *in vitro* using enzymes, such as exonucleases. Initial methods at sequencing directly ds DNA with the chain termination method were usually not very successful. Sanger had developed his method employing ss DNA as template, isolated from the bacteriophage ϕX174 (Sanger *et al*., 1977). Also at this time only a few laboratories specialized in the chemical synthesis of oligodeoxynucleotides, the primers needed for Sanger sequencing.

2.3.3 Messing M13 vectors

However, the problem of preparation of ss DNA templates for sequencing was solved due to the efforts of Joachim Messing, who pioneered the development of vectors for cloning based upon filamentous phage M13 (Messing, 1983, 1991; Messing and Vieira, 1982; Messing *et al*., 1977, 1981; Norrander *et al*., 1983; Vieira and Messing, 1988; Yanisch-Perron *et al*., 1985). The only region of the M13 genome available for inserting foreign DNA is in the intergenic region, which also contains the origins of replication. The rest of the genome is wall-to-wall phage genes, with virtually no non-coding regions between genes. Insertion into any other part of the viral genome would inactivate a phage gene and be non-productive. Messing was successful in cloning a 759 bp *Hin*dII DNA fragment, encoding the amino terminal region of the *lac*Z gene, into a *Bsu*I site in the intergenic region of M13. The resultant phage, termed M13mp1, was viable (Messing *et al*., 1977). Chemical mutagenesis modified the *lac*Z region creating a unique *Eco*RI restriction enzyme site spanning the fifth to seventh codons of the β-galactosidase gene producing the vector M13mp2 (Gronenborn and Messing, 1978). This allowed foreign DNA fragments to be cloned into the *lac*Z region.

Normal β-galactosidase is a tetramer enzyme capable of hydrolysing the colorless indicator molecule X-gal releasing a deep blue indole dye moiety. The *lac*Z region cloned into the M13mp vectors codes for the first 146 amino acid residues (the so-called α-peptide) of the β-galactosidase enzyme. This polypeptide is sufficient to form tetramers with the defective monomer enzyme lacZΔM15. The *lac*ZΔM15 mutation deletes residues 11 to 41 of the β-galactosidase protein and in *E. coli* strains possessing this mutation, the monomer enzyme is incapable of forming tetramers and does not yield blue colonies on X-gal containing agar plates. Infection of *lac*ZΔM15 strains of

E. coli, such as JM101, with a parental M13mp series of phage vector allows expression of the phage encoded α-peptide, which can complement the mutation and thus result in the production of blue plaques. Cloning of foreign DNA into the *Eco*RI site disrupts the *lac*Z sequence, resulting in failure to produce a functional α-peptide, thus there is no complementation of the ΔM15 mutation. This means that recombinant phage produce clear plaques on indicator plates, whereas the parental vector produces blue plaques.

These vectors provide a direct visualization method allowing differentiation of plaques on an agar plate into recombinant or parental phage. Previous cloning vectors such as pBR322 allowed for selection of bacteria on antibiotic plates, but did not distinguish between parental and recombinant clones. Messing went on to modify the M13mp2 vector by introducing polylinker sequences containing unique restriction enzyme sites into the *lac*Z region creating the mp7, mp8/9, mp10/11 and mp18/19 series (Messing, 1983; Messing and Vieira, 1982; Messing *et al.*, 1981; Yanisch-Perron *et al.*, 1985). This system has been adapted for use in plasmid vector cloning systems (Vieira and Messing, 1982), and others have constructed a range of derivative vectors, and many of these are available commercially. Very few molecular biologists will not have used a blue/white cloning-screening system based upon those developed by Messing.

Others have also taken the original M13 vectors and modified them to make them more versatile, for example the M13tg vectors of Kieny *et al.* (1983). By making sure that the introduced polylinker sequences maintain the reading frame of the *lac*Z sequence, the α-peptide synthesized by the parental vector will always complement the ΔM15 mutation in strain JM101, and thus produce blue plaques. The longer the polylinker, the more useful restriction enzyme sites are included in the vectors. The downside of this is the observation that the blue color produced with these newer vectors is paler compared to the original M13mp2 vector, which produces a deep blue color. This in not surprising as the extra DNA sequences translate to extra amino acids in the α-peptide which may adopt a less favorable interaction with the β-galactosidase resulting in a poorer hydrolysis of the X-gal indicator.

Due to the structure of the filamentous phage particles there is no theoretical limit to the size of phage particles. Cloning of a 7000 bp fragment into the 7229 bp M13mp8 genome will produce a 14 000 base ss DNA genome, which will be packaged into a phage particle approximately twice as long as parental M13mp8 virions. All that is required to package a larger genome is more gVIIIp to form the filamentous tube surrounding the ss DNA. In practice such large inserts are unstable in M13mp vectors, and the practical size limit for cloning fragments into M13 vectors is about 3000 bp, although larger fragments can be stable. This is probably a consequence of the actual position in the intergenic region that the original *lac*Z fragment was cloned. Barnes has developed another series of M13 vectors by cloning the *lac*Z sequences into a different site in the intergenic region (Barnes and Bevan, 1983). They were able to stably clone DNA fragments up to 14 000 bp in their M13 vector.

The availability of M13 derived cloning vectors meant that it was now easy to produce ss DNA molecules, the ideal substrate templates for Sanger chain termination sequencing (Messing et al., 1981; Sanger *et al.*, 1980). M13 vectors are ideal for sub-cloning smaller DNA fragments for sequence analysis. The availability of the Messing vectors, and cheap oligonucleotide primers, lead to a rapid explosion of sequence data generation in the 1980s and early 1990s. There is still an exponential increase in new DNA sequence acquisition, although improvements in technology, mainly brought about by the polymerase chain reaction (PCR), have meant the use of M13 vectors for sequence analysis is not as popular as it once was. Having said that, the best sequence data generated with the latest generation of automated fluorescent sequencers is that produced utilizing a M13 ss DNA template.

Even with the latest generation of automated instruments, the sequence data determined in each sequence reaction is in the region 500–1000 bases. Thus, to sequence a large cloned DNA fragment of 10 000 bp, or even herpesviral (~200 000 bp) or bacterial (several million bp) genomes requires strategies to break down the target DNA into fragments suitable for individual sequencing (see Jones, 1996). Computer programs are then used to compile the sequence data into the final complete sequence of the DNA. Given the length of sequence obtainable with automated instruments M13 vectors are ideal for cloning DNA fragments for sequence analysis. After a 5-hour growth period which can yield as much as 10^{12} phage particles per ml, ~5 μg of ss DNA is produced, which is ample DNA for several sequence reactions (Messing and Bankier, 1989). Simple centrifugation removes the bacterial cells from the culture, leaving the phage particles in the supernatant. The phage particles are recovered by precipitation with polyethylene glycol and sodium chloride, and the ss DNA released by treatment with phenol (Messing and Bankier, 1989; Schreier and Cortese, 1979) or heating in the presence of a detergent such as Triton X100 (Chissoe *et al.*, 1991).

The easiest method for randomly fragmenting a DNA molecule is by sonication, and various protocols have been developed (Deininger, 1983). This method is still used today for fragmenting whole bacterial genomes (Fleischmann *et al.*, 1995), and BAC and cosmid library clones. This has culminated in the publication of the sequence of human chromosome 22 (Dunham *et al.*, 1999) and 21 (Hattori *et al.*, 2000). It has also been used in the whole genome sequencing of *Drosophila melanogaster* (Myers *et al.*, 2000). Excellent protocols and strategies have been published and reviewed (Andersson *et al.*, 1994; Bankier and Barrell, 1989; Bankier *et al.*, 1987; Chissoe *et al.*, 1991; Messing and Bankier, 1989; Messing *et al.*, 1981). Today there is no reason why researchers should be working with cloned DNA molecules that have not had their complete nucleotide sequence determined. Sequencing is relatively easy and extremely fast, particularly with the recent capillary electrophoresis instruments. A single ABI 3700 sequencer can determine approximately 400 bases of data from each of ~960 clones in just 24 hours, a staggering total of ~400 000 bases (see the website: http://www.pebio.com/ga/3700/).

One disadvantage of working with M13 is because the DNA is single stranded; it is only possible to determine the sequence of one end of a cloned DNA fragment without further *in vitro* and *in vivo* manipulations to re-orientate the insert in relation to the universal sequence priming site. These manipulations are not trivial to perform. However, Blattner developed the Janus phage vectors to overcome this problem (Burland *et al.*, 1993). The Janus vector contains the λ att recombination sequences, so when the phage infects the appropriate host strain the DNA insert is inverted relative to the cloning site, and subsequent isolation of phage produces ss DNA with the insert in the opposite orientation. Blattner used this method to help in completing the sequence of the 4 639 221 base pair sequence of *Escherichia coli* K-12 (Blattner *et al.*, 1997). Employment of PCR and flanking M13 vector primers allows for an easy method for preparing ds insert DNA for sequence analysis starting with a ss M13 DNA template.

Although many laboratories now use ds plasmid DNA vectors for sequencing projects, mainly because of the ability to sequence inserts from either end, phage λ has made a comeback. Whole genome sequence analysis (Bult *et al.*, 1996) is a strategy whereby the complete genomic DNA is shotgun cloned for sequencing, without any prior cloning or mapping of clones. Computers then assemble the individual sequence gel reads. There can be problems in trying to assemble data from large genomes compared to assembly of data from smaller cloned DNA fragments. To aid in assembly, a large insert library of (~15–20 kbp) is constructed in λ vectors. Generating sequences of ~400 bases from each end produces a scaffold of data upon which the random shotgun data can be assembled. The λ end data gives sequence reads that are a known distance apart, ~15–20 kbp, and this aids dramatically the assembly of contigs into a complete single continuous sequence.

Although the predominant use of filamentous phage vectors is the production of ss DNA for sequence analysis, it is also used as a template for site-directed mutagenesis of cloned genes (Zoller and Smith, 1983). This method allows you to change any nucleotide residue of a gene into any other. The true era of 'genetic engineering' was here. Utilizing M13 to clone and alter genes has lead to the rapid understanding of the functioning of genes and proteins. The function of any amino acid residue in a protein can be studied by the ability to change it into any of the other 19 naturally occurring residues. Today manipulation of genes and mutagenesis strategies generally rely upon PCR, rather than filamentous phage methods.

2.3.4 Filamid vectors

The realization that only the ~500 bp intergenic region is required for packaging ss DNA into a phage particle led to the development of a new kind of hybrid phage–plasmid cloning vector, the filamids. Although filamentous phage vectors are extremely useful vectors, they are not as versatile as the more traditional plasmid vectors such as pBR322 or the pUC series (Vieira and Messing, 1982) for manipulation of cloned DNA fragments. Dotto and colleagues cloned the intergenic region of phage f1 into the *Eco*RI site of pBR322 (Dotto and Horiuchi, 1981; Dotto *et al.*, 1981) and showed that if plasmid containing cells were infected with f1 phage, the virion capsids secreted into the culture medium contained a mixture of f1 and plasmid ss DNA. Thus, ss DNA of the plasmid could very easily be obtained for DNA sequence analysis. Plasmid vectors with filamentous phage origins of replication have been variously called filamids (filamentous plasmids), phagemids (phage plasmids), phasmids (phage plasmids) or even plages (plasmid phages).

Dente *et al.* (1983) expanded the versatility of such vectors by cloning the f1 *ori* region into pUC plasmids to create the pEMBL series. Since then many variants on plasmids with dual origins of replication have been developed. Growth of plasmid-containing bacterial cells results in plasmid replication as ds DNA. Upon infection with a helper phage, which provides all the phage replication proteins in *trans*, phage particles are produced containing one strand of the plasmid DNA. The actual strand of plasmid DNA that is packaged into phage particles depends upon the orientation of the f1 *ori* sequences. To cover all eventualities there are four pEMBL vectors depending upon the orientation of the f1 *ori* and the polylinker sequence relative to each other, pEMBL8(+), 8(−), 9(+) and 9(−). Many commercial companies now market a wide range of filamid vectors containing a versatile array of features. Stratagene market a very good filamid vector, equivalent to the pEMBL series, called Bluescript (Alting-Mees and Short, 1989; Alting-Mees *et al.*, 1992).

Addition of a helper phage to a bacterial cell containing a filamid vector results in packaging of the filamid DNA and the helper phage DNA. Thus, two types of phage are produced. Providing that apart from the f1 *ori* sequences there are no other sequences in common then the helper phage DNA is an unobtrusive partner. In general the ratio of the two types of ss DNA isolated is in relation to their size, thus a 3.5 kbp filamid infected with a 7 kbp helper phage will yield a roughly 2:1 molar ratio of filamid to helper phage DNA. Davies and Hutchison (1991) have developed an extremely useful helper phage for the production of ss filamid DNA, M13CO7. Their helper phage has an antibiotic resistance marker cloned into the phage *ori* region. Thus, although the phage is viable its origin is sub-optimal for the efficient production of phage particles, whereas the filamid vector has a perfect wild-type filamentous origin of replication. When M13CO7 is used to infect filamid containing bacterial cells, the phage proteins replicate almost exclusively the filamid DNA such that the phage particles produced contained virtually only ss filamid DNA, with only extremely small amounts of helper phage DNA (<5%).

2.3.5 Composite vectors

Combining origins of replication of different phage and plasmid systems allows for the development of some very interesting and extremely useful cloning vectors. Lambda ZAP (Short and Sorge, 1992; Short *et al.*, 1988) overcomes one of the major problems of working with λ insertion vectors such as λgt10 and gt11; that of re-cloning the insert out of λ into a more useful plasmid vector. λZAP contains the complete filamid vector pBluescriptII (Alting-Mees *et al.*, 1992) cloned into phage λ. The filamid sequences were linearized at the filamentous origin of replication, effectively separating the functions of initiation of filamentous phage DNA replication from the termination signals, and then cloned into a composite λ vector derived from λ L47.1 and λgt11 (Short and Sorge, 1992) (*Figure 2.1* and *2.3*). Infection of an *E. coli* cell with a filamentous helper phage, which is already infected with λZAP, provides all the *trans* functions for replication of the filamid sequences and packaging into a filamentous phage particle. The net result is efficient excision of the filamid sequences from the lambda vector. This elegant system utilizes the best of all three vector systems (λ, M13 and plasmid). *In vitro* packaging of λ sequences allows the efficient production of recombinant libraries containing many clones. The 'insert' sequences are excised from the λ arms as a filamentous phage particle, which can infect fresh *E. coli* cells yielding

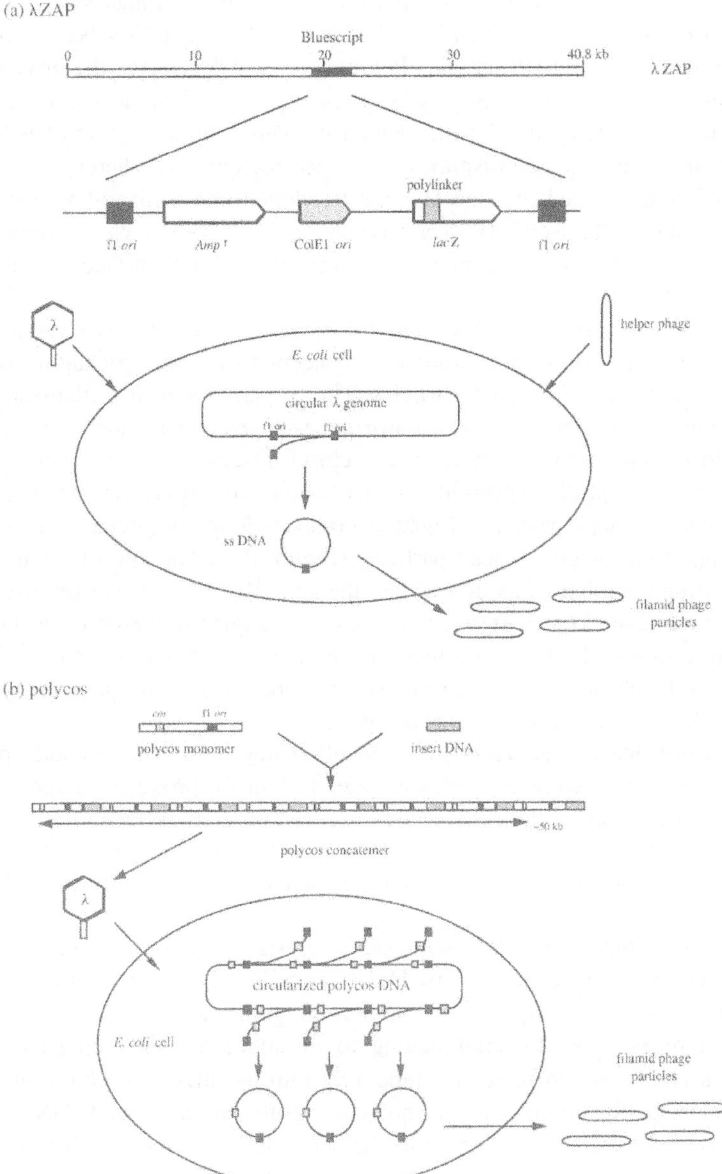

Figure 2.3. Combination vectors. (a) λZAP system. After infection of an *E.coli* cell, the λ genome circularizes. Addition of helper filamentous phage provides all the *trans* proteins which act on the f1 origin producing ss filamid DNA which is packaged into infectious filamentous phage particles. This concept of two f1 origins yielding ss filamids is exploited in the Davies and Hutchison Tn3 deletion sequencing strategy (Davies and Hutchison, 1991). (b) The polycos system. Ligation of the polycos vector to insert DNA generates a concatemer which can package DNA into λ phage heads *in vitro*. Once inside a bacterial cell the DNA circularizes and starts to express the filamentous phage genes resulting in ss DNA replication and production of filamentous phage.

colonies containing easily manipulated plasmid sequences. A similar system of cloning DNA into a plasmid vector, which is embedded in a λ phage, has been exploited by others. For example, Burmeister and Lehrach (1988) developed the Jekyll vectors, which allowed for excision from λ by homologous recombination due to short, direct repeat sequences flanking the plasmid sequences. This λ-filamid system has been used to create a filamentous phage display system (see section 2.4) whereby sequences for surface display are cloned into the f1 gene III cloned into a filamid vector, which is itself cloned into λ; the SurfZAP vector (Hogrefe *et al.*, 1993). This system has been used to create libraries expressing Fab antibody on the surface of filamentous bacteriophage particles.

Another useful system, also developed at Stratagene, is the polycos system (Alting-Mees and Short, 1993). This is similar in concept to *in vitro* packaging of cosmid sequences into infectious phage λ particles. The polycos vector is a filamentous phage vector containing the ampicillin resistance genes together with the λ *cos* sequences (identical to the *cos* sequence in cosmid vectors). Inserts are cloned into a modified M13 gene III such that the expressed gene III fusion will display the encoded sequence on the surface of phage particles. Ligation produces long concatemers suitable for *in vitro* packaging into λ phage head particles (*Figure 2.3*). The *cos* sites are ~7–8 kbp apart depending upon the DNA ligated into the gene III region. However, there will be *cos* sequences ~42–48 kbp apart, which allow for packaging into λ particles. Once these particles are introduced into *E. coli* then the circularized DNA is a head to tail tandem array of ~6 M13 phage sequences. Expression of the encoded phage proteins resolves itself into 6 M13 phage particles, exactly like the situation with λZAP, but without the need for helper phage. Yet again the high efficiency of *in vitro* lambda packaging combined with *in vivo* excision yields a novel system for producing large libraries of filamentous phage display clones. This system could possibly be taken further replacing the M13 vector with a filamid sequence. Once packaged into λ particles and used to infect *E. coli*, subsequent infection with a M13 helper phage should rescue the filamid clones.

An elegant method for DNA sequence analysis using filamid vectors and the transposon Tn3 has been developed by Davies and Hutchison (1991), and as described above exploits two f1 *ori* sequences being present on the same circular DNA molecule. The method relies upon bacterial mating to introduce a Tn3 conjugative plasmid, containing a f1 *ori*, which integrates randomly into the filamid vector containing the gene of interest. Resolving the co-integrate results in a filamid DNA molecule containing two f1 origins. Infection with a helper phage provides gIIp that nicks at both f1 origins in the same molecule, resulting in production of two filamid molecules (see *Figure 2.3*). Only one of these molecules contains the original antibiotic marker on the filamid, kanamycin resistance, and the ampicillin resistance marker introduced with the mini-Tn3. These clones are selected on Amp/Kan agar plates. The net result is a deletion of a portion of the cloned gene between the site of Tn3 integration and the f1 *ori* of the filamid vector. The deletion can be sequenced using a universal primer binding to Tn3 sequences remaining in ss DNA. The beauty of this method is the simplicity of manipulations. All that is required is to clone the target DNA of interest into the filamid vector, and then carry out extremely simple bacterial matings to produce thousands of random deletion clones for sequence analysis. The individual deletion clones are rescued as ss DNA using helper phage, and cover the whole region of the target gene.

2.4 Phage display systems

2.4.1 Fusion phages–the beginnings

George Smith can be considered the founding father of phage display (Smith, 1985) when he used filamentous phage f1 to clone DNA fragments from plasmid pAN4 containing the *Eco*RI endonuclease and methylase genes. He showed that foreign sequences could form fusion proteins with the phage encoded gIIIp protein, which are able to produce viable phage particles. The foreign fusion sequences were displayed on the surface of the virion and were accessible to antibody raised against the fusion partner. Such fusion phage could be enriched over a 1000-fold over normal f1 phage by antibody derived affinity selection. He went on to develop better bacteriophage fusion vectors, based upon fd, whereby foreign DNA could be cloned into the *gIIIp* gene at the amino terminus of the mature gIIIp (*Figure 2.4*) (Parmley and Smith, 1988). This paper showed proof of principle of two essential attributes that are essential for phage display systems. Firstly, it established that quite large, as well as small, fusion partners at the amino terminus of gIIIp were stable and that the fusion-gIIIp could be assembled into infectious phage particles. Although the yield of phage particles is less than that obtained with wild-type filamentous phages, the yield is adequate. It thus appeared that any peptide sequence could be expressed and displayed on the surface of a filamentous phage, without affecting the viability of the bacteriophage. This is truly amazing when you consider that the function of gIIIp is to bind to the surface of *E. coli* F^+ cells and initiate phage infection.

Secondly, it introduced the concept of 'biopanning'. This is whereby a ligand specific for the fusion partner displayed on the phage surface, is used in an immobilized form to bind the fusion phage. Any phages displaying other fusions are not selected and are removed by washing the immobilized ligand. The selected phages are then eluted, and because the phages are still infectious, they can be used to re-infect bacterial cells and produce further phage. Depending upon the ligand used, there will be some non-specific selection of irrelevant phage. The selected phages are submitted to several rounds of binding, washing, elution and re-growth; 'biopanning' results in enrichment and selection for phage displaying a fusion partner recognized by the immobilized ligand. This enrichment and selection can be up to 10^6. In other words one specific clone can be isolated from a mixture of 1 million other phage. Yet again this is a fantastic concept. As mentioned above, screening a λgt10 or gt11 library for a clone of interest requires plating out nearly 10^6 phage onto 10–20 150 mm plates (see section 2.2.2) and is an extremely arduous undertaking. However, with a phage display system, a single 1.5 ml tube, containing 10^9 different bacteriophage, can be sampled and the required phage isolated by 3–4 rounds of simple biopanning.

Scott and Smith (1990) utilized phage display to construct a peptide display library whereby virtually every possible hexapeptide sequence was displayed on the surface of

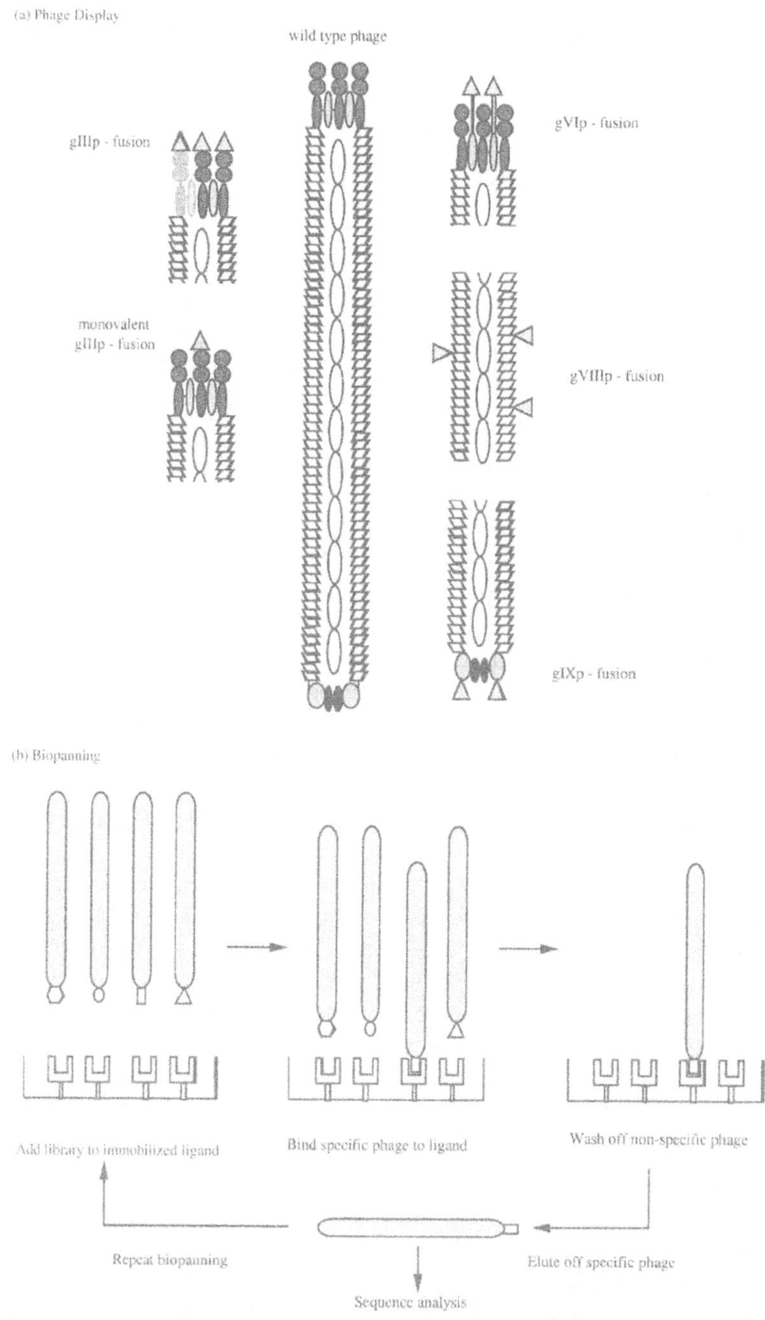

Figure 2.4. Phage display vector systems. (a) The various possible modes of displaying peptides and proteins fused to filamentous phage capsid proteins. (b) The basic concept behind biopanning is shown. Several rounds of biopanning, 3 to 4, are performed, then the final eluted phage are analysed by DNA sequencing.

fd as a gIIIp fusion. Their library contained ~4×10^7 different hexapeptide sequences (this represents 69% of the theoretical 6.4×10^7 hexapeptide sequences), and was constructed by cloning synthetic oligonucleotide duplexes with a central degenerate region encoding a hexapeptide sequence, $(NNK)_6$; where N represents an equal mixture of the 4 bases A, G, C and T, and K stands for an equal mixture of G and T. Thus, NNK represents 32 of the possible 64 codons, including a codon for all 20 amino acids and the amber termination codon (TAG). Due to the degeneracy of the genetic code, $(NNK)_6$ potential encodes 32^6 (~10^9) hexamer codons, and some sequences will be present more than others. For example, the sequence Met_6 can only be encoded by the sequence $(ATG)_6$, whereas there are three possible codons for Arg (CGT, CGG and AGG) and thus there are 729 possible DNA sequences encoding Arg_6 compared to only 1 sequence for Met_6. To allow for access to the displayed hexapeptides on the phage surface, flanking the random core sequence are flexible structureless linkers composed of glycine and alanine residues.

Scott and Smith (1990) used their library to select for hexapeptide sequences recognized by a monoclonal antibody raised against the sequence DFLEKI, an epitope of the protein myohemerythrin. After three rounds of biopanning selected phage were plated and individual phage analysed by DNA sequence analysis to reveal the encoded hexapeptide region. The sequence of the hexapeptide epitopes selected by the monoclonal antibody encoded the following sequences; DFLEWL (isolated 16 times), DFLEML and DFLAWL (both isolated once). As can be seen the 'epitope' isolated closely matches the real epitope that generated the monoclonal antibody.

This peptide epitope display library illustrates a third essential feature of phage display systems. Isolation of a particular phage, by virtue of its interaction of the surface displayed peptide, contains the gene sequence which encodes that peptide. Thus, the peptide is physically linked to the gene that encodes it. George Smith is happy to send his libraries to any group to help their research (http://www.biosci.missouri.edu/smithgp/).

At the same time Devlin *et al.* (1990) and Cwirla *et al.* (1990) constructed similar peptide display phage libraries. Cwirla *et al.* (1990) biopanned their hexapeptide library, consisting of 3×10^8 independent clones, against a monoclonal antibody against the peptide hormone β-endorphin. Most clones isolated started with the sequence YGG, which agrees with the actual β-endorphin amino terminal sequence, YGGFL. Devlin and colleagues (1990) constructed a library of random 15-mers, $(NNS)_{15}$, and screened for peptide sequences capable of binding to streptavidin. From the library of 2×10^7 phage clones were isolated which contained a common sequence motif, HPQF, responsible for streptavidin binding. Thus, even though the 15-mer library contained only a very small fraction of the possible 3×10^{19} 15-mer peptide sequences, it contains a large array of overlapping hexamers, heptamers, etc.

2.4.2 Gene gIIIp systems

Since the first studies described above, many groups have developed and expanded the display of peptides and proteins as amino terminal fusions to gIIIp on the surface of filamentous bacteriophages. The examples discussed below are just a small fraction of the published work. As explained above, any peptide sequence appears to be able to be displayed on the tip of phage particles, and thus can be used to study its interaction with other proteins. The potential is endless.

A very nice example of the use of the Smith hexapeptide library is that of Stephen and Lane (1992). Lane and colleagues have isolated a monoclonal antibody, PAb240, which recognizes a normally cryptic epitope of the protein p53 that is exposed in mutant p53 proteins. Biopanning the phage library against PAb240 isolated phage which allowed identification of the cryptic epitope, RHSVV. Screening the genomic sequence databases with the epitope sequence identified the *Xenopus laevis* protein TFIIIA as containing the same peptide sequence, and subsequent Western blotting of *Xenopus* oocyte extracts with PAb240 detected the TFIIIA protein.

The MHC class II molecules are highly polymorphic membrane glycoproteins that display peptides for recognition by T cells. Screening of peptide epitope libraries against purified class II molecules reveals the sequence of peptides capable of binding into the peptide-groove of class II molecules (Hammer *et al.*, 1992, 1993). These studies help understand the interaction of the polymorphic class II molecules with peptide fragments that are displayed on the surface of cells. A novel use of peptide display libraries may well be their use in targeting particular cells. Once cell-specific binding peptides have been identified, these peptide sequences can then be used for imaging cells in whole animals (Ladner, 1999; Pasqualini, 1999). Yip *et al.* (1999) have injected phage into mice to show the feasibility of *in vivo* biopanning for selection of phage displaying Fab antibodies.

In the original Smith peptide display libraries (Scott and Smith, 1990) the hexapeptide sequence was connected by a short flexible linker. In effect the peptide could adopt an infinite variety of conformations. By incorporation of flanking cysteine residues which result in a disulphide bond formation, the peptide sequence is constrained in cyclic loop conformation. Koivunen *et al.* (1995) constructed constrained cyclic peptide libraries (CX_5C and CX_6C) and isolated phage selectively binding to the cell surface integrin receptor. The tri-peptide sequence, RGD, appears to specify binding to $\alpha_5\beta_1$ integrin receptors. A display library of constrained peptides, XCX_4CX, was screened against streptavidin (McLafferty *et al.*, 1993), and identified the peptide sequence HPQF. The sequence is identical to that obtained by Devlin *et al.* (1990).

Peptide display libraries have been used to isolate peptide sequences, which mimic non-protein antigens, such as carbohydrates (Hoess *et al.*, 1993; Oldenburg *et al.*, 1992). Oldenburg and colleagues screened an octapeptide display library with the carbohydrate-binding protein concanavalin A and isolated phage with the consensus sequence YPY (Oldenburg *et al.*, 1992). A 12-mer peptide containing the consensus sequence was found to inhibit precipitation of the carbohydrate moiety α-glucan dextran 1355 by concanavalin A. The monoclonal antibody B3 recognizes the $Lewis^Y$ carbohydrate antigen, and biopanning a surface octamer display library isolated phage encoding the peptide sequence APWLYGPA.

These studies clearly showed that peptides can mimic carbohydrates, and together with the results of finding streptavidin-binding peptides, suggest that peptides are capable of mimicking virtually any chemical ligand. One major benefit is that comprehensive filamentous display libraries, containing mixtures of constrained peptides, flexible peptides, hexamers to 30-mers, etc. can be made relatively easily, and this vast array can be biopanned against any ligand. It does not matter what type of display peptide is selected in the rounds of biopanning, as the sequence of the gene encoding the displayed peptide will reveal exactly what the sequence is and its context.

The above studies have utilized peptide display libraries whereby potentially every possible short peptide sequence is present in the libraries for screening. It is possible to display larger amino acid sequences fused to gIIIp.

Roberts *et al.* (1992) displayed the 58 residue bovine pancreatic trypsin inhibitor fused to gIIIp, Corey *et al.* (1993) have displayed trypsin protease, and Matthews and Wells (1993) have displayed human growth hormone with protease linker sequences. The work by Matthews and Wells (1993) was to study the subtilisin protease cleavage site. They constructed a filamid vector expressing the human growth hormone sequence fused to a library of sequences of the format $GPGGX_5GGPG$ or $GPAAX_5AAPG$ in frame with gIIIp. Infection of *E. coli* containing the filamid vector with a helper phage produces phage particles that display just one gIIIp fusion protein molecule per virion. The remaining gIIIp protein is made up of normal wild-type gIIIp molecules provided by the helper phage, thus the phage particles are monovalent. This contrasts with the situation with filamentous phage vectors where all the gIIIp molecules are the fusion protein; there is no wild-type gIIIp present. The phage particles were bound to a solid support using an immobilized antibody against the human growth hormone sequence. Treatment of the bound phage with subtilisin released phage that possessed a protease-sensitive cleavage site, whereas phage that were still bound had a sequence that was resistant to subtilisin cleavage. Thus, it is possible to sample in the library the most sensitive and most resistant sequences.

Atwell and Wells (1999) have used phage display to select for improved enzyme activities in subtiligase. Subtiligase is a mutant subtilisin protease that catalyses the ligation of peptides. Utilizing a biotinylated peptide substrate, which allowed for capture of phage with immobilized neutravidin, a library of mutant subtiligases ($>10^9$) was screened for improved ligation activity. The screen relied upon the activity of subtiligase to ligate the biotinylated peptide to itself, such that ligase-active phage were able to be bound to the solid support.

Displaying on the surface of phages a repertoire of protein variants allows for the selection of those sequences with desired properties. The selected sequences can then undergo further rounds of mutagenesis of the whole displayed sequence or just a particular region, and offers an extremely powerful methodology for the selection of proteins with improved or new specificities and/or properties.

Applying phage display to enzymes is more difficult, as the reaction products diffuse away from the enzyme. Demartis *et al.* (1999) have elegantly overcome this problem. The enzyme under study is expressed as a fusion with a calmodulin-gIIIp fusion. The substrate for the enzyme is linked to a calmodulin-binding peptide, which is bound to the calmodulin-fusion. The enzyme then converts the substrate to product which is bound to the phage particle. Antibody specific to the reaction product can then be used to biopan phage with functional enzyme displayed.

A major use of gIIIp system has been the surface display of antibody fragments, Fabs and single chain Fv fragments (scFv) (*Figure 2.5*). This work has been pioneered by Winter and colleagues. They were among the first to show the potential of phage display

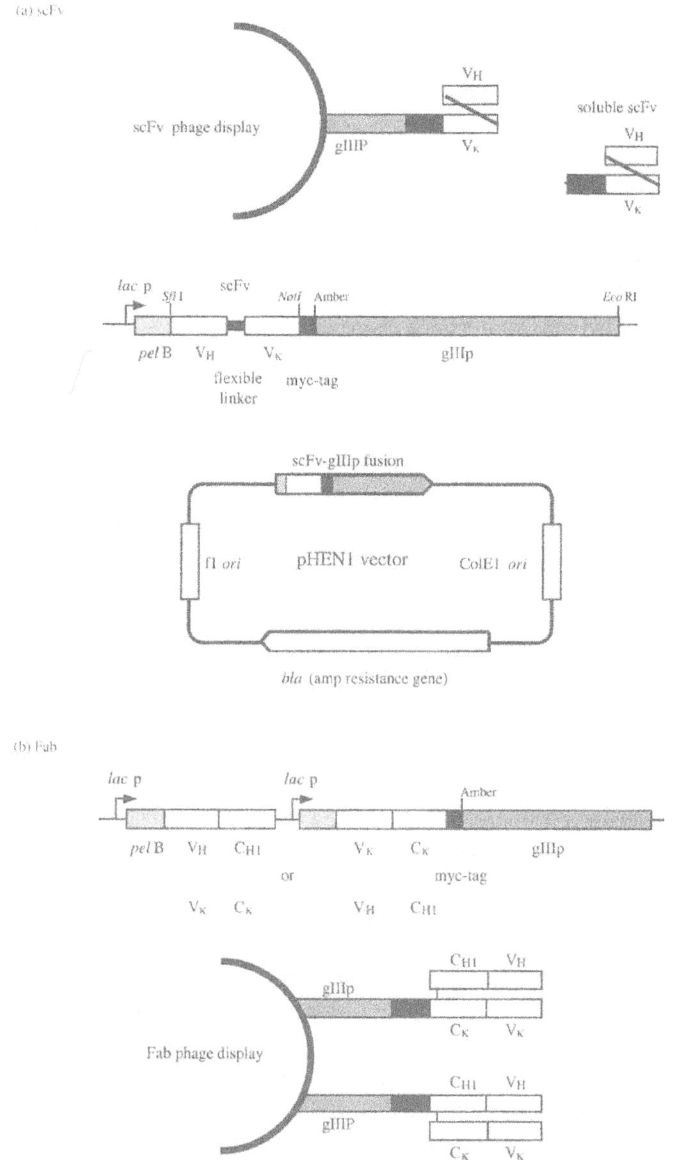

Figure 2.5. scFv antibody display. (a) The pHEN1 vector is shown with a cloned scFV construct. This is shown expanded above the circular map in more detail. The flexible linker is composed of $(Gly_4Ser)_3$ and links the V_H to the V_κ. Expression in an amber suppressor strain of *E. coli* results in read through from the scFV sequences into the gIIIp resulting in display of the scFv on the surface of phage particles. Once a particular clone had been isolated, infection of wild-type *E. coli* results in secretion of the scFv alone as a soluble molecule. (b) Fab display utilizes similar vectors producing two proteins, either V_H linked to gIIIp or V_κ, and these associate with V_κ or V_H, respectively and are displayed on the phage surface. The two subunits of the Fab are covalently linked by disulfide bridge formation between the C domains.

for the isolation of antigen specific antibodies (Clackson *et al.*, 1991; McCafferty *et al.*, 1990). Barbas *et al.* (1991) also showed the feasibility of antibody display using filamentous phage display.

The whole range of immunological processes involved in the development of the antibody response has been beautifully duplicated with filamentous phage display (Marks *et al.*, 1992). mRNA from peripheral blood lymphocytes can be reverse transcribed and PCR amplification yields the whole repertoire of heavy and light chain variable domain segments which can be manipulated *in vitro* to yield scFv genes. In scFv a flexible peptide sequence, $(Gly_4Ser)_3$, links into a single peptide chain the heavy and light V sequences (*Figure 2.5*). However, it is possible to assemble multi-subunit proteins on the surface of phage. Hoogenboom *et al.* (1991) expressed either the heavy or light chain fused to gIIIp, and the complementary light or heavy chain V segment was expressed independently, with subsequent secretion into the bacterial periplasm. This resulted in the formation of heterodimeric Fab fragments on the phage surface. Also, by introducing amber mutations (a UAG termination codon) between the antibody and the gIIIp sequences then the Fab can be displayed on the phage surface or produced as a soluble molecule depending whether a amber suppressor or non-suppressor *E. coli* strain is used for phage growth (Hoogenboom *et al.*, 1991).

The need for a purified protein ligand for immunization is no longer necessary, as Chowdhury and Pastan (1999) have successfully utilized DNA immunization. Purification of a protein ligand can be a very time-consuming process and in many cases nearly impossible, particularly for mammalian surface membrane-expressed proteins. However, in many cases isolation of the gene can be easily achieved and subsequent expression *in vivo* can yield the fully processed and mature protein. Chowdhury and Pastan (1999) expressed mesothelin cDNA in mice, and subsequently constructed a scFv phage library from mouse splenic mRNA and biopanned against mesothelin-positive cells. They thus produced mesothelin-specific antibody phage without the need to purify the protein itself.

The process of prior immunization can be bypassed altogether with phage antibody display. Traditionally for the isolation of monoclonal antibodies the particular antigen is first used to immunize a mouse, with subsequent cell fusion and screening for a specific monoclonal antibody. This is a very time-consuming procedure and not without its pitfalls. Winter and colleagues have eliminated the need for immunization by generating large scFv libraries from unimmunized donors (Marks *et al.*, 1991). Diverse libraries of immunoglobulin heavy and light chains, both kappa and lambda, were prepared by PCR and randomly combined to produce a combinatorial library of $>10^7$ scFVs. Rare scFv phage could be selected after four rounds of growth and biopanning. The need for mRNA or rearranged V genes from unimmunized donors as a source is unnecessary, as they have shown that germline V_H gene segments can be manipulated to produce scFv libraries from which specific antibodies can be isolated (Hoogenboom and Winter, 1992). These studies clearly showed that specific antibody fragments can be produced entirely *in vitro*. Generating scFv libraries from 50 donors has led to the production of a combined library of 4×10^9 antibody phage (Little *et al.*, 1999).

As with peptide display libraries, scFv phage libraries once made can be screened for any antibody specificity; there is no need for immunization and preparation of specific libraries for each ligand under study. This is clearly revolutionary technology (Griffiths and Duncan, 1998). Even if the bacterial expressed scFv does not have the necessary characteristics, once an scFv has been isolated, the gene sequence can be manipulated to construct a 'proper' full length antibody gene which can be transfected into suitable mammalian cells to synthesize the natural immunoglobulin molecule.

2.4.3 Gene gVIIIp systems

Whilst the predominant filamentous virion protein to be utilized for surface display is gIIIp, the major structural protein gVIIIp has also been developed. The mature processed protein is only 50 amino acids in length (see section 2.3.1). Greenwood *et al.* (1991) expressed peptides from the major surface antigen of the malaria parasite *Plasmodium falciparum*, Markland *et al.* (1991) expressed bovine pancreatic trypsin inhibitor, and Felici and colleagues have constructed peptide libraries utilizing gVIIIp (Felici *et al.*, 1991; Luzzago *et al.*, 1993). The problem of utilizing gVIIIp is that it does not tolerate large inserts. It would appear that the maximum size of displayed peptide is ~6 amino acid residues. This is applicable to vectors whereby the gVIIIp gene is modified such that all copies of gVIIIp are displaying peptide fusions. Given the structure of the filamentous virion it is not surprising that modification of all ~3000 gVIIIp molecules prevents correct assembly of a viable phage particle. However, by constructing filamid vectors which express the gVIIIp fusion, and infecting cells with a helper phage overcomes the size problem. The helper phage provides normal gVIIIp and together with the fusion-gVIIIp yield infectious phage particles. As the fusion protein gene is cloned on a filamid vector, the ss DNA genome will be packaged thus physically linking the gene to the surface-displayed peptide. In general, it appears that a relatively low number of fusions are displayed on the phage surface; the predominant protein is the wild-type gVIIIp. This leads to multivalent display of a peptide sequence, with a single molecule at any particular location on the surface of the phage. Greenwood *et al.* (1991) reported obtaining between 10% and 30% of the gVIIIp-fusion sequences on phage particles. Thus, by controlling expression of the gVIIIp fusions it is possible to alter the number of displayed fusions giving flexibility on the absolute number displayed per virion.

Markland *et al.* (1991) have constructed filamid gVIIIp fusions, but have also constructed M13 phage vectors that have the gVIIIp fusion cloned into the M13 genome. Thus, no helper phage is required as the phage has both the wild-type gVIIIp and gVIIIp-fusion genes. To eliminate homology between the two copies of the gVIIIp genes, a different signal-leader sequence was used. Markland *et al.* (1991) used the bacterial alkaline phosphatase signal peptide sequence. Winter and colleagues have used this *pel*B signal sequence for their gIIIp scFv fusions (Hoogenboom *et al.*, 1991; McCafferty *et al.*, 1990).

The gVIIIp fusion display system has also been used to display scFv antibody fusions, and by utilizing either a filamid/helper phage or phage (with wild-type and fusion genes) it is possible to modulate the valency of the antibody displayed (Chappel *et al.*, 1998; Kang *et al.*, 1991).

Not to be left behind in the phage technology, workers on T-cells have displayed functional αβ single chain T-cell receptor molecules on phages (Weidanz *et al.*, 1998).

They engineered a three domain single chain T-cell receptor (scTCR), displayed it on the surface of phage as a gVIIIp fusion, and showed that it was functional.

2.4.4 cDNA display

A major drawback to the utilization of the gIIIp and gVIIIp fusion systems for the display of cDNA libraries is the fact that the fusions have to be at the amino terminus of the phage proteins. In general construction of cDNA libraries yields partial cDNA fragments, which are usually truncated in the middle of the gene's orf; the cDNAs lack the amino terminus and the initiator codon. For construction of cDNA expression libraries there is a necessity for fusion at the carboxy terminus of the fusion partner. However, the ingenuity of molecular biologists has lead to several solutions to this problem in phage display.

Crameri and Suter (1993) utilized the tight interaction of the leucine zipper regions of the proteins *jun* and *fos*. The leucine zipper domain of *jun* (jzip) was fused to the amino terminus of gIIIp such that it is displayed on the phage surface. The *fos* zipper (fzip) was cloned such that it could accept carboxy terminal in-frame cDNA inserts, onto the same filamid vector creating pJuFo. Both zipper fusions had *pel*B signal sequences so that the proteins would be exported to the periplasmic space. There the two proteins interact resulting in display of the cDNA sequences at the tip of the phage particle (*Figure 2.6*). To prevent dissociation of the zipper after release of the phage into the growth media, cysteine residues were engineered either side of the zipper regions. This results in the formation of disulphide bridges chemically linking the zippers together, and as with filamid vectors the gene encoding for the cDNA will be present inside the phage particle. The pJuFo vector can thus be used to construct cDNA expression libraries that can be simply and efficiently screened by biopanning (Crameri *et al.*, 1994). Palzkill *et al.* (1998) utilized a modified pJuFo vector to express genomic fragments of *E. coli* to analyse the complete bacterial genome against various ligands.

The filamentous phage particle also contains the minor proteins gVIp, gVIIp and gIXp. The gVIp protein has been used to express carboxy terminal fusion proteins for display on the phage surface (Fransen *et al.*, 1999; Hufton *et al.*, 1999; Jespers *et al.*, 1996). This overcomes the slightly complicated strategy employed with the pJuFo vector system.

Gao *et al.* (1999) have utilized gVIIp and gIXp to display antibody sequences. These two phage proteins are present at the tip of the phage, opposite to that of gIIIp and gVIp, and are closely associated together. As with gIIIp and gVIIIp, the fusions are at the amino terminus of the virion proteins.

The yeast 2-hybrid system (Fields and Song, 1989) is a technology for studying protein–protein interactions. At present there is no real equivalent based on *E. coli*. However, filamentous phages offer possible solutions. Gramatikoff *et al.* (1994) utilized gIIIp to study protein–protein interactions. In this system gIIIp is split into two independent genes. The amino terminal part of gIIIp, necessary for binding to the F pilus, was fused to fzip, Np3-fzip, and the carboxy terminus of gIIIp was fused to jzip, Njzip-p3C. Infectious phage particles can only be produced if the zipper partners associate. Thus, it is possible to make libraries of cDNA fused to Np3, Np3-cDNA, on a

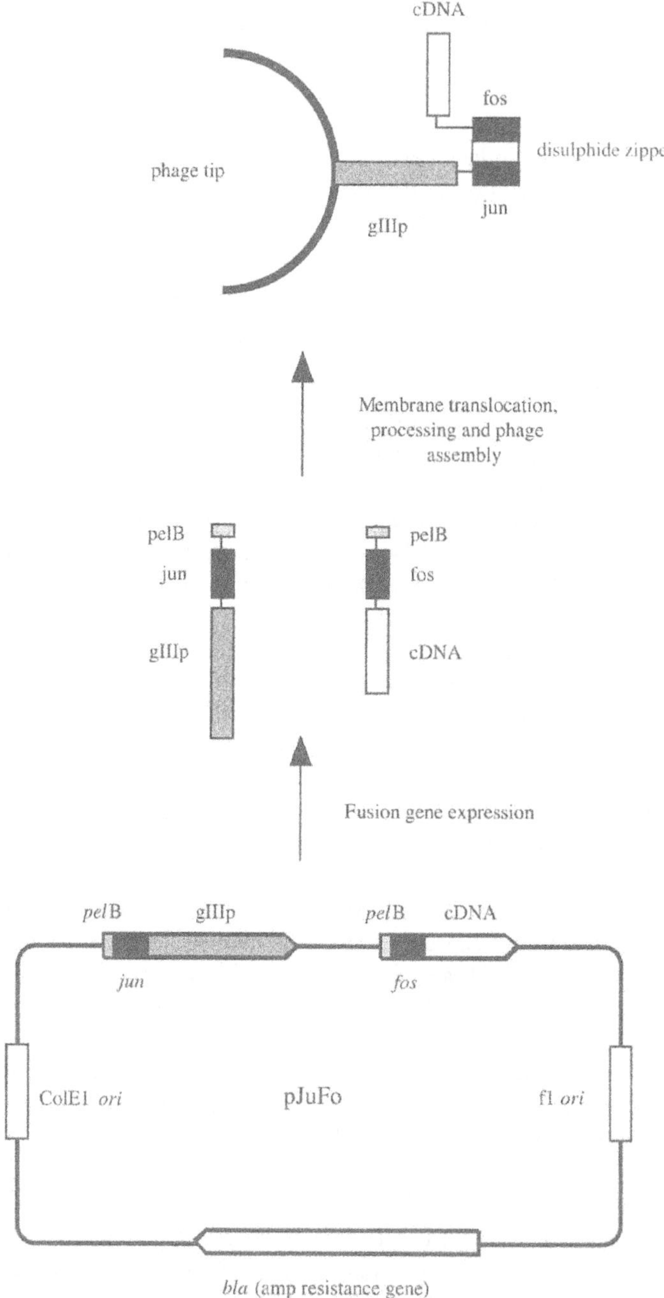

Figure 2.6. pJuFo vector system. At the bottom is a map of the vector. Expression produces a jun-gIIIp and a fos-cDNA fusions. These proteins associate in the cell membrane due to the leucine zipper activity of fos and jun. The result is display of the cDNA sequences on the surface of a phage particle.

vector expressing a bait ligand fused to p3C, bait-p3C. Only if a cDNA produces a fusion which associates with the bait will an infectious phage particle be produced. Malmborg *et al.* (1997) have developed a similar approach. They constructed fusion proteins with the *E. coli* pilin gene *tra*A, such that the fusion protein is displayed upon the pilus of the bacterial cell. Expression of 13 to 15 amino acid long peptide fusions abolishes the infection of such *E. coli* by wild-type filamentous phages. They then expressed scFv-gIIIp fusion phage which recognized the pilin-displayed peptides. These phage bound to the peptides due to antibody–peptide recognition and subsequently entered the cells. These two studies showed that it is possible to manipulate the interactions and structure of gIIIp to study protein–protein interactions.

2.4.5 λ, T4 and T7 systems

Although most of the work on peptide/protein phage display has used filamentous phages, work has commenced with the larger bacteriophages λ and T7 (Dunn, 1996b). Work has naturally been slower due to the larger genomes involved and hence the slightly more difficult DNA manipulations required to generate the display vectors. However, the successes with these phages would seem to indicate that possibly any bacteriophage could be suitably adapted for surface display of foreign peptides.

Bacteriophage λ has been modified to display peptides on its surface. The virion proteins manipulated include; the small head protein gD (Mikawa *et al.*, 1996; Sternberg and Hoess, 1995) and the pV tail protein (Dunn, 1995, 1996c; Maruyama *et al.*, 1994). Santini *et al.* (1998) have constructed an HCV cDNA expression library fused to the carboxy terminus of λ gD. However, the development of λ display for peptide or protein libraries still has to be fully exploited.

Similarly, bacteriophage T4 has been studied for surface display (Efimov *et al.*, 1995; Ren *et al.*, 1996). Ren *et al.* (1996) utilized the small outer capsid protein, SOC, for their T4 system.

Bacteriophage T7 has been developed commercially by Novagen as an alternative to filamentous phage display, employing the capsid p10 protein (Rosenberg *et al.*, 1996). Zozulya *et al.* (1999) have used the T7 system to rapidly identify protein interactions with epidermal growth factor receptor (EGFR). Due to the rapid growth and lysis of bacteria by T7, it is possible to go through three rounds of biopanning and PCR amplification of the target insert in just 2 days.

2.5 Phage gene delivery vectors

2.5.1 Gene delivery systems

In recent years there has been much research on the development of vectors for the delivery of recombinant genes to mammalian cells. At present these gene therapy or gene delivery vector (GDV) systems fall into two camps, synthetic systems and mammalian virus systems. Viruses are highly evolved vehicles for gaining entry of their genomes into host cells. However, the use of mammalian viruses, like retroviruses and adenoviruses, are not ideal in many aspects. For example, retroviruses can integrate at random into the host genome and there are size constraints on the size of target genes that can be packaged into a retrovirus particle, usually less than 7000 bp. Adenoviruses do not integrate their DNA into the host genome, and thus the target gene may not be stably maintained in the host cells. With mammalian viruses it is expensive to grow them in tissue culture systems, and there can be problems with the host immune response. There is also the possibility of replication competent recombinant viruses being inadvertently produced and escaping containment to infect other hosts.

What about bacteriophages as GDVs? The obstacle to the use of phages to deliver genes to mammalian cells is the obvious one; phages do not possess surface ligands to enabling binding to mammalian cells and subsequent internalization. Bacteriophages only infect their specific bacterial host. Also, phages have evolved to utilize bacterial gene control systems, not mammalian ones.

In 1979, Richardson and colleagues attempted to infect the western grey kangaroo (*Macropus fuliginosus*) which is deficient for galactose-1 phosphate uridyl transferase (GPUT), with a lambda phage containing the bacterial enzyme (Richardson *et al.*, 1979). Intravenously injected phage particles were cleared by the reticuloendothelial system or were inactivated by the immune system 4 days later. There was no evidence for expression of the bacterial GPUT. As lambda phage is not taken up by mammalian cells and in general bacterial transcription units are not expressed in eukaryotic cells, it is not surprising that this early attempt at phage gene therapy failed.

2.5.2 Phage transfection

Several groups of workers have utilized successfully intact phage particles of λ clones containing human genes to transfect directly into mammalian cell lines in culture (Okayama and Berg, 1985; Stambrook *et al.*, 1984). These methods effectively employed calcium phosphate co-precipitation of the phage particles onto the cells, with subsequent uptake. Ishiura *et al.* (1989) have also used packaged cosmids to introduce genes into mammalian cells.

Yokoyama-Kobayashi and Kato showed that filamentous phage carrying a urokinase expression system can be transfected into COS-7 cells in the presence of DEAE-dextran (Yokoyama-Kobayashi and Kato, 1993). They have also shown that recombinant filamentous phage, carrying the selectable marker gene blasticidin S deaminase, under control of the mouse sarcoma virus long terminal repeat (LTR) promoter, can transfect monkey COS7 cells when mixed with a lipopolyamine (Yokoyama-Kobayashi and Kato, 1994). They compared phage particles with naked double stranded DNA, and found they required 20-fold less phage particles to obtain similar numbers of blasticidin S-resistant cells. This effect is presumably due to degradation of the naked ds DNA by nucleases in the culture media, thus requiring large amounts of ds DNA, whereas the ss DNA is fully protected within the phage particle. Yokoyama-Kobayashi and Kato's experiments clearly showed that filamentous phage particles could be used to transfect ss DNA into mammalian cells. How the phage particles enter the nucleus, how the ss DNA is released from the phage and how it is converted to ds DNA is not understood at present. However, the net result was that phage particles were able to gain entry, and once inside the cell expressed the encoded mammalian gene.

The above studies effectively employed techniques normally used to transfect naked DNA into cells. The advantage they possess is the fact the DNA in the phage particles is protected from degradation in the culture medium.

2.5.3 Phage targeting

Hart and colleagues have shown that expression of the integrin binding sequence, Arg-Gly-Asp (RGD) on the surface of filamentous bacteriophages, allows the cell binding and internalization of the phage particles (Hart *et al.*, 1994). The peptide was expressed as a fusion with the phage gVIIIp protein; approximately 10% of the 3000 copies of the major phage coat protein were the RGD-fusion protein. The RGD sequence was displayed in a restricted loop conformation on the surface of the phage due to two flanking cysteine residues. The phage DNA did not contain a suitable marker gene, so it was not possible to assess whether the ss DNA genome once inside the cell could have been expressed. Dunn (1996a) has used a cyclized RGD sequence displayed on the tail tube major subunit (gVp) proteins of bacteriophage λ to target mammalian cells.

This concept has been taken further by Larocca *et al.* (1998, 1999). In 1998 they reported the targeted delivery of a filamentous phage particle to COS-1 cells. They chemically attached mammalian targeting proteins to the surface of phage particles containing a filamid vector containing a reporter gene. The reporter genes encoded were either β-galactosidase or green fluorescent protein (GFP), both controlled by the cytomegalovirus (CMV) immediate early gene promoter. The targeting ligands were basic fibroblast growth factor (FGF2), transferrin and epidermal growth factor (EGF), and the best targeting was seen with the FGF2 ligand. These studies were taken a step further by cloning the FGF2 coding sequences in frame with the M13 *gIIIp* gene in a recombinant M13 vector (Larocca *et al.*, 1999), resulting in display of FGF2 sequences on the surface of the phage particles. These phages were able to transfect COS-1 cells and express a phage-encoded GFP reporter gene.

Larocca and colleagues have recently shown that it is possible to select for targeting specific phages by transfection of mammalian cells with phage displaying libraries (Kassner *et al.*, 1999). They demonstrated that EGF-phage could be selected from libraries of peptide display phage by their ability to transduce EGF receptor bearing COS-1 cells. The EGF-displaying phages were enriched 1 million-fold by four rounds of selection. Their data clearly indicate the possibility of applying phage gene delivery to select novel cell-specific DNA-targeting ligands and also to select for genetically altered phage that are specifically designed and evolved for use as gene therapy vectors.

The above studies have shown that it is feasible to develop bacteriophage particles that display on their surface mammalian targeting sequences, which result in uptake of the virions, and subsequent expression of encoded marker genes. Although the actual mechanism whereby the particles are taken up and how the DNA reaches the nucleus remains to be elucidated, the proof of principle has been established.

Bacteriophage GDV particles thus offer the potential for development into successful gene therapy vehicles. Phages can be propagated in simple cheap media, can be grown in extremely large quantities, and can be purified to homogeneity. Costs for large scale production will be much less than those necessary for tissue culture production of mammalian viral GDVs, such as retroviruses or adenoviruses. The size of foreign DNA accommodated by phage GDVs is much larger than that possible with mammalian viral systems. Bacteriophage λ can accommodate large inserts of foreign DNA, up to 20 kbp. The development of an *in vivo* λ packaging system in *E. coli* for cosmids would allow for human genomic DNA of 45 kbp to be packaged into a phage-GDV. Extending this to include a P1/T4 bacteriophage packaging system could allow the delivery of ~100 kbp of mammalian DNA. It will thus be possible to include large stretches of upstream and downstream sequences flanking the gene of interest, potentially containing all of the natural regulatory sequences necessary for regulated expression of the target gene.

Due to the fact that the targeting display sequences are encoded by prokaryotic gene constructs means that the genes will not be expressed in eukaryotic cells. The GDV is thus a single hit vector; it can get into a mammalian cell and deliver its payload of DNA, but will be unable to replicate and leave the cell. There exists the possibility that such recombinant phages could infect normal gut flora *E. coli*. Replication and production of phage GDVs can be prevented by incorporating non-sense termination codons in phage replicative and targeting genes, such that the phage will only be able to grow on laboratory suppresser strains of *E. coli*. The construction of *in vivo* packaging bacterial strains that only require *cis* packaging sequences on the vector DNA, with all the virion capsid proteins provided in *trans*, would also eliminate the possibility of further GDV production in bacterial gut flora.

The potential is here to exploit bacteriophage GDVs for specific cell targeting with any recombinant gene. When the problems of inefficient homologous recombination of GDV DNA into host chromosomes are solved, with subsequent natural regulation of the introduced genes, a near perfect GDV system will be available to the medical profession. Until that time, there is plenty to do; there is an abundance of bacteriophages to exploit apart from M13 and λ.

2.6 Bacteriophages as antibiotics

Briefly reported in this section is the use of bacteriophages as 'antibiotics' to kill bacteria causing infections in man and other animals (Carlton, 1999). This does not quite come under the topics of this book, namely genetically engineered viruses, as virtually all the work has used naturally occurring bacteriophages or culture-adapted phages. However, the potential to engineer bacteriophages to improve their bacterial killing potential remains an exciting future growth area, given the emergence of bacteria that are resistant to all the traditional antibiotics available to the medical profession (Lederberg, 1996). Much of this work has been performed by workers in Russia and Poland, probably due to the expense of manufacturing antibiotics compared with the cheaper methods to produce bacteriophages.

For example, Slopek and colleagues treated 138 patients with bacteriophages and good results were obtained in 129 (Slopek *et al.*, 1983). Between 1981 and 1986 they treated 550 cases of suppurative bacterial infections, and in only four cases was the treatment considered to be ineffective (Slopek *et al.*, 1987). They have also studied the immunological consequences of phage treatment (Kucharewicz-Krukowska and Slopek, 1987; Weber-Dabrowska *et al.*, 1987).

In Britain, Smith and Huggins (1982, 1983) have carried out studies on the use of phages in systemic *E. coli* infections in mice and diarrhoeal disease in young calves. They found that phage could protect mice from lethal doses of *E. coli*, whereas all mice died who were just given the pathogenic bacteria. Also, in a study in calves with diarrhoea, they found the phage were able to reduce the severity of the infection and prevent death (Smith *et al.*, 1987). Barrow and Soothill (1997) have recently reviewed current work in the West.

The following points should be borne in mind when considering phage therapy (Kutter, 1997).

(i) The phages will replicate in the bacterial hosts, but when all the host cells have been killed, then the phages will gradually be eliminated, thus they are self-replicating and self-limiting.
(ii) There have been few reported cases of side effects.
(iii) They can be targeted more specifically to particular bacteria, more so than conventional antibiotics. They can easily be manipulated in the laboratory to develop new activities by selection on various bacterial hosts, as well as being genetically modified *in vitro* using genetic engineering.
(iv) They could be used by people with allergies to conventional antibiotics.
(v) Phage therapy could be used in conjunction with antibiotics.
(vi) Phages could be used prophylactically to help prevent spread of bacterial disease.

2.7 Conclusion

In conclusion, although in recent years work on bacteriophages has not been seen as 'trendy' or the 'in-thing' to do, there is clearly great potential in the development of phage vector systems for efficacy in many aspects of disease. The production of engineered phages for medical use, whether it be for production of monoclonal antibodies, gene therapy or a new generation of 'antibiotics' will bring about a renewed interest in these 'filterable entities'.

References

Allday MJ, Jones MD. (1987) Rapid processing of nitrocellulose filter lifts of bacteriophage lambda libraries. *Nucleic Acids Research* **15:** 10592.

Alting-Mees MA, Short JM. (1989) pBluescript II: gene mapping vectors. *Nucleic Acids Res.* **17:** 9494.

Alting-Mees MA, Short JM. (1993) Polycos vectors: a system for packaging filamentous phage and phagemid vectors using lambda phage packaging extracts. *Gene* **137:** 93–100.

Alting-Mees MA, Sorge JA, Short JM. (1992) pBluescriptII: Multifunctional Cloning and Mapping Vectors. *Methods Enzymol.* **216:** 483–495.

Andersson B, Povinelli CM, Wentland MA, Shen Y, Muzny DM, Gibbs RA. (1994) Adaptor-based Uracil DNA Glycosylase Cloning simplifies Shotgun Library Construction for Large-scale Sequencing. *Anal. Biochem.* **218:** 300–308.

Atwell S, Wells JA. (1999) Selection for improved subtiligases by phage display. *Proc. Natl. Acad. Sci. USA* **96:** 9497–9502.

Bankier AT, Barrell BG. (1989) Sequencing single-stranded DNA using the chain-termination method. In: *Nucleic Acids Sequencing: a Practical Approach.* (eds CJ Howe and ES Ward), pp 37–78. Oxford University Press, Oxford.

Bankier AT, Weston KM, Barrell BG. (1987) Random cloning and sequencing by the M13/dideoxynucleotide chain termination method. *Methods Enzymol.* **155:** 51–93.

Barbas CF, Kang AS, Lerner RA, Benkovic SJ. (1991) Assembly of combinatorial antibody libraries on phage surfaces: the gene III site. *Proc. Natl. Acad. Sci. USA* **88:** 7978–7982.

Barnes WM, Bevan M. (1983) Kilo-sequencing: an ordered strategy for rapid DNA sequence data. *Nucleic Acids Res.* **11:** 349–368.

Barrow PA, Soothill JS. (1997) Bacteriophage therapy and prophylaxis: rediscovery and renewed assessment of the potential. *Trends Microbiol.* **5:** 268–271.

Beck E, Sommer R, Auerswald EA, Kurz C, Zink B, Osterburg G, Schaller H, Sugimoto K, Sugisaki H, Okamoto T, Takanami M. (1978) Nucleotide sequence of bacteriophage fd DNA. *Nucleic Acids Res.* **5:** 4495–4503.

Becker A, Gold M. (1975) Isolation of the bacteriophage lambda A-gene protein. *Proc. Natl. Acad. Sci. USA* **72:** 581–585.

Benton WD, Davis RW. (1977) Screening λgt recombinant clones by hybridization to single plaques *in situ*. *Science* **196:** 180–182.

Blattner FR, Williams BG, Blechl AE *et al.* (1977) Charon phages: safer derivatives of bacteriophage lambda for DNA cloning. *Science* **196:** 161–169.

Blattner FR, Plunkett G, Bloch CA *et al.* (1997) The complete genome sequence of *Escherichia coli* K-12. *Science* **277:** 1453–1474.

Bult CJ, White O, Olsen GJ *et al.* (1996) Complete genome sequence of the methanogenic archaeon, *Methanococcus jannaschii*. *Science* **273:** 1058–1073.

Burland V, Daniels DL, Plunkett G, Blattner FR. (1993) Genome sequencing on both strands: the Janus strategy. *Nucleic Acids Res.* **21:** 3385–3390.

Burmeister M, Lehrach H. (1988) Jekyll, a family of phage-plasmid shuttle vectors. *Gene* **73:** 245–250.

Carlton RM. (1999) Phage therapy: past history and future prospects. *Arch. Immunol. Ther. Exp. (Warsz)* **47:** 267–274.

Chappel JA, He M, Kang AS. (1998) Modulation of antibody display on M13 filamentous phage. *J. Immunol. Methods* **221:** 25–34.

Chatellier J, Hartley O, Griffiths AD, Fersht AR, Winter G, Riechmann L. (1999) Interdomain interactions within the gene 3 protein of filamentous phage. *FEBS Lett.* **463:** 371–374.

Chissoe SL, Wang Y-F, Clifton SW *et al.* (1991) Strategies for Rapid and Accurate DNA Sequencing. *Methods: Comp. to Meth. Enzymol.* **3:** 55–65.

Chowdhury PS, Pastan I. (1999) Analysis of cloned Fvs from a phage display library indicates that DNA immunization can mimic antibody response generated by cell immunizations. *J. Immunol. Methods* **231:** 83–91.

Clackson T, Hoogenboom HR, Griffiths AD, Winter G. (1991) Making antibody fragments using phage display libraries. *Nature* **352:** 624–628.

Corey DR, Shiau AK, Yang Q, Janowski BA, Craik CS. (1993) Trypsin display on the surface of bacteriophage. *Gene* **128:** 129–134.

Coulson A, Waterston R, Kiff J, Sulston J, Kohara Y. (1988) Genome linking with yeast artificial chromosomes. *Nature* **335:** 184–186.

Crameri R, Suter M. (1993) Display of biologically active proteins on the surface of filamentous phages: a cDNA cloning system for selection of functional gene products linked to the genetic information responsible for their production. *Gene* **137:** 69–75.

Crameri R, Jaussi R, Menz G, Blaser K. (1994) Display of expression products of cDNA libraries on phage surfaces. A versatile screening system for selective isolation of genes by specific gene-product/ligand interaction. *Eur. J. Biochem.* **226:** 53–58.

Cwirla SE, Peters EA, Barrett RW, Dower WJ. (1990) Peptides on phage: A vast library of peptides for identifying ligands. *Proc. Natl. Acad. Sci. USA* **87:** 6378–6382.

Davies CJ, Hutchison III CA. (1991) A directed DNA sequencing strategy based upon Tn3 transposon mutagenesis: application to the ADE1 locus on *Saccharomyces cerevisiae* chromosome I. *Nucleic Acids Res.* **19:** 5731–5738.

Deininger PL. (1983) Random subcloning of sonicated DNA: application to shotgun DNA sequence analysis. *Anal. Biochem.* **129:** 216–223.

Demartis S, Huber A, Viti F *et al.* (1999) A strategy for the isolation of catalytic activities from repertoires of enzymes displayed on phage. *J. Mol. Biol.* **286:** 617–633.

Deng LW, Malik P, Perham RN. (1999) Interaction of the globular domains of pIII protein of filamentous bacteriophage fd with the F-pilus of *Escherichia coli. Virology* **253:** 271–277.

Denhardt DT, Dressler D, Ray DS. (eds) (1978) *The Single-Stranded DNA Phages.* Cold Spring Harbor Laboratory, New York.

Dente L, Cesareni G, Cortese R. (1983) pEMBL: a new family of single stranded plasmids. *Nucleic Acids Res.* **11:** 1645–1655.

Devlin JJ, Panganiban LC, Devlin PE. (1990) Random peptide libraries: a source of specific protein binding molecules. *Science* **249:** 404–406.

D'Herelle F. (1917) Sur un microbe invisible antagoniste des bac. Dysent Èriques. *Compt. Rend. Acad. Sci. Paris* **165:** 373.

D'Herelle F, Twort FW, Bordet J, Gratia A. (1922) Discussion on the Bacteriophage (Bacteriolysin). From the Ninetieth Annual Meeting of the British Medical Association, Glasgow, July, 1922. *British Medical Journal* **2:** 289–297.

Dotto GP, Horiuchi K. (1981) Replication of a plasmid containing two origins of bacteriophage. *J. Mol. Biol.* **153:** 169–176.

Dotto GP, Enea V, Zinder ND. (1981) Functional analysis of bacteriophage f1 intergenic region. *Virology* **114:** 463–473.

Dunham I, Shimizu N, Roe BA *et al.* (1999) The DNA sequence of human chromosome 22. *Nature* **402:** 489–495.

Dunn IS. (1995) Assembly of functional bacteriophage lambda virions incorporating C-terminal peptide or protein fusions with the major tail protein. *J. Mol. Biol.* **248:** 497–506.

Dunn IS. (1996a) Mammalian cell binding and transfection mediated by surface-modified bacteriophage lambda. *Biochimie* **78:** 856–861.

Dunn IS. (1996b) Phage display of proteins. *Curr. Opin. Biotechnol.* **7:** 547–553.

Dunn IS. (1996c) Total modification of the bacteriophage lambda tail tube major subunit protein with foreign peptides. *Gene* **183:** 15–21.

Efimov VP, Nepulev IV, Mesyanzhinov VV. (1995) Bacteriophage T4 as a surface display vector. *Virus Genes* **10:** 172–177.

Felici F, Castagnoli L, Musacchio A, Jappelli R, Cesareni G. (1991) Selection of antibody ligands from a large library of oligopeptides expressed on a multivalent exposition vector. *J. Mol. Biol.* **222:** 301–310.

Fields S, Song O. (1989) A novel genetic system to detect protein–protein interactions. *Nature* **340:** 245–246.

Fleischmann RD, Adams MD, White, O *et al.* (1995) Whole-genome random sequencing and assembly of *Haemophilus influenzae* Rd. *Science* **269:** 496–512.

Fransen M, Van-Veldhoven PP, Subramani S. (1999) Identification of peroxisomal proteins by using M13 phage protein VI phage display: molecular evidence that mammalian peroxisomes contain a 2,4-dienoyl-CoA reductase. *Biochem. J.* **340:** 561–568.

Frischauf AM, Lehrach H, Poustka A, Murray N. (1983) Lambda replacement vectors carrying polylinker sequences. *J. Mol. Biol.* **170:** 827–842.

Gao C, Mao S, Lo CH, Wirsching P, Lerner RA, Janda KD. (1999) Making artificial antibodies: a format for phage display of combinatorial heterodimeric arrays. *Proc. Natl. Acad. Sci. USA* **96:** 6025–6030.

Gottesman M. (1999) Bacteriophage lambda: the untold story. *J. Mol. Biol.* **293:** 177–180.

Gramatikoff K, Georgiev O, Schaffner W. (1994) Direct interaction rescue, a novel filamentous phage technique to study protein–protein interactions. *Nucleic Acids Res.* **22:** 5761–5762.

Greenwood J, Willis AE, Perham RN. (1991) Multiple display of foreign peptides on a filamentous bacteriophage. Peptides from *Plasmodium falciparum* Circumsporozoite protein as antigens. *J. Mol. Biol.* **220:** 821–827.

Griffiths AD, Duncan AR. (1998) Strategies for selection of antibodies by phage display. *Curr. Opin. Biotechnol.* **9:** 102–108.

Gronenborn B, Messing J. (1978) Methylation of single-stranded DNA in vitro introduces new restriction endonuclease cleavage sites. *Nature* **272:** 375–377.

Hammer J, Takacs B, Sinigaglia F. (1992) Identification of a motif for HLA-DR1 binding peptides using M13 display libraries. *J. Exp. Med.* **176:** 1007–1013.

Hammer J, Valsasnini P, Tolba K, Bolin D, Higelin J, Takacs B, Sinigaglia F. (1993) Promiscuous and allele-specific anchors in HLA-DR-binding peptides. *Cell* **74:** 197–203.

Hart SL, Knight AM, Harbottle RP *et al.* (1994) Cell binding and internalization by filamentous phage displaying a cyclic Arg-Gly-Asp-containing peptide. *J. Biol. Chem.* **269:** 12468–12474.

Hattori M, Fujiyama A, Taylor TD *et al.* (2000) The DNA sequence of human chromosome 21. *Nature* **405:** 311–319.

Hendrix RW, Roberts JW, Stahl FW, Weisberg RA (eds.) (1983) *Lambda II.* Cold Spring Harbor Laboratory, New York.

Hill DF, Petersen GB. (1982) Nucleotide sequence of bacteriophage f1 DNA. *J. Virol.* **44:** 32–46.

Hoess R, Brinkmann U, Handel T, Pastan I. (1993) Identification of a peptide which binds to the carbohydrate-specific monoclonal antibody B3. *Gene* **128:** 43–49.

Hofschneider PH. (1963) Untersuchungen über "kleine" *E. coli* Bakteriophagen 1. and 2. Mitteilung. *Z. Naturforsch.* **18b:** 203.

Hogrefe HH, Amberg JR, Hay BN, Sorge JA, Shopes B. (1993) Cloning in a bacteriophage lambda vector for the display of binding proteins on filamentous phage. *Gene* **137:** 85–91.

Hohn B, Hohn T. (1974) Activity of empty, headlike particles for packaging of DNA of bacteriophage lambda in vitro. *Proc. Natl. Acad. Sci. USA* **71:** 2372–2376.

Holliger P, Riechmann L, Williams RL. (1999) Crystal structure of the two N-terminal domains of g3p from filamentous phage fd at 1.9 A: evidence for conformational lability. *J. Mol. Biol.* **288:** 649–657.

Hoogenboom HR, Winter G. (1992) By-passing immunisation. Human antibodies from synthetic repertoires of germline VH gene segments rearranged in vitro. *J. Mol. Biol.* **227:** 381–388.

Hoogenboom HR, Griffiths AD, Johnson KS, Chiswell DJ, Hudson P, Winter G. (1991) Multi-subunit proteins on the surface of filamentous phage: methodologies for displaying antibody (Fab) heavy and light chains. *Nucleic Acids Res.* **19:** 4133–4137.

Hufton SE, Moerkerk PT, Meulemans EV, de Bruine A, Arends JW, Hoogenboom HR. (1999) Phage display of cDNA repertoires: the pVI display system and its applications for the selection of immunogenic ligands. *J. Immunol. Methods* **231:** 39–51.

Huynh TV, Young RA, Davis RW. (1985) Constructing and screening cDNA libraries in λgt10 and λgt11. In: *DNA Cloning Volume I, a practical approach.* (DM Glover, ed.), pp. 49–78, IRL Press, Oxford.

Ishiura M, Ohashi H, Uchida T, Okada Y. (1989) Phage particle-mediated gene transfer of recombinant cosmids to cultured mammalian cells. *Gene* **82:** 281–289.

Jespers LS, De-Keyser A, Stanssens PE. (1996) LambdaZLG6: a phage lambda vector for high-efficiency cloning and surface expression of cDNA libraries on filamentous phage. *Gene* **173:** 179–181.

Jones MD. (1996) Manual and automated DNA sequencing. In: *Clinical Gene Analysis and Manipulation.* (eds JAZ Jankowski, JM Polak), pp 303–331, Cambridge University Press, Cambridge.

Kang AS, Barbas CF, Janda KK, Benkovic SJ, Lewrner RA. (1991) Linkage of recognition and replication functions by assembling combinatorial antibody Fab libraries along phage surfaces. *Proc. Natl. Acad. Sci. USA* **88:** 4363–4366.

Karn J, Matthes HWD, Gait MJ, Brenner S. (1984) A new selective phage cloning vector, lambda 2001, with sites for *Xba*I, *Bam*HI, *Hind*III, *Eco*RI, *Sst*I *and Xho*I. *Gene* **32:** 217–224.

Kassner PD, Burg MA, Baird A, Larocca D. (1999) Genetic selection of phage engineered for receptor-mediated gene transfer to mammalian cells. *Biochem. Biophys. Res. Comm.* **264:** 921–928.

Kieny MP, Lathe RF, Lecocq JP. (1983) New versatile cloning and sequencing vectors based on bacteriophage M13. *Gene* **26:** 91–99.

Koivunen E, Wang B, Ruoslahti E. (1995) Phage libraries displaying cyclic peptides with different ring sizes: ligand specificities of the RGD-directed integrins. *Bio/Technology* **13:** 265–270.

Kucharewicz-Krukowska A, Slopek S. (1987) Immunogenic effect of bacteriophage in patients subjected to phage therapy. *Arch. Immunol. Ther. Exp. (Warsz)* **37:** 553–561.

Kutter E. (1997) Phage therapy: bacteriophages as antibiotics. Web page address: http://www.evergreen.edu/user/T4/PhageTherapy/Phagethea.html

Ladner RC. (1999) Polypeptides from phage display. A superior source of *in vivo* imaging agents. *Quart. J. Nuclear Med.* **43:** 119–124.

Larocca D, Witte A, Johnson W, Pierce GF, Baird A. (1998) Targeting bacteriophage to mammalian cell surface receptors for gene delivery. *Hum. Gene Therapy* **9:** 2393–2399.

Larocca D, Kassner PD, Witte A, Ladner RC, Pierce GF, Baird A. (1999) Gene transfer to mammalian cells using genetically targeted filamentous bacteriophage. *FASEB J.* **13:** 727–734.

Lederberg EM. (1951) Lysogenicity in *E. coli* K-12. *Genetics* **36:** 560.

Lederberg J. (1996) Smaller fleas ... ad infinitum: therapeutic bacteriophage redux. *Proc. Natl. Acad. Sci. USA* **93:** 3167–3168.

Little M, Welschof M, Braunagel M *et al.* (1999) Generation of a large complex antibody library from multiple donors. *J. Immunol. Methods* **231:** 3–9.

Loeb T. (1960) Isolation of a bacteriophage specific for the F^+ and Hfr mating types of *Escherichia coli* K12. *Science* **131:** 932.

Luzzago A, Felici F, Tramontano A, Pessi A, Cortese R. (1993) Mimicking of discontinuous epitopes by phage-displayed peptides, I. Epitope mapping of human ferritin using a phage library of constrained peptides. *Gene* **128:** 51–57.

Malmborg A-C, Söderlind E, Frost L, Borrebaeck CAK. (1997) Selective phage infection mediated by epitope expression on F pilus. *J. Mol. Biol.* **273:** 544–551.

Markland W, Roberts BL, Saxena MJ, Guterman SK, Ladner RC. (1991) Design, construction of a multicopy display vector using fusions to the major coat protein of bacteriophage M13. *Gene* **109:** 13–19.

Marks JD, Hoogenboom HR, Bonnert TP, McCafferty J, Griffiths AD, Winter G. (1991) By-passing immunization. Human antibodies from V-gene libraries displayed on phage. *J. Mol. Biol.* **222:** 581–597.

Marks JD, Hoogenboom HR, Griffiths, AD, Winter G. (1992) Molecular evolution of proteins on filamentous phage. Mimicking the strategy of the immune system. *J. Biol. Chem.* **267:** 16007–16010.

Marvin DA, Hoffman-Berling H. (1963) Physical and chemical properties of two new small bacteriophages. *Nature* **197:** 517.

Maruyama IN, Maruyama HI, Brenner S. (1994) Lambda foo: a lambda phage vector for the expression of foreign proteins. *Proc. Natl. Acad. Sci. USA* **91:** 8273–8277.

Matthews DJ, Wells JA. (1993) Substrate phage: selection of protease substrates by monovalent phage display. *Science* **260:** 1113–1117.

McCafferty J, Griffiths AD, Winter G, Chiswell DJ. (1990) Phage antibodies: filamentous phage displaying antibody variable domains. *Nature* **348:** 552–554.

McLafferty MA, Kent RB, Ladner RC, Markland W. (1993) M13 bacteriophage displaying disulfide-constrained microproteins. *Gene* **128:** 29–36.

Messing J. (1983) New M13 vectors for cloning. *Methods Enzymol.* **101:** 20–78.

Messing J. (1991) Cloning in M13 phage or how to use biology at its best. *Gene* **100:** 3–12.

Messing J, Bankier AT. (1989) The use of single-stranded phage in DNA sequencing. In: (eds. C.J. Howe and E.S. Ward), *Nucleic Acids Sequencing: a Practical Approach,* pp 1–36, Oxford University Press, Oxford.

Messing J, Vieira J. (1982) A new pair of M13 vectors for selecting either strand of a double-digest restriction fragment. *Gene* **19:** 269–276.

Messing J, Gronenborn B, Müller-Hill B, Hofschneider PH. (1977) Filamentous coliphage M13 as a cloning vehicle: insertion of a *Hin*dII fragment of the *lac* regulatory region in the M13 replicative form in vitro. *Proc. Natl. Acad. Sci. USA* **74:** 3642–3646.

Messing J, Crea R, Seeburg PH. (1981) A system for shotgun DNA sequencing. *Nucleic Acids Res.* **9:** 309–321.

Mikawa YG, Maruyama IN, Brenner S. (1996) Surface display of proteins on bacteriophage lambda heads. *J. Mol. Biol.* **262:** 21–30.

Myers EW, Sutton GG, Delcher AL *et al.* (2000) A whole-genome assembly of *Drosophila. Science* **287:** 2196–2204.

Norrander J, Kempe T, Messing J. (1983) Construction of improved M13 vectors by oligonucleotide-directed mutagenesis. *Gene* **26:** 101–106.

Okayama H, Berg P. (1985) Bacteriophage lambda vector for transducing a cDNA clone library into mammalian cells. *Mol. Cell. Biol.* **5:** 1136–1142.

Oldenburg KR, Loganathan D, Goldstein IJ, Schultz PG, Gallop MA. (1992) Peptide ligands for a sugar-binding protein isolated from a random peptide library. *Proc. Natl. Acad. Sci. USA* **89:** 5393–5397.

Palzkill T, Huang W, Weinstock GM. (1998) Mapping protein–ligand interactions using whole genome phage display libraries. *Gene* **221:** 79–83.

Parmley SF, Smith GP. (1988) Antibody-selectable filamentous fd phage vectors: affinity purification of target genes. *Gene* **73:** 305–318.

Pasqualini R. (1999) Vascular targeting with phage peptide libraries. *Quart. J. Nuclear Med.* **43:** 159–162.

Ren ZJ, Lewis GK, Wingfield PT, Lockem EG, Steven AC, Black LW. (1996) Phage display of intact domains at high copy number: a system based on SOC, the small outer capsid protein of bacteriophage T4. *Protein Sci.* **5:** 1833–1843.

Richardson BJ, Inglis B, Poole WE, Rolfe B. (1979) Galactose-1 phosphate uridyl transferase deficiency in the western grey kangaroo (*Macropus fuliginosus*; marsupialia): a model system for gene therapy studies. *Aust. J. Exp. Biol. Med. Sci.* **57:** 43–49.

Roberts BL, Markland W, Siranosian K, Saxena MJ, Guterman SK, Ladner RC. (1992) Protease inhibitor display M13 phage: selection of high-affinity neutrophil elastase inhibitors. *Gene* **121:** 9–15.

Rosenberg A, Griffin G, Studier FW, McCormick M, Berg J, Mierendorf R. (1996) T7Select phage display system: a powerful new protein display system based on bacteriophage T7. *inNovations* **6:** 1–6.

Sanger F, Nicklen S, Coulson AR. (1977) DNA sequencing with chain-terminating inhibitors. *Proc. Natl. Acad. Sci. USA* **74:** 5463–5467.

Sanger F, Coulson AR, Barrell BG, Smith AJ, Roe BA. (1980) Cloning in single-stranded bacteriophage as an aid to rapid DNA sequencing. *J. Mol. Biol.* **143:** 161–178.

Sanger F, Coulson AR, Hong GF, Hill DF, Petersen GB. (1982) Nucleotide sequence of bacteriophage lambda DNA. *J. Mol. Biol.* **162:** 729–773.

Santini C, Brennan D, Mennuni C, Hoess RH, Nicosia A, Cortese R, Luzzago A. (1998) Efficient display of an HCV cDNA expression library as C-terminal fusion to the capsid protein D of bacteriophage lambda. *J Mol. Biol.* **282:** 125–135.

Schreier PH, Cortese R. (1979) A fast and simple method for sequencing DNA cloned in the single-stranded bacteriophage M13. *J. Mol. Biol.* **129:** 169–172.

Scott JK, Smith GP. (1990) Searching for peptide ligands with an epitope library. *Science* **249:** 386–390.

Shizuya H, Birren B, Kim UJ, Mancino V, Slepak T, Tachiiri Y, Simon M. (1992) Cloning and stable maintenance of 300-kilobase-pair fragments of human DNA in *Escherichia coli* using an F-factor-based vector. *Proc. Natl. Acad. Sci. USA* **89:** 8794–8797.

Short JM, Sorge J. (1992) *In vivo* excision properties of bacteriophage λ ZAP expression vectors. *Methods Enzymol.* **216:** 495–508.

Short JM, Fernandez JM, Sorge J, Huse WD. (1988) Lambda ZAP: a bacteriophage lambda expression vector with in vivo excision properties. *Nucleic Acids Res.* **16:** 7583–7600.

Slopek S, Durlakowa I, Weber-Dabrowska B, Kucharewicz-Krukowska A, Dabrowski M, Bisikiewicz R. (1983) Results of bacteriophage treatment of suppurative bacterial infections. I. General evaluation of the results. *Arch. Immunol. Ther. Exp. (Warsz)* **31:** 267–291.

Slopek S, Weber-Dabrowska B, Dabrowski M, Kucharewicz-Krukowska A. (1987) Results of bacteriophage treatment of suppurative bacterial infections in the years 1981–1986. *Arch. Immunol. Ther. Exp. (Warsz)* **35:** 569–583.

Smith GP. (1985) Filamentous fusion phage: novel expression vectors that display cloned antigens on the virion surface. *Science* **228:** 1315–1317.

Smith HW, Huggins RB. (1982) Successful treatment of experimental *E. coli* infections in mice using phage: its general superiority over antibiotics. *J. Gen. Microbiology* **128:** 307–318.

Smith HW, Huggins RB. (1983) Effectiveness of phages in treating experimental *E. coli* Diarrhoea in calves, piglets and lambs. *J. Gen. Microbiology* **129:** 2659–2675.

Smith HW, Huggins RB, Shaw KM. (1987) The control of experimental *E. coli* diarrhea in calves by means of bacteriophage. *J. Gen. Microbiology* **133:** 1111–1126.

Stambrook PJ, Dush MK, Trill JJ, Tischfield JA. (1984) Cloning of a functional human adenine phosphoribosyltransferase (APRT) gene: identification of a restriction fragment length polymorphism and preliminary analysis of DNAs from APRT-deficient families and cell mutants. *Som. Cell Mol. Gen.* **10:** 359–367.

Stephen CW, Lane DP. (1992) Mutant conformations of p53. Precise epitope mapping using a filamentous phage epitope library. *J. Mol. Biol.* **225:** 577–583.

Sternberg N, Hoess RH. (1995) Display of peptides and proteins on the surface of bacteriophage lambda. *Proc. Natl. Acad. Sci. USA* **92:** 1609–1613.

Twort FW. (1915) An investigation on the nature of ultramicroscopic viruses. *Lancet* **II:** 1241.

van Wezenbeck PMGF, Hulsebos TJM, Schoenmakers JGG. (1980) Nucleotide sequence of the filamentous bacteriophage M13 DNA genome: comparison with phage fd. *Gene* **11:** 129–148.

Vieira J, Messing J. (1982) The pUC plasmids, an M13mp7 derived system for insertion mutagenesis and sequencing with synthetic universal primers. *Gene* **19:** 259–268.

Vieira J, Messing J. (1988) Production of single-stranded plasmid DNA. *Methods Enzymol.* **153:** 3–11.

Watson JD. (1970) *The Double Helix.* Penguin Books, London.

Weber-Dabrowska B, Dabrowski M, Slopek S. (1987) Studies on bacteriophage penetration in patients subjected to phage therapy. *Arch. Immunol. Ther. Exp. (Warsz)* **35:** 563–568.

Weidanz JA, Card KF, Edwards A, Perlstein E, Wong HC. (1998) Display of functional alphabeta single-chain T-cell receptor molecules on the surface of bacteriophage. *J. Immunol. Methods* **221:** 59–76.

Yanisch-Perron C, Vieira J, Messing J. (1985) Improved M13 phage cloning vectors and host strains: nucleotide sequences of the M13mp18 and pUC19 vectors. *Gene* **33:** 103–119.

Yip YL, Hawkins NJ, Smith G, Ward RL. (1999) Biodistribution of filamentous phage-Fab in nude mice. *J. Imm. Methods* **225:** 171–178.

Yokoyama-Kobayashi M, Kato S. (1993) Recombinant f1 phage particles can transfect monkey COS-7 cells by DEAE dextran method. *Biochem. Biophys. Res. Comm.* **192:** 935–939.

Yokoyama-Kobayashi M, Kato S. (1994) Recombinant f1 phage-mediated transfection of mammalian cells using lipopolyamine technique. *Anal. Biochem.* **223:** 130–134.

Zinder ND, Valentine RC, Roger M, Stoeckenius W. (1963) f1, a rod-shaped male-specific bacteriophage that contains DNA. *Virology* **20:** 638.

Zoller M, Smith M. (1983) Oligonucleotide-directed mutagenesis of DNA fragments cloned into M13 vectors. *Methods Enzymol.* **100:** 468–500.

Zozulya S, Lioubin M, Hill RJ, Abram C, Gishizky ML. (1999) Mapping signal transduction pathways by phage display. *Nature Biotechnol.* **17:** 1193–1198.

Chapter 3

Insect viruses

Ian M. Jones

3.1 Introduction

3.1.1 Viruses as expression vectors

Viruses use the protein expression machinery of the host cell in order to produce progeny virions. With the exception of viral latency, most viruses accomplish this in a relatively short period of time and increase virus numbers several thousand fold in the process. Generally speaking, viruses are also efficient in their use of the genetic information required to complete the life cycle. The sequences of viral genomes usually reveal little redundant coding capacity and genes are more often than not under strict temporal control producing the proteins required for the early part of the life cycle (nucleic acid replication) before the proteins required for later virus assembly (virion structural proteins). These properties have evolved to ensure efficient and timely production of viral components but they also make viruses ideal candidates as vectors for the expression of heterologous proteins. Temporal control of gene expression can be used as a natural induction step to ensure that an inserted gene is activated at a unique stage of the virus life cycle often when the virus infection has shut down much of the host cell protein expression. In addition, the use of virus late promoters, evolved to produce large quantities of protein for the formation of new virus particles, often leads to high level synthesis of the gene product of interest. As most viruses are ultimately lytic, the expression of gene products that would otherwise be toxic to cell growth is also allowed, at least for a period of time.

In the early development of viruses as expression vectors, the efficient use of coding capacity in virus genomes represented one of the few limitations to be overcome. Early and late gene expression vectors based on SV40 for example required that the early or late gene functions replaced by the heterologous gene, influenza HA, be provided by a helper virus grown at the same time (Gething and Sambrook, 1981). Such 'helper dependent' virus expression systems, whilst not preventing virus exploitation for heterologous gene expression, do introduce limitations. Degeneration of the recombinant viral stock by outgrowth of the helper or by recombination may lead to an apparent loss of protein expression over time and careful maintenance of recombinant virus stocks is essential.

With sequence information available for larger viruses however, the size and complexity of the genome provides ample opportunities for insertion of additional genes without loss or alteration of an existing function. In addition, viruses whose life

Genetically Engineered Viruses: Development and Applications, edited by C.J.A. Ring and E.D. Blair

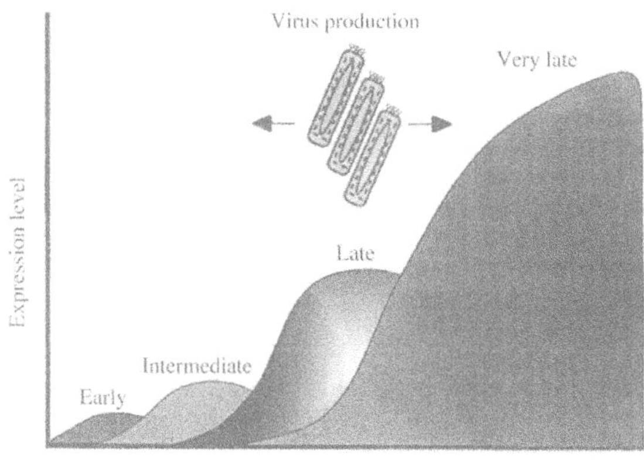

Figure 3.1. A schematic representation of the temporal control of baculovirus gene expression. On entering cells immediate early genes are transcribed to produce products that co-ordinate the later cascades of viral gene expression. The very late genes are involved in polyhedrin inclusion body formation. They have no effect on the virus yield in tissue culture which is completed before maximum very late gene expression is reached.

cycles involve more than one host, or whose replication requires distinct phases of host involvement, often encode proteins whose role is redundant when virus growth is restricted to tissue culture. Redundant genes can be readily exchanged for heterologous genes without loss of virus infectivity and the heterologous protein product is then expressed *in lieu* of the original viral protein in a truly helper-independent expression system. The exploitation of the most popular insect viruses for heterologous protein expression (the baculoviruses) is an example of such a system. The genomes of these viruses are large (>130 kbp) (Possee and Rohrmann, 1997) and they contain a number of genes that are not required for virus growth in cell culture. Moreover, there is strict temporal control of gene expression and, in addition to early and late genes, a set of 'very late' genes are transcribed after virus replication and virus assembly are complete (*Figure 3.1*). The characteristics of the very late gene promoters are known and have formed the basis for the development of most available virus expression vectors although other promoters, active at earlier times in the life cycle, have also been used. As very late gene expression occurs after virus production the addition of a heterologous gene does not affect virus yield and the recombinant virus grows to high titre and is normally genetically stable.

3.1.2 Insect viruses used as expression vectors

By far the most exploited group of insect viruses for the purposes of heterologous gene expression are the baculoviruses (below) but they are not the only insect viruses that have been used experimentally as expression vectors. Entomopoxviruses (EPVs), large DNA viruses that are morphologically and genetically similar to the vertebrate pox viruses, have been isolated from at least three orders of insects (Arif, 1995) and have a wide host range offering the possibility of expression in many cell types. During the virus life cycle they produce spheroidin, a protein that assembles into a paracrystalline matrix in the cytoplasm of infected cells and encapsulates virus particles. The gene encoding spheriodin, *sph*, has been sequenced for a number of viruses and has formed the basis of expression vector development for *Amsacta moorei* EPV (Palmer *et al.*, 1995). The viruses have also been suggested as non-replicating gene delivery vehicles for the expression of proteins in vertebrate cells (Li *et al.*, 1997). Much smaller insect viruses, the Densoviruses, have also been manipulated as expression vectors (Corsini *et al.*, 1996). The cloned copy of the densovirus genome is infectious and their replication strategy suggests them as candidates for the generation of non-lytic constitutive expression systems in insect cells (Afanasiev *et al.*, 1994; Giraud *et al.*, 1992). Expression systems based on either entomopoxviruses or densoviruses are specialist in their application however and neither is well developed. Whilst they may indicate areas of future insect virus exploitation, the use of baculoviruses and cells lines transfected with vectors based on baculoviral promoters, remains the most popular method of foreign protein expression in insect cells.

3.2 The baculovirus expression system

3.2.1 Background

Two baculoviruses have been widely used as expression vectors, *Autographa californica* multiple nuclear polyhedrosis virus (AcMNPV) and *Bombyx mori* nuclear polyhedrosis virus (BmNPV). The use of the *Bombyx* (Silkworm) virus allows the recombinant virus to be used to infect silkworm larval hosts for which there are automated rearing facilities and a long history of cultivation (Choudary *et al.*, 1995). It remains the lesser of the two viruses used however and by far the majority of virus manipulation and development of expression vectors has been done with AcMNPV. The virus was first studied extensively as a method of biological crop control (below) and was originally cultured, as were many insect viruses, by passage from insect to insect of the polyhedrin inclusion bodies (PIBs) from which the virus name derives. PIBs are produced by the polyhedrin protein which is synthesized at very high levels in the late stages of virus infection and serves to occlude many virus particles in a crystalline protein array. In the natural infection, AcNPV is transmitted horizontally via release from the decaying insect and deposition onto solid surfaces. By encapsidating virions the polyhedra endow the virus with two important features for long-term survival. They protect viruses from harsh environmental conditions such as desiccation and exposure to UV light and they provide an excellent delivery vehicle for eventual onward infection by ensuring that the ingesting larva is 'hit' by many virus particles at one time. The PIB dissolves readily in the alkaline mid gut of the insect and the infection cycle renews. In keeping with a role in natural virus transmission the polyhedrin protein serves no useful purpose when the virus is cultured in insect cell culture and viruses that have the polyhedrin gene deleted or replaced, whilst now sensitive to inactivation in the field, grow with equivalent kinetics and to similar titres to the wild type in cell culture.

The polyhedrin gene is only one of several very late baculovirus genes and although the majority of baculovirus recombinants use both the polyhedrin transcription signals and the polyhedrin locus as the basis for recombinant formation, a second gene, *p10*, has also been exploited to provide signals for heterologous gene expression. Like the polyhedrin protein, p10, a 10 kDa phosphoprotein is produced in vast amounts at the end of the virus life cycle, probably because it has a structural role in the assembly of the polyhedra (Gross *et al.*, 1994). The strength of the p10 transcription signals are on a par with those of the polyhedrin gene and have formed the basis of the development of P10-based transfer vectors (Heldens *et al.*, 1997; Martens *et al.*, 1995; Tomita *et al.*, 1995). In addition, vectors designed for multiple protein expression have been based on the combined use of P10 and polyhedrin promoters so avoiding the vector instability associated with duplication of the polyhedrin promoters. Alternating use of the polyhedrin and P10 promoters has proven a successful route to the generation of dual (Weyer and Possee, 1991), triple and quadruple promoter expression vectors (Belyaev and Roy, 1993). In most cases however these vectors remain destined for the polyhedrin locus of the recombinant virus as the flanking sequences required for recombination are those which surround the polyhedrin gene. Vectors based on the P10 promoter that include the surrounding P10 sequences allow integration at the P10 locus and, consequently, can be used to make recombinant viruses that continue to produce polyhedra, a useful property if infection of insects is the desired goal (Chaabihi *et al.*, 1997; Heldens *et al.*, 1997; Martens *et al.*, 1995).

3.2.2 Baculovirus insecticides

Extensive use has been made both of wild-type baculovirus and, latterly, genetically modified baculoviruses, as insect control agents. The active development of baculoviruses as biological pesticides has been ongoing for more than 30 years and they have been used locally in certain parts of the world for far longer. In the last decade research has centred on genetic improvements to the virus to counter perceived deficiencies that were holding back their widespread use and with the safety of any released virus in the environment. Several naturally occurring baculoviruses of *Helicoverpa zea, Lymantria dispar, Orgyia pseudotsuga* and *Neodiprion sertifer* have been used in field tests and although the result of virus application was generally successful, problems in the production and application of the virus (produced in infected larvae as PIBs) limited commercialization and a more widespread use. In addition, a major deficiency of insect pesticides when compared to chemical pesticides, speed of kill, was clearly illustrated (Cory and Bishop, 1997; Entwistle and Evan, 1985). A typical baculovirus infection can take up to 7 days to kill the host and as the main area of insect viral pesticides has been foliar application to prevent crop damage, a faster speed of kill, or at least cessation of feeding, is clearly necessary to minimize crop damage after application. The focus of genetic manipulation in the field of biological control has been to improve this key parameter. Two strategies have been pursued, the deletion of viral genes that appear to prolong the virus life cycle (evolved to improve final virus yield) so shortening the time to host death; and addition of foreign genes that encode proteins that might hasten host death. In the former, AcMNPV carrying deletions in the gene encoding ecdysteroid glucosyltransferase (*egt*) kill host up to 30% faster than the wild-type virus (O'Reilly and Miller, 1991). The mechanism of enhanced killing is not clear although it appears linked to the disruption of the normal hormonal balance of the insect host. More importantly, insects infected with *egt*$^{-ve}$ viruses consume up to 40% less food than wild-type viruses and their action in the field is competitive with *Bacillus thuringensis* toxin (Bt), a widely used biocontrol agent.

Several genes have been amended to the wild-type virus genome in attempts to improve speed of kill but in the main strategies aimed at altering insect hormonal balances have met with only limited success (Black *et al.*, 1997). Direct improvement has been seen however with viruses engineered to express insect specific neurotoxins during the virus infection. When the scorpion toxin *AaIT* gene is incorporated into AcMNPV and the virus used to infect insects they develop contractile paralysis that prevents them from continued feeding (Stewart *et al.*, 1991). Field trials with the virus showed a significant reduction in the amount of crop damage associated with pest infestation when compared to the pest treated with wild-type virus (Cory et al., 1994).

Whilst genetic alteration of AcMNPV (and other baculoviruses) appears a plausible route to the development of successful faster-acting viral pesticides it formally introduces foreign genes to the environment and requires extensive risk assessment before widespread release can be considered. Despite its promise therefore it is likely to be some time before commercially viable virus-based pesticides are available on a wide scale.

3.2.3 Expression vectors

There are now a great variety of expression vectors for the production of recombinant baculoviruses. Most use the sequences that surround the polyhedrin locus to enable the recombination event that places the gene of interest in the baculovirus genome. A system using the p10 flanking sequences has been described (Chaabihi *et al.*, 1997) and it is clear from gene knockout experiments that a number of other sites in the viral genome could be engineered for the insertion of foreign genes driven by a variety of promoters. Recombination between the transfer vector can occur in insect cells (Kitts and Possee, 1993), the 'classical' approach to recombinant formation or in bacterial (Luckow *et al.*, 1993) or yeast cells (Patel *et al.*, 1992). Direct ligation of genes into pre-engineered viral genomic DNA can also be used (Ernst *et al.*, 1994) although it is not particularly efficient as it includes no counterselection for background viruses.

As host cell protein expression is already being shut down by the time the polyhedrin or P10 promoters are maximally active, protein modifications such as glycosylation may not function as well late in infection as they do at earlier times. To overcome this it is possible to use promoters from genes that are active earlier. The basic protein promoter, which is a late rather than very late promoter, has been used in this way (Chazenbalk and Rapoport, 1995) with a consequential increase in the bioactivity of the expressed protein when compared to the same protein expressed under polyhedrin control. A promoter from the immediate early genes has also been used, specifically to provide a glycosylation enzyme that will modify glycoproteins made from the polyhedrin locus (Jarvis and Finn, 1996). The levels of protein expressed are inevitably lower however and the use of the polyhedrin promoter for high level expression remains predominant. The requirements for high-level expression are met by all current vectors, academic and commercial, and derive from mapping experiments that examined the levels of expression obtained when the length of the polyhedrin leader sequence was sequentially reduced (Matsuura *et al.*, 1987). The full untranslated leader sequence from the transcription start site to the A of the ATG codon marking the start of the polyhedrin coding region was required for full activity. Polylinkers and expressed tags since introduced to provide more convenience in vector usage have all been introduced downstream of the leader and sometimes well into the polyhedrin coding sequence itself (the ATG having been inactivated by mutagenesis). A typical single promoter vector with a variety of the many modifications available is shown in *Figure 3.2.* The vector backbone is generally a small multicopy plasmid that may have single strand capability (e.g. Livingstone and Jones, 1989) although the prevalence of mutagenesis procedures based on double stranded DNA makes this feature now largely redundant. The polyhedrin promoter occurs about 100 bp upstream of the restriction sites used for insertion and the polyhedrin leader sequence to at least the position of the original polyhedrin ATG is preserved. Thereafter, a number of modifications may exist to assist in the purification of the expressed protein or expression in a desired cell compartment (e.g. secretion). The precise tags used vary with the development of the expression system (*Figure 3.2* shows polyhistidine and glutathione-S-transferase as two commonly used examples). Secretion of proteins from infected insect cells is best achieved by the use of insect signal sequences, although most mammalian signals are cleaved efficiently by the insect cell secretion mechanism. The signal from the major baculovirus glycoprotein gp67, a classical type I membrane glycoprotein, has been successfully used in this regard (Wang *et al.*, 1995) as have selected human (Jarvis *et al.*, 1993;

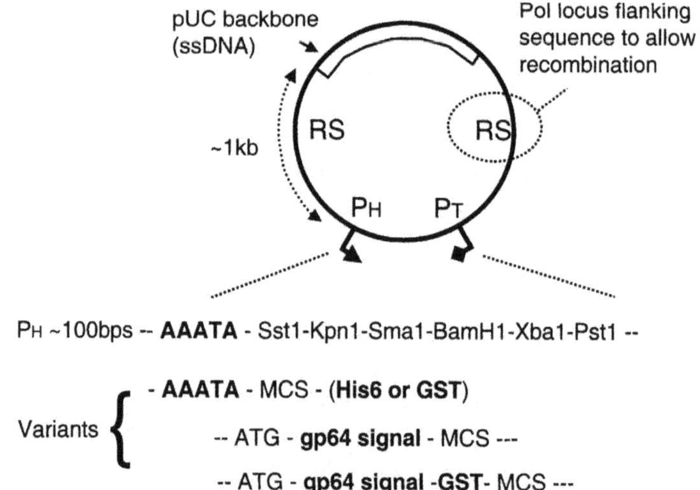

Figure 3.2. A typical baculovirus transfer vector. The signals for transcription, here shown as the polyhedrin promoter (Ph) and terminator (Pt), are flanked on either side by about 1 kb by DNA required for recombination (RS) into the polyhedrin locus of the viral genome. Alternate methods of integration, such as transposition in *E. coli* are discussed in the text. Multiple cloning sites (MCS) such as the one shown are common in many vectors. The full untranslated leader RNA to include the AAAT sequence in front of the ATG initiator must be present for maximum expression. A variety of fusion tags are available in vectors designed for both cytoplasmic expression and secretion.

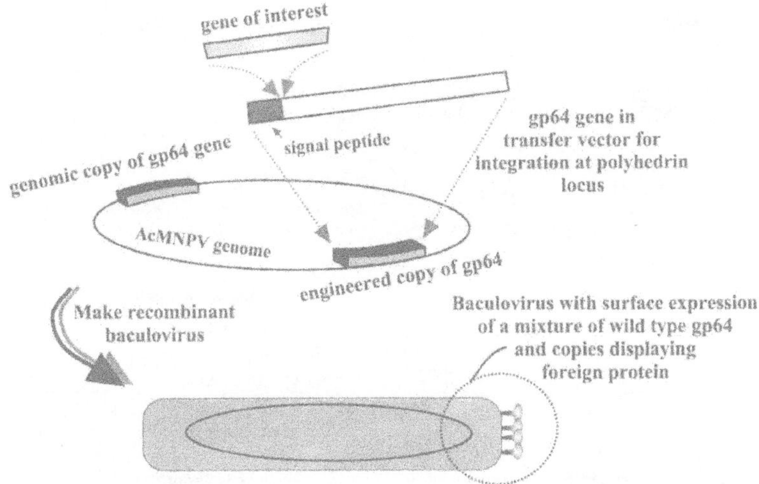

Figure 3.3. A specialized form of expression vector is that designed for display of the foreign protein on the virus surface. The major baculovirus glycoprotein gp64 provides the signals for secretion and for anchoring the fusion to virion membrane. As gp64 is essential for infection a second copy destined for the polyhedrin locus is used for display purposes.

Kuhn and Zipfel, 1995) and bacterial (Allet *et al.*, 1997) protein signal sequences.

When the vector contains a specific fusion tag, a second copy of the major baculovirus glycoprotein gp64, downstream of the signal peptide, introduced coding regions, in frame with both the signal and the gp64 sequence are displayed on the virus surface (Boublik, 1995); Grabherr, 1997; Mottershead *et al.*, 1997). As the construction of recombinant viruses is generally a one-by-one process the opportunities to develop the technology to allow the display of eukaryotic libraries have been limited. However, a limited library of ca.10^4 variants of an antibody epitope has been constructed and an improved binding variant directly selected (Ernst *et al.*, 1998). The display of a foreign protein fragment on the surface of the virus has also proven an efficient way to raise and screen monoclonal antibodies without having to purify the protein fragment concerned (Lindley *et al.*, 2000).

3.2.4 Host cell involvement

Insect (Lepidopteran) cells are typical higher eukaryotic cells and carry out all the post-translational modifications found in eukaryotic proteins and governed by a variety of signal sequences. Thus, signal sequences are recognized and lead to protein translocation into the lumen of the ER; phosphorylation sites are phosphorylated, glycosylation sites have glycans added and many cellular protease processing sites are recognized and cleaved. Expression of mammalian proteins therefore in any of the widely used insect cell lines (*Sf*9 and *Sf*21 from *Spodoptera frugipurda* and High 5 from *Trichoplusia ni*) generally leads to a protein that is processed as it would be in a mammalian cell and that frequently attains full biological activity (*Figure 3.4*). The exception of the precise glycan

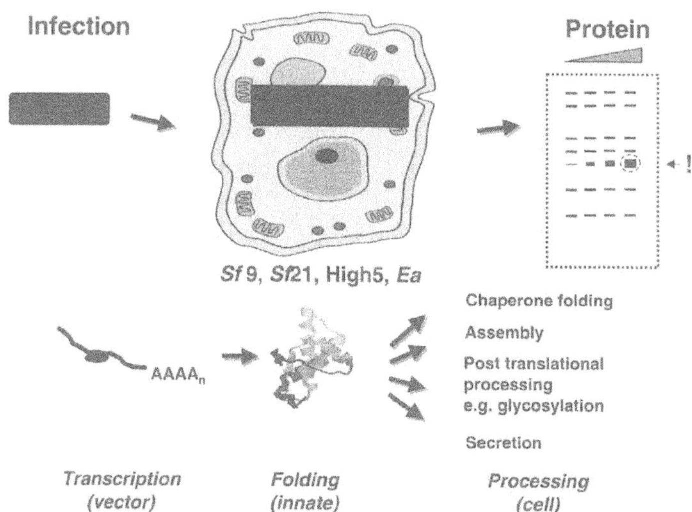

Figure 3.4. Cellular post-translational processing may limit the yield of any protein, particularly glycoproteins. In many cases where expression is high, the processing of the expressed protein within the insect cell is of little concern (upper). For glycoproteins and other highly processed proteins however many post-translational processing steps may be required, any one of which could reduce yield and/or bioactivity.

pattern and some specific proteolytic cleavages are discussed below but in the main, the default expectation for an average protein expressed using an insect virus expression system would be for a high yield of soluble protein with full or near-full specific activity. Where uncertainties lie these are in the processing of the protein after translation (*Figure 3.4*). For cytosolic proteins with simple domain structures there are few barriers to high level expression but for larger proteins, and particularly for glycoproteins, the yield of biologically active product can be much less than expected given the strength of the promoter in use. Compare for example the yield of 1–2 milligrams per litre for the HIV envelope glycoprotein (Morikawa *et al.*, 1990) with the several hundred milligrams per litre of the polyhedrin protein itself (O'Reilly *et al.*, 1992). The discrepancy undoubtedly lies in the post-transitional processing of these proteins and their innate biochemical properties. For example, by forming a crystal the polyhedrin protein accumulates in a non-accessible form that limits proteolysis. The opposite is true for a complex glycoprotein where long folding times and passage through several cellular compartments increase the possibility for proteolysis. In addition, the levels of initial synthesis, dictated by the level of mRNA available, may simply swamp the quality control aspects of glycoprotein synthesis. Interaction with folding chaperones such as BiP or calnexin (e.g. Kim and Arvan, 1995) can effectively limit the amount of the final folded form. In recognition of these limitations some studies have modified the insect cell secretory pathway to express the proteins involved in folding in order to maximize the level of fully folded glycoprotein possible (Hsu *et al.*, 1994; Lenhard and Reilander, 1997; Whiteley *et al.*, 1997). Engineering of the host cell in addition to engineering of the baculovirus itself is likely to feature in the future development of the expression system.

3.2.5 Known limitations

The glycosylation pattern of glycoproteins derived from insect cells is significantly different from that of mammalian cells. The canonical triplet recognized by the N-glycosylation machinery, Asn-X-Ser/Thr is the same but as glycosylation is a function of the processing enzymes present in the host cell, the glycans attached to the polypeptide backbone are different. As in mammalian glycosylation, the initial step is the addition of a high mannose glycan precursor to the asparagine residue. As the protein proceeds through the endoplasmic reticulum, the glycan is trimmed by resident glycosidases to produce a structure with only three mannose residues linked via two N-acetyl glucosamine residues to the polypeptide although fucose is also often present (Butters *et al.*, 1998; Jarvis and Finn, 1995). In mammalian cells the tri-mannose core is rebuilt to a variety of branched glycans depending on the complement of glycotransferases present in the expressing cell. In insect cells however the rebuild is much less complex. A good deal of the attached glycans remain as the tri-mannose core and whilst some may be extended, mannose is by far the predominant sugar (*Figure 3.5*). As a result, insect expressed glycosylated proteins always have a high mannose content and are usually smaller than the equivalent protein produced in mammalian cells, although this is only evident if the level of glycosylation is high. If glycan identity is important for protein function then expression in insect cells can pose difficulties (e.g. Rosa *et al.*, 1996) but it is rare. More usefully, the glycan composition can be used to advantage as all insect glycoproteins can be purified away from the cytosolic components by mannose specific lectin chromatography.

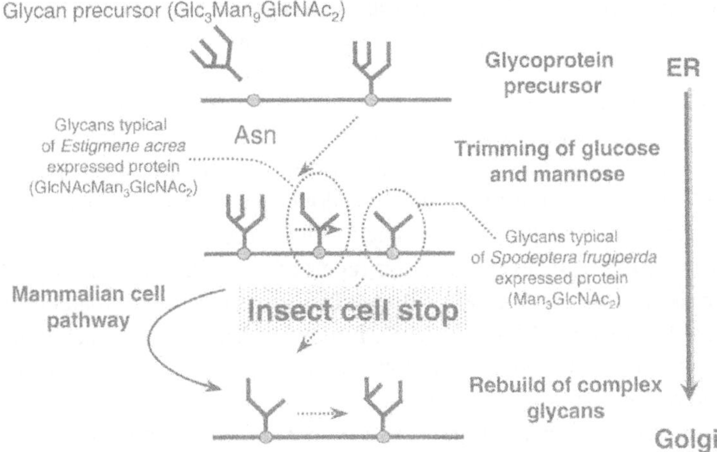

Figure 3.5. Glycosylation in insect cells follows the same initial pathway as any other higher eukaryotic cell and the glycan precursor is reduced by sequential endoglycosidase steps that occur during protein migration through the ER. In contrast to mammalian cells however, there is little or no rebuild of the glycan structure once the tri-mannose core has been produced. *Estigmene acrea* cells fail to cleave the final GlcNAc residue from the core leaving a glycoprotein that is slightly more glycosylated than the same protein expressed in *Spodoptera* cells. Where glycan levels or identity are required for bioactivity, expression of the recombinant protein in *Ea*4 cells may offer advantages.

Cell lines derived from *Estigmene acrea* (Ea) appear not to possess the enzyme N-acetyl-β-glucosaminidase which is responsible, during glycoprotein processing in *Spodoptera* cells, for trimming the last GlcNAc residue to yield the tri-mannose core (Wagner *et al.*, 1996). As a consequence glycosylated proteins expressed in Ea cells have glycan profiles that more closely resemble the glycan content of mammalian-derived proteins (Ogonah *et al.*, 1996). Purposeful reconstruction of the glycosylation pathways in insect cells to allow mammalian-like glycan addition have also been reported by the transfection of cDNAs that encode mammalian processing enzymes such as β1,4-galactosyltransferase. When recombinant viruses expressing glycoproteins were used to infect these cells the recovered, expressed protein contained galactose (Hollister *et al.*, 1998) in contrast to the glycans obtained when unmodified cell lines were used (Jarvis and Finn, 1995).

Certain proteolytic cleavages represent a second limitation for the production of fully functional proteins in insect cells. The cleavage of prohormones, some proteases and a number of viral glycoproteins is poor following expression in insect cells. The cleavage sequence in each case, basic-x-basic-basic, is the hallmark of a family of protease convertases, typified by the convertase furin, that are thought to be golgi resident enzymes. The observation that the provision of furin by co-expression allowed the authentic cleavage to take place (e.g. Morikawa *et al.*, 1993) suggested that this enzyme is absent from insect cells but more recent studies show that the insect cell homologue exists and exhibits a similar substrate specificity (Cieplik *et al.*, 1998). Low, rather than absent, activity is therefore the reason for poor cleavage of overexpressed glycoproteins.

The expression of the human malaria parasite *Plasmodium falciparum* merozoite surface protein 1 (MSP-1) resulted in only intracellular antigen expression following expression using recombinant baculoviruses (Kedees *et al.*, 2000). The lack of cell surface expression was attributed to failure of the insect cell to add a glycosylphosphatidyl-inositol (GPI) anchor to the protein following expression. The lack of a GPI anchor was confirmed by expression of *Toxoplasma gondii* surface antigen SAG1 which was also not modified by a GPI-anchor addition (Azzouz *et al.*, 2000). The authors went on to show that biosynthesis of the early GPI intermediate GlcNH(2)-PI is the key step blocked in baculovirus-infected cells and concluded that the baculovirus system was not appropriate for the expression of GPI-anchored proteins (Azzouz *et al.*, 2000). However, an earlier report of the expression of the GPI-anchored human complement inhibitor CD59 reported cell surface expression and at least partial GPI presence (Davies and Morgan, 1993). It may be therefore that baculovirus infected cells are inefficient at some post-translational processing steps and that this is compounded by high protein expression levels.

An uncharacterized limitation that has recently become apparent is the availability or not of host molecules that may interact with the expressed recombinant protein. Whilst not affecting protein expression *per se*, deficiencies in the ability of the insect cell cytosol to mimic the mammalian environment were apparent in studies of the transport of retroviral capsids (Parker and Hunter, 2000). It is not yet known if this represents a true deficiency in the protein complement of insect compared to mammalian cells or if, as in the furin case, there is simply too little of the functional homologue to ensure efficient interaction.

In contrast to the lack of certain host proteins, unwanted proteins produced as part of the virus life cycle can affect recombinant protein expression. To aid lysis of infected insects in the wild, baculoviruses encode a chitinase related to the chitinase of *Serratia marcescens* (Hawtin *et al.*, 1995). The protein has an endoplasmic reticulum retention signal at the C-terminus and is found in large amounts in the ER and golgi late in infection (Thomas *et al.*, 1998). The Chitinase minus (chi −ve) viruses are viable and when used as the basis of recombinant generation, the level of secreted proteins such as enterokinase is enhanced when compared to chi +ve recombinants (Possee – personal communication). The virus also encodes a protease of the cathepsin family with substrate specificity for dibasic sequences whose activity is greatest at acidic pHs (Slack *et al.*, 1995). As insect cell media is acidic this can mean that recombinant proteins with the substrate cleavage site are sensitive to degradation although this can be reduced by the addition of inhibitors such as leupeptin and E64 to the media during the final stage of expression. At least one commercial source of baculovirus reagents offers a genomic DNA that has had the cathepsin gene removed.

3.2.6 New directions

The finding that baculoviruses can enter mammalian cells (Boyce and Bucher, 1996; Hofmann *et al.*, 1995; Shoji *et al.*, 1997) has led to the development of new vectors for mammalian gene transfer. Entry into the cell is mediated by the major baculovirus surface glycoprotein gp64 and may involve electrostatic interactions with heparin sulfate (or similar molecules) on the cell surface (Duisit *et al.*, 1999) although if the subsequent fusion of viral and host membranes is gp64 mediated is not known. The viral late and very late promoters so widely used in insect cell expression systems are dependent on factors produced earlier in infection acting *in trans* for activity. They are not active in mammalian cells but if the recombinant baculovirus is constructed to contain a gene under control of a promoter active in mammalian cells then entry of the virus is followed by gene expression. The virus appears able to enter a wide range of cells including primary human cells (Condreay *et al.*, 1999) and expression in most cell types can be enhanced by treatment with butyrate, consistent with the baculovirus genome being assembled into histone complexes. The efficiency of mammalian cell entry has been improved by pseudotyping the virus with vesicular stomatitis virus G glycoprotein which appears to improve the movement of the virus from the entry vesicle to the cytoplasm proper (Barsoum *et al.*, 1997).

If the recombinant baculovirus also contains a selectable marker then permanent cell lines can be established in which at least 12 kb of baculoviral DNA is inserted into the host genome (Condreay *et al.*, 1999). In essence the virus genome is acting as a large backbone plasmid which, when compared to plasmid transfection protocols, is being delivered to the cell far more efficiently by the virus than it would be as DNA alone. Whether the baculovirus DNA is transcriptionally active when integrated, either through the inherent activity of the immediate early promoters or by transcripts that originate outside the inserted sequence is not clear and would have consequences for the development of the virus for delivery in the clinical setting. In this arena however the susceptibility of the virus to inactivation by human complement would also need to be addressed. The virus is rapidly inactivated by the classical complement pathway but removal of a key complement component (C3) by treatment with cobra venom factor provides full protection (Hofmann and Strauss, 1998). Addition of a known complement inhibitor protein CR1 also provided some protection (Hofmann *et al.*, 1999). Baculoviruses engineered to resist such inactivation as well as to incorporate molecules for cell-specific targeting may be available in the future.

3.2.7 Isolated viral promoters

Ever since the development of baculoviruses as tools for protein expression (Smith *et al.*, 1983) there have been attempts to generate permanent cell lines based on viral promoters that express recombinant proteins. Late or very late promoters have essentially no activity in isolated form but the immediate early (IE) promoters of the virus are recognized and transcribed by host polymerases (Jarvis *et al.*, 1990) in the absence of other viral proteins. They make good candidates for the development of permanent insect cell lines but their use, particularly the use of IE1, was hampered by low expression levels until recently. By linking constitutive promoters to insect virus enhancer elements, the level of transcription can be boosted a thousand fold making permanent insect cell lines expressing relatively good levels of protein, although not at the level expected for baculovirus expressed proteins, a feasibility (Lu *et al.*, 1997). Unfortunately, vectors based on insect viral promoters interfere with transcription by incoming recombinant viruses (Jarvis, 1993) so the use of permanent cell lines made in this way to provide helper functions for the products expressed by recombinant viruses is still limited. An independent review of the development of permanent insect cell lines including those based on viral promoters has been published recently (Pfeifer, 1998).

References

Afanasiev BN, Kozlov YV, Carlson JO, Beaty BJ. (1994) Densovirus of Aedes aegypti as an expression vector in mosquito cells. *Exp. Parasitol.* **79**: 322–339.

Allet B, Bernard AR, Hochmann A, Rohrbach E, Graber P, Magnenat E, Mazzei GJ, Bernasconi L. (1997) A bacterial single peptide directs efficient secretion of eukaryotic proteins in the baculovirus expression system. *Protein Expr. Purif.* **9**: 61–68.

Arif BM. (1995) Recent advances in the molecular biology of entomopoxviruses. *J. Gen. Virol.* **76**: 1–13.

Azzouz N, Kedees MH, Gerold P, Becker S, Dubremetz JF, Klenk HD, Eckert V, Schwarz RT. (2000) An early step of glycosylphosphatidyl-inositol anchor biosynthesis is abolished in lepidopteran insect cells following baculovirus infection. *Glycobiology* **10**: 177–183.

Barsoum J, Brown R, McKee M, Boyce FM. (1997) Efficient transduction of mammalian cells by a recombinant baculovirus having the vesicular stomatitis virus G glycoprotein. *Hum. Gene Ther.* **8**: 2011–2018.

Belyaev AS, Roy P. (1993) Development of baculovirus triple and quadruple expression vectors: co-expression of three or four bluetongue virus proteins and the synthesis of bluetongue virus-like particles in insect cells. *Nucleic. Acids Res.* **21**: 1219–1223.

Black BC, Brennam LA, Dierks PM, Gard IE. (1997) The baculoviruses. In: *The Viruses.* L.K. Miller (ed), pp. 341–381, Plenum Press, New York, London.

Boublik Y, Di Bonito P, Jones IM. (1995) Eukaryotic virus display: engineering the major surface glycoprotein of the Autographa californica nuclear polyhedrosis virus (AcNPV) for the presentation of foreign proteins on the virus surface *Biotechnology (NY),* **13**: 1079–1084.

Boyce FM, Bucher NLR. (1996) Baculovirus-mediated gene transfer into mammalian cells. *Proc. Natl Acad. Sci. USA* **93**: 2348–2352.

Butters TD, Yudkin B, Jacob GS, Jones IM. (1998) Structural characterisation of the N-linked oligosaccharides derived from HIVgp120 expressed in lepidopteran cells. *Glycoconjugate Journal* **15**: 83–88.

Chaabihi H, Cetre C, Berne A. (1997) A new vector for efficient generation of p10-single-late-promoter recombinant baculoviruses. *J .Virol. Methods* **63**: 1–7.

Chazenbalk GD, Rapoport B. (1995) Expression of the extracellular domain of the thyrotopin receptor in the baculovirus system using a promoter active earlier than the polyhedrin promoter. Implications for the expression of fuctional highly glycosylated proteins. *J. Biol. Chem.* **270**: 1543–1549.

Choudary PV, Kamita SG, Maeda S. (1995) Expression of foreign genes in *Bombyx mori* larvae using baculovirus vectors. *Methods Mol. Biol.* **39**: 243–264.

Cieplik M, Klenk HD, Garten W. (1998) Identification and characterization of spodoptera frugiperda furin: a thermostable subtilisin-like endopeptidase. *Biol. Chem.* **379**: 1433–1440.

Condreay JP, Witherspoon SM, Clay WC, Kost TA. (1999) Transient and stable gene expression in mammalian cells transduced with a recombinant baculovirus vector. *Proc. Natl Acad. Sci. USA* **96**: 127–132.

Corsini J, Afanasiev B, Maxwell IH, Carlson JO. (1996) Autonomous parvovirus and densovirus gene vectors. *Adv. Virus Res.* **47**: 303–351.

Cory JS, Bishop DH. (1997) Use of baculoviruses as biological insecticides. *Mol. Biotechnol.* **7**: 303–313.

Cory JS, Hirst ML, Williams T, Hails RS, Goulson D, Green BM, Carty TM, Possee RD, Cayley PJ, Bishop DHL. (1994) Field trials of a genetically improved baculovirus insecticide. *Nature* **270**: 138.

Davies A, Morgan BP. (1993) Expression of the glycosylphosphatidylinositol-linked complement-inhibiting protein CD59 antigen in insect cells using a baculovirus vector. *Biochem. J.* **295**: 889–896.

Duisit G, Saleun S, Douthe S, Barsoum J, Chadeuf G, Moullier P. (1999) Baculovirus vector requires electrostatic interactions including heparan sulfate for efficient gene transfer in mammalian cells [In Process Citation]. *J. Gene Med.* **1**: 93–102.

Entwistle PF, Evan HF. (1985) Viral control. In: LI Gilbert, GA Kerkut, (eds), *Comprehensive Insect Physiology, Biochemistry and Pharmacology,* vol 12. Pergamon, Oxford, pp 347–412.

Ernst, WJ, Grabherr RM, Katinger HW. (1994) Direct cloning into the Autographa californica nuclear polyhedrosis virus for generation of recombinant baculoviruses. *Nucleic Acids Res.* **22**: 2855–2856.

Ernst W, Grabherr R, Wegner D, Borth N, Grassauer A, Katinger H. (1998) Baculovirus surface display: construction and screening of a eukaryotic epitope library. *Nucleic Acids Res.* **26**: 1718–1723.

Gething MJ, Sambrook J. (1981) Cell-surface expression of influenza haemagglutinin from a cloned DNA copy of the RNA gene. *Nature* **293**: 620–625.

Giraud C, Devauchelle G, Bergoin M. (1992) The densovirus of Junonia coenia (Jc DNV) as an insect cell expression vector. *Virology* **186**: 207–218.

Grabherr R. *et al.* (1997) Expression of foreign proteins on the surface of *Autographa californica* nuclear polyhedrosis virus. *Biotechniques,* **22**: 730–753.

Gross CH, Russell RL, Rohrmann GF. (1994) Orgyia pseudotsugata baculovirus p10 and polyhedron envelope protein genes: analysis of their relative expression levels and role in polyhedron structure. *J. Gen. Virol.* **75**: 1115–1123.

Hawtin RE, Arnold K, Ayres MD *et al.* (1995) Identification and preliminary characterization of a chitinase gene in the Autographa californica nuclear polyhedrosis virus genome. *Virology* **212**: 673–685.

Heldens JG, Kester HA, Zuidema D, Vlak JM. (1997) Generation of a p10-based baculovirus expression vector in yeast with infectivity for insect larvae and insect cells. *J. Virol. Methods* **68**: 57–63.

Hofmann C, Strauss M. (1998) Baculovirus-mediated gene transfer in the presence of human serum or blood facilitated by inhibition of the complement system. *Gene Therapy* **5**: 531–536.

Hofmann C, Sandig V, Jennings G, Rudolph M, Schlag P, Strauss M. (1995) Efficient gene transfer into human hepatocytes by baculovirus vectors. *Proc. Natl Acad. Sci. USA* **92**: 10099–10103.

Hofmann C, Huser A, Lehnert W, Strauss M. (1999) Protection of baculovirus-vectors against complement-mediated inactivation by recombinant soluble complement receptor type 1. *Biol. Chem.* **380**: 393–395.

Hollister JR, Shaper JH, Jarvis DL. (1998) Stable expression of mammalian beta 1,4-galactosyltransferase extends the N-glycosylation pathway in insect cells [In Process Citation]. *Glycobiology* **8**: 473–480.

Hsu TA, Eiden JJ, Bourgare P, Meo T, Betenbaugh MJ. (1994) Effects of co-expressing chaperone BiP on functional antibody production in the baculovirus system. *Protein Expr. Purif.* **5**: 595–603.

Jarvis DL. (1993) Effects of baculovirus infection on IE1-mediated foreign gene expression in stably transformed insect cells. *J. Virol.* **67**: 2583–2591.

Jarvis DL, Finn EE. (1995) Biochemical analysis of the N-glycosylation pathway in baculovirus-infected lepidopteran insect cells. *Virology* **212**: 500–511.

Jarvis DL, Finn EE. (1996) Modifying the insect cell N-glycosylation pathway with immediate early baculovirus expression vectors [In Process Citation]. *Nat. Biotechnol.* **14**: 1288–1292.

Jarvis DL, Fleming JA, Kovacs GR, Summers MD, Guarino LA. (1990) Use of early baculovirus promoters for continuous expression and efficient processing of foreign gene products in stably transformed lepidopteran cells. *Biotechnology (NY)* **8**: 950–955.

Jarvis DL, Summers MD, Garcia Jr, A, Bohlmeyer DA. (1993) Influence of different signal peptides and prosequences on expression and secretion of human tissue plasminogen activator in the baculovirus system. *J. Biol. Chem.* **268**: 16754–16762.

Kedees MH, Gerold P, Azzouz N, Blaschke T, Shams-Eldin H, Muhlberger E, Holder AA, Klenk HD, Schwarz RT, Eckert V. (2000) Processing and localisation of a GPI-anchored Plasmodium falciparum surface protein expressed by the baculovirus system [In Process Citation]. *Eur. J. Cell. Biol.* **79**: 52–61.

Kim PS, Arvan P. (1995) Calnexin and BiP act as sequential molecular chaperones during thyroglobulin folding in the endoplasmic reticulum. *J. Cell. Biol.* **128**: 29–38.

Kitts PA, Possee RD. (1993) A method for producing recombinant baculovirus expression vectors at high frequency. *Biotechniques* **14**: 810–817.

Kuhn S, Zipfel PF. (1995) The baculovirus expression vector pBSV-8His directs secretion of histidine-tagged proteins. *Gene* **162**: 225–229.

Lenhard T, Reilander H. (1997) Engineering the folding pathway of insect cells: generation of a stably transformed insect line showing improved folding of a recombinant membrane protein. *Biochem. Biophys. Res. Commun.* **238**: 823–830.

Li Y, Hall RL, Moyer RW. (1997) Transient, nonlethal expression of genes in vertebrate cells by recombinant entomopoxviruses. *J. Virol.* **71**: 9557–9562.

Lindley KM, Su JL, Hodges PK, Wisely GB, Bledsoe RK, Condreay JP, Winegar DA, Hutchins JT, Kost TA. (2000) Production of monoclonal antibodies using recombinant baculovirus display gp64-fusion proteins. *J. Immunol. Methods* **234**: 123–135.

Livingstone C, Jones IM. (1989) Baculovirus expression vectors with single-strand capability. *Nucl. Acids Res.* **17**: 2366.

Lu M, Farrell PJ, Johnson R, Iatrou K. (1997) A baculovirus (Bombyx mori nuclear polyhedrosis virus) repeat element functions as a powerful constitutive enhancer in transfected insect cells. *J. Biol. Chem.* **272**: 30724–30728.

Luckow VA, Lee SC, Barry GF, Olins PO. (1993) Efficient generation of infectious recombinant baculoviruses by site-specific transposon-mediated insertion of foreign genes into a baculovirus genome propagated in *Escherichia coli*. *J. Virol.* **67**: 4566–4579.

Martens JW, van-Oers MM, van-de-Bilt BD, Oudshoorn P, Vlak JM. (1995) Development of a baculovirus vector that facilitates the generation of p10-based recombinants. *J. Virol. Methods* **52**: 15–19.

Matsuura Y, Possee RD, Overton HA, Bishop DH. (1987) Baculovirus expression vectors: the requirements for high level expression of proteins, including glycoproteins. *J. Gen. Virol.* **68**: 1233–1250.

Morikawa Y, Overton HA, Moore JP, Wilkinson AJ, Brady RL, Lewis SJ, Jones IM. (1990) Expression of HIV-1 gp120 human soluble CD4 by recombinant baculoviruses and their interaction in vitro. *AIDS Res. Hum. Retro.* **6**: 765–773.

Morikawa Y, Barsov E, Jones IM. (1993) Legitimate and illegitimate cleavage of human immunodeficiency virus glycoproteins by furin. *J. Virol.* **67**: 3601–3604.

Mottershead D, van der Linden I, von Bonsdorff CH, Keinanen K, Oker-Blom C. (1997) Baculoviral diplay of the green fluorescent protein and rubella virus envelope proteins. *Biochem. Biophys. Res. Cummun.* **238**: 717–722.

Ogonah OW, Freedman RB, Jenkins N, Patel K, Rooney BC. (1996) Isolation and characterisation of an insect cell line able to perform complex N-linked glycosylation on recombinant proteins. *Nature Biotechnol.* **14**: 197–202.

O'Reilly DR, Miller LK. (1991) Improvement of a baculovirus pesticide by deletion of the *egt* gene. *Biotechnology* **9**: 1086–1089.

O'Reilly DR, Miller LK, Luckow VA. (1992) Baculovirus expression vectors: a laboratory manual. WH Freeman and Company, New York, 347 pp.

Palmer CP, Miller DP, Marlow SA, Wilson LE, Lawrie AM, King LA. (1995) Genetic modification of an entomopoxvirus: deletion of the spheroidin gene does not affect virus replication in vitro. *J. Gen. Virol.* **76**: 15–23.

Parker SD, Hunter E. (2000) A cell-line-specific defect in the intracellular transport and release of assembled retroviral capsids. *J. Virol.* **74**: 784–795.

Patel G, Nasmyth K, Jones N. (1992) A new method for the isolation of recombinant baculovirus. *Nucl. Acids Res.* **20**: 97–104.

Pfeifer TA. (1998) Expression of heterologous proteins in stable insect cell culture. *Curr. Opin. Biotechnol.* **9**: 518–521.

Possee RD, Rohrmann GF. (1997) Baculovirus genome organisation and evolution. In: (LK Miller, ed), The Baculoviruses. Plenum Press, New York, pp 109–140.

Rosa D, Campagnoli S, Moretto C, Guenzi E, Cousens L, Chin M, Dong C, Weiner AJ, Lau JYN, Choc Q-L, Chien D, Pileri P, Houghton M, Abrignani S. (1996) A quantitative test to estimate neutralising antibodies to the hepatitis C virus: cytofluorimetric assessment of envelope glycoprotein 2 binding to target cells. *Proc. Natl Acad. Sci. USA* **93**: 1759–1763.

Shoji I, Aizaki H, Tani H, Ishii K, Chiba T, Saito I, Miyamura T, Matsuura Y. (1997) Efficient gene transfer into various mammalian cells, including non-hepatic cells, by baculovirus vectors. *J. Gen. Virol.* **78**: 2657–2664.

Slack JM, Kuzio J, Faulkner P. (1995) Characterization of v-cath, a cathepsin L-like proteinase expressed by the baculovirus Autographa californica multiple nuclear polyhedrosis virus. *J. Gen. Virol.* **76**: 1091–1098.

Smith GE, Summers MD, Frazer MJ. (1983) Production of human beta-interferon in insect cells infected with a baculovirus expression vector. *Mol. Cell. Biol.* **3**: 2156–2165.

Stewart LMD, Hirst M, Ferber ML, Merryweather AT, Cayley PJ, Possee RD. (1991) Construction of an improved baculovirus insecticide containing an insect-specific toxin gene. *Nature* **352**: 85–88.

Thomas CJ, Brown HL, Hawes CR, Lee BY, Min MK, King LA, Possee RD. (1998) Localization of a baculovirus-induced chitinase in the insect cell endoplasmic reticulum. *J. Virol.* **72**: 10207–10212.

Tomita S, Kanaya T, Kobayashi J, Imanishi S. (1995) Isolation of p10 gene from Bombyx mori nuclear polyhedrosis virus and study of its promoter activity in recombinant baculovirus vector system. *Cytotechnology* **17**: 65–70.

Wagner R, Geyer H, Geyer R, Klenk HD. (1996) N-acetyl-beta-glucosaminidase accounts for differences in glycosylation of influenza virus hemagglutinin expressed in insect cells from a baculovirus vector. *J. Virol.* **70**: 4103–4109.

Wang YH, Davies AH, Jones IM. (1995) Expression and purification of glutathione-S-transferase tagged HIV-1 gp120; no evidence for an interaction with CD26. *Virology* **208**: 142–146.

Weyer U, Possee RD. (1991) A baculovirus dual expression vector derived from the Autographa californica nuclear polyhedrosis virus polyhedrin and p10 promoters: co-expression of two influenza virus genes in insect cells. *J. Gen. Virol.* **72**: 2967–2974.

Whiteley EM, Hsu TA, Betenbaugh MJ. (1997) Thioredoxin domain non-equivalence and anti-chaperone activity of protein disulfide isomerase mutants in vivo. *J. Biol. Chem.* **272**: 22556–22563.

Chapter 4

Plant viruses

Christophe Lacomme, Greg P. Pogue, T. Michael A. Wilson and Simon Santa Cruz

4.1 Introduction

Progress in the understanding of plant virus gene regulation has provided a diverse collection of tools that are routinely exploited in both basic and applied plant molecular biology. Regulatory elements have been derived from both RNA and DNA plant viruses that have found widespread application in transgenic plant technology (Mushegian and Shepherd, 1995). For example the cauliflower mosaic virus (CaMV) 35S promoter and the tobacco mosaic virus (TMV) Ω translational enhancer are routinely used to direct and regulate transgene expression. In addition to their utility as 'toolboxes' in providing sequences of value in plant genetic engineering plant viruses also offer the possibility of providing convenient and readily manipulated vectors for the rapid and high-level expression of foreign genes in plants. To date the most widespread application of plant virus-based vectors has been in the area of basic research and less attention has focused on the potential of virus-based vectors in commercial biotechnology. In part the emphasis on viral vectors as research tools reflects the successes achieved in the development of stable transformation technologies for plants. Techniques for *Agrobacterium*-mediated transformation and more recently biolistic introduction of foreign genes that are effective in a wide range of agronomically important species have been developed over the last 15 years. The products of these novel molecular breeding technologies are now commercially available and include agronomically valuable traits, such as herbicide tolerance, disease resistance and pest resistance (Birch, 1997). In addition to agronomic improvement transgenic plants also have the potential for exploitation as bioreactors for the bulk production of desirable proteins (for reviews see: Fischer *et al.*, 1999a; 1999b; Porta and Lomonossoff, 1996).

Despite the many benefits of stable plant transformation there are circumstances where high-level transient gene expression, using virus-based vectors, may be of more use than expression from stably integrated transgenes. In basic research applications the ability to introduce foreign genes to mature, differentiated tissue avoids problems associated with toxicity of foreign gene products during regeneration of transgenic plants. In addition viral vectors can be prepared and tested *in planta* in a matter of weeks whereas stable transformation and the subsequent regeneration of transformants can be a lengthy process that is likely to take several months to provide primary transformants and even longer for production of progeny transgenic lines. Thus, even where the ultimate goal is stable transformants, the use of transient gene expression *via*

Genetically Engineered Viruses: Development and Applications, edited by C.J.A. Ring and E.D. Blair

a virus-based vector can allow rapid screening and evaluation of gene function, allowing selection of optimal transgenes before embarking on the time-consuming process of preparing transgenic plant lines. In terms of transient gene expression systems viruses can also be compared with plasmid-based, non-integrative, strategies for foreign gene expression. Plasmid-based gene expression has been of particular use for promoter mapping, in which expression levels of reporter genes can be correlated to the structure of the promoter. However, for protein expression studies this approach is limited by the cell autonomous nature of gene expression, compared with viruses that encode their own intercellular transport functions, and the relatively low levels of foreign gene expression compared with autonomously replicating virus-based vectors. In addition to applications in basic research the extremely high level of gene expression achievable with certain viral vectors can also provide commercial advantages where plants are being used as bioreactors for the over-production of biologically active proteins. Ongoing progress in developing reliable viral vectors, which give rapid systemic infections, is likely to open up a variety of opportunities for commercial production of recombinant proteins as well as providing a powerful tool for gene discovery.

The use of plant viruses as gene expression tools has been reviewed in some detail previously (Lacomme *et al.,* 1998; Lomonossoff and Johnson, 1995; Palmer and Rybicki, 1997; Porta and Lomonossoff, 1996; Scholthof *et al.*, 1996). The objective of this chapter is to summarize the various virus-based vector technologies that have been developed and discuss the applications for which different types of viral vector have been exploited. The literature describing virus-based gene expression in plants has increased substantially over the last few years and it is intended to highlight the applications of virus-based vectors that have proven most successful with particular reference to recent developments.

4.2 General properties of plant viruses

A common feature of both DNA and RNA viruses is their efficient use of a limited amount of genetic material to direct their amplification and propagation. For plant viruses a complement of core genes, encoding functions associated with replication and intercellular transport are basic prerequisites for pathogenicity. In the case of the majority of plant viruses all the necessary functions for replication, spread and transmission to new hosts are encoded by genomes of 10 kb or less and a variety of expression strategies have evolved to direct efficient production and regulation of viral gene products (Drugeon *et al.*, 1999).

Other virus-encoded functions, at least when considering the construction of a vector, can be considered as either core or peripheral depending on the particular virus in question. Thus the coat protein (CP) of some plant viruses is indispensable for cell-to-cell movement whereas for other viruses encapsidation is a prerequisite for systemic but not local movement. For potyviruses, and other viruses belonging to the picornavirus superfamily, virus-encoded proteolytic activities are essential for processing of the viral polyprotein and are thus indispensable for viability of the virus. In contrast other viral gene products, such as those encoding proteins involved in the suppression of gene silencing (see Section 4.4.4), will be of greater or lesser importance depending upon the specific application for which the vector is to be used. Finally, viruses that are transmitted by either arthropod, nematode or fungal vectors usually encode proteins required for transmission, such genes are frequently dispensable for other aspects of the virus lifecycle and can be replaced (or deleted) without adversely affecting viral replication or movement. The following section briefly outlines the salient features of genome structure and function for the three types of plant virus that have been tested as gene expression vectors.

4.2.1 Structure, organization and replication of viral genomes

Double-stranded DNA viruses. Caulimoviruses are pararetroviruses that have narrow host-ranges and infect only dicotyledonous plants. Replication of the ~8 kb double-stranded (ds) DNA genome occurs by reverse transcription of the genome-length, terminally redundant, RNA. Replication of CaMV, the best studied of all the caulimoviruses, has been intensively investigated and has been described in detail elsewhere (Covey, 1991). The circular dsDNA genome contains gaps at specific sites, one in the α-strand (the strand used for transcription) and two in the complementary strand. There are two phases in the replication cycle, in the first, the dsDNA of the infecting particle moves to the cell nucleus, where the gaps are covalently closed to form a fully dsDNA and virion DNA is supercoiled presumably by a host topoisomerase. These minichromosomes form the template used by the host enzyme, DNA-dependent RNA polymerase II for RNA transcription. The RNAs migrate to the cytoplasm for the second phase of the replication cycle. The genome-length polycistronic 35S RNA with a terminal redundancy of 180 nucleotides serves as a template for viral replication by a virus-encoded reverse transcriptase synthesizing the DNA minus strand. Following synthesis of the negative strand DNA the 35S RNA template is degraded by an RNase H activity and synthesis of the plus strand DNA then occurs. Progeny viral dsDNA is then packaged in icosahedral particles.

Single-stranded DNA viruses. Geminiviruses are plant viruses with a genome that comprises either one or two single-stranded (ss), circular DNAs that encapsidate in virions which appear as twinned (geminate) icosahedra. Three groups of geminiviruses are currently recognized (Palmer and Rybicki, 1997). The first, the mastrevirus which includes maize streak virus (MSV) and wheat dwarf virus (WDV), have monopartite genomes and are transmitted between plants by whiteflies. The mastreviruses have narrow host ranges that are restricted to monocotyledenous plants. The sole member of the second group (curtovirus) is beet curly top virus, which has a monopartite genome, is leafhopper transmitted, and infects dicotyledonous plants from several dozen families. Members of the third group of geminiviruses, the begomovirus, which includes tomato golden mosaic virus (TGMV) and African cassava mosaic virus (ACMV), are whitefly-transmitted viruses with bipartite genomes that infect dicotyledonous plants.

Replication of geminiviruses occurs in the nucleus and requires the concerted action of virus-encoded and cellular enzymes (reviewed by Timmermans *et al.*, 1994). Replication is believed to occur by a rolling-circle mechanism similar to that employed by M13 bacteriophage. The proposed sequence of events in geminivirus replication, following virus infection, and transport of the ssDNA genome to the host cell nucleus, are as follows: (i) host-encoded enzymes convert the ssDNA into the ds replicative form (RF), followed by (ii) replication of the RF and (iii) production of viral ssDNA. Both complementary strand synthesis at the beginning of the viral replication cycle and replication of the RF intermediate depend only on host functions. Synthesis of progeny, genome sense, ssDNA occurs *via* a rolling-circle replication mechanism requiring endonuclease, helicase and ligase activities all of which are encoded by viral proteins. The two genomic DNAs found in the bipartite geminiviruses, are termed A and B. Only the A component is capable of autonomous replication in plant cells, whereas the B component provides proteins necessary for intracellular, intercellular and systemic movement. The single genomic DNA of the monopartite geminiviruses can be regarded as equivalent to the DNA A component of the bipartite viruses, to which small genes required for intercellular movement have been added. The autonomously replicating geminivirus genomes range in size from approximately 2.6–3 kb and encode between four and six proteins in both the virion-encapsidated strand and the complementary strand.

Single-stranded, positive sense RNA viruses. A total of 28 groups of plant viruses with positive sense ssRNA genomes have been established that together represent more than 70% of the known plant viruses. Replication of these viruses occurs in two main stages: (i) translation of replication associated protein(s) allowing synthesis of a complementary (negative strand) RNA using the genomic positive-strand RNA as a template; (ii) synthesis of progeny positive-strand RNA (and for members of the alphavirus superfamily subgenomic mRNAs) using the negative-stranded RNA as a template. At least one virus-encoded protein, an RNA-dependent RNA polymerase (RdRp), is required to catalyse RNA synthesis directed by the RNA template. To enable specific and efficient amplification of viral RNA, the virus encoded polymerase enzymes form complexes with host-encoded factors and with other virus-encoded proteins that interact with the *cis*-acting regulatory elements on viral RNAs. Several models of replication, involving different structural intermediates, have been the subject of recent reviews (Buck, 1996, 1999) and will not be discussed here.

In general the 5′-terminal open reading frame (orf) of RNA viruses is directly translated from the genomic RNA and, as with animal RNA viruses, a variety of strategies are exploited to direct translation of internal orfs carried on the viral genome. Mechanisms exploited to ensure translation of all virus-encoded gene products include: (i) synthesis of subgenomic RNAs; (ii) translation of polyproteins that are subsequently processed into individual mature proteins; (iii) multipartite genomes; (iv) read-through proteins and (v) *trans*frame proteins. Both plant and animal ssRNA viruses are classified into supergroups (alpha-, carmo-, sobemo-, picorna- and corona-like viruses) based on genome organization and gene expression strategies (reviewed by Drugeon *et al.*, 1999; Maia *et al.*, 1996).

4.2.2 Movement of plant viruses

For effective colonization of their hosts and thus for most applications of virus-based vectors, a key requirement is the ability of the virus to move, both between cells and organs, within the infected plant. Intercellular movement of plant viruses presents problems not encountered by animal viruses and as a consequence plant viruses have evolved unique strategies for both cell-to-cell and long-distance transport.

The idea that viruses move through plants in two distinct modes was first proposed over 60 years ago (Samuel, 1934) and it is now firmly established that plant viruses move from cell to cell and over long distances by exploiting and modifying pre-existing pathways for macromolecular transport within cells, between cells, and between organs. Because of the importance of efficient cell-to-cell movement in developing a viable plant virus-based vector a brief overview of virus movement processes is presented below, however, for a more comprehensive discussion of this topic the reader should refer to several recent reviews (Carrington *et al.*, 1996; Ding, 1998; Ghoshroy *et al.*, 1997; Santa Cruz, 1999).

Cell-to-cell movement. The cell-to-cell movement of plant viruses requires the transport of newly synthesized genomes to and through plasmodesmata, the small membrane-lined channels that interconnect plant cells. However, both viruses and viral nucleic acids are substantially larger than the functional channel presented by plasmodesmata, which must therefore be modified before viral movement can occur. To facilitate this process plant viruses encode functions specifically required for transport and movement proteins (MPs), with dedicated functions in intercellular transport have been identified in most families of plant viruses. The process of cell-to-cell movement can be considered as involving three distinct steps: (i) transfer of newly synthesized genomes from sites of replication (nucleus or cytoplasm) to the intracellular transport machinery; (ii) directed transport of genomes to plasmodesmata; and (iii) transit through plasmodesmata. Two distinct mechanisms of viral intercellular transport have been well documented. In one strategy, exemplified by TMV, the viral nucleic acid associates with a virus-encoded MP that is able to transport the nucleic acid to and through the plasmodesmata by transiently increasing the plasmodesmal pore size. In the second strategy, exemplified by cowpea mosaic virus (CPMV), the virus-encoded MP assembles into a proteinaceous tubule that is inserted through the cell wall and which provides a conduit *via* which intact virus particles are transported to adjacent cells. For other viruses, including some that have been widely used as vectors such as potato virus X (PVX) and tobacco etch virus (TEV), the precise mechanism of movement is still poorly understood.

Long-distance movement. At the organismal level, the movement pathway for a systemically infecting virus involves traversing several cell types to gain access to the phloem, which serves as the long-distance transport pathway. Phloem-dependent movement requires that the virus be able to enter and exit sieve elements, most probably *via* their associated companion cells. The sieve elements, which serve as the conducting elements of the phloem, allow rapid virus movement (centimetres/hour) by bulk flow to tissues that are sinks for photoassimilate. The unloading of viruses in sink tissues requires exit from sieve elements into companion cells where viral replication and accumulation can occur. Subsequent cell-to-cell movement then allows newly invaded leaves to become fully infected. Two critical points along the long-distance movement pathway are the entry into and exit from sieve elements. Plasmodesmata connecting a sieve element with its supporting companion cell possess a unique morphology and have gating capacities that differ from those between mesophyll cells. Phloem entry and/or exit appears to present a 'bottle-neck' to virus movement and the process may involve a novel set of viral and host factors (Carrington *et al.*, 1996). It is a common experience that viral genomes modified for foreign gene expression are either completely debilitated or severely restricted in their long-distance transport function when compared to the progenitor, wild-type virus.

4.3 Development of plant virus-based vectors

4.3.1 General considerations

Historically, interest in the potential of virus-based vectors for plant transformation dates back to the early 1980s when the lack of robust genetic transformation systems for plants was recognized as an impediment to basic studies in plant molecular biology (Grill, 1983; Howell, 1982; van Vloten-Doting, 1983). With the isolation and characterization of DNA viruses the earliest attempts to derive autonomously replicating gene expression vectors were based on the use of members of the two groups of plant DNA viruses, the caulimoviruses and geminiviruses. Despite significant early successes in directing foreign gene expression *via* DNA virus-based vectors a number of factors have hindered the widespread development and utilization of DNA viruses as vectors. Recombination, leading to the deletion of foreign genes, is generally found to occur rapidly from DNA virus-based vectors due to limitations imposed on the size of foreign inserts due to packaging or other unidentified constraints (Scholthof *et al.*, 1996). The cloning of full-length cDNA copies of plant RNA viruses from which infectious transcripts could be synthesized (Ahlquist *et al.*, 1984; Dawson *et al.*, 1986; Meshi *et al.*, 1986) paved the way for the development of modified clones that could serve as vectors for gene expression. A chronology of the evolution of plant virus-based vectors is presented in *Figure 4.1*.

Ultimately the engineering of an effective vector for gene delivery will depend to a large extent on the level of understanding of the virus concerned and for obvious reasons the plant viruses that have been adapted for use as gene expression vectors are those viruses that are best characterized at the molecular level. Nevertheless, a detailed knowledge of the molecular biology of any given virus is no guarantee that a useful vector can be derived from an infectious cDNA clone. Over and above the requirements for an infectious clone (or clones in the case of viruses with multipartite genomes) the choice of a particular virus for development of an expression vector will depend on several features including:

(i) Facile genetic manipulation of the viral genome *via* a full-length infectious clone.
(ii) Ease of mechanical infection (with either RNA or DNA inocula).
(iii) Genetic stability of the modified viral genome.
(iv) Optimized viral functions (high-level of replication, efficient promoter(s) for foreign gene expression, efficiently translated mRNA, rapid local and systemic movement).

1984:	CaMV ORF II replacement (DHFR)	Brisson *et al.*
1986:	BMV CP replacement (CAT)	French *et al.*
1987:	TMV CP replacement (CAT)	Takamatsu *et al.*
1989:	TMV gene insertion (CAT)	Dawson *et al.*
1990:	TMV CP carboxy-terminal peptide presentation	Takamatsu *et al.*
1991:	TMV/ORSV insertion (NPTII & DHFR)	Donson *et al.*
1993:	CPMV peptide presentation (FMDV VP1) TMV CP carboxy-terminal readthrough (ACEI) TMV/ORSV insertion (α–trichosanthin)	Usha *et al.* Hamamoto *et al.* Kumagai *et al.*
1994:	CyRSv DI complementation (TBSV & TAV CP)	Burgyan *et al.*
1996:	PVX amino-terminal CP fusions (GFP) BaMV satellite complementation (CAT)	Santa Cruz *et al.* Lin *et al.*
1997:	CPMV-mediated protection (MEV VP2)	Dalsgaard *et al.*
1998:	TMV-based functional screening of cDNA library TMV-based expression of MAb (colon cancer antigen) PVX CP-scFv fusion	Karrer *et al.* Verch *et al.* Smolenska *et al.*
1999:	TMV scFv-mediated immunotherapy (NHL)	McCormick *et al.*

Figure 4.1. Highlights in the development of plant virus-based expression factors.

(v) No size constraint for genome packaging.
(vi) The availability of hypovirulent virus strains (causing mild symptoms).
(vii) Wide host range.
(viii) No restriction to specific cell or tissue types.

For any given virus the overall genome organization and gene expression strategies, a knowledge of which genes are essential and which, if any, are dispensable, and the requirement for *cis*-acting regulatory sequences will determine the possible strategies that could be used for foreign gene expression (*Figure 4.2*). The remainder of this section discusses different strategies that have been developed for foreign gene expression from plant viral vectors.

4.3.2 Gene replacement vectors

The earliest attempts to construct plant virus-based vectors used a gene replacement strategy in which a viral gene, which is not essential for replication, is replaced by the gene for prospective expression. In the case of CaMV the gene II product, which is responsible for aphid transmission, was replaced by the sequence encoding either dihydrofolate reductase (DHFR) or human α-D-interferon (Brisson *et al.*, 1984; De Zoeten *et al.*, 1989). These studies in plant virus-based vector development took advantage of the genome composition of CaMV (dsDNA virus) that made cloning and manipulation easier in the absence of reverse transcription techniques. Taking account

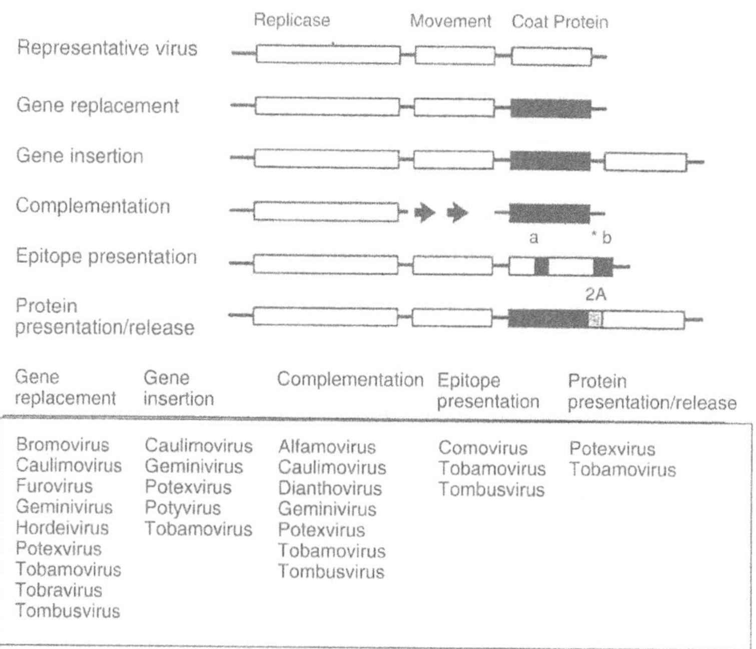

Figure 4.2. Strategies used to express a foreign gene (black box) by different pant viruses. White boxes indicate native viral genes. The epitope presentation method (a) indicates a translation fusion of a small peptide sequence inside the coat protein gene or (b) translational readthrough of an amber stop codon (*) at the 3′-end.

of all the non-essential CaMV sequence that could be deleted places a limit on insert size of approximately 1 kb. However, in addition to the size constraint on the inserted sequence, a further consideration is the genomic instability of genetically engineered CaMV that results in rapid loss of inserted sequence with a consequent reduction in foreign protein accumulation.

The ssDNA geminiviruses have also been investigated as potential gene replacement vectors and, at least in theory, they have several characteristics that make them suitable candidates for the development of expression vectors: (i) their replication mode, which involves a double-stranded DNA intermediate, should be less error prone than replication of either caulimoviruses, which replicate *via* a reverse transcribed RNA intermediate, or RNA viruses; (ii) the high copy number of template reached in the nuclei of infected cells leading to high levels of viral gene expression and (iii) the geminiviruses infect a wide range of crop species including several agronomically important monocotyledonous crops. The finding that the CP of ACMV is not required for systemic infection of tobacco suggested that this gene could be replaced with a foreign gene. Manual inoculation of *Nicotiana benthamiana* with a recombinant ACMV DNA A, where the CP was replaced by CAT, together with a wild-type DNA B resulted in symptoms similar to those induced by the wild-type virus and gave rise to high-level expression of CAT (80 U mg^{-1} soluble protein, Ward *et al.*, 1988). However, although in some cases geminivirus DNA is infectious when mechanically inoculated onto plants, many geminiviruses are not infectious by this route and can only be introduced either by insect feeding or, as discussed below, agroinoculation.

Agroinoculation involves T-DNA-mediated delivery of viral genomes into plant cells. To achieve this multimeric copies of the viral DNA genome are inserted between the T-DNA borders of a binary vector allowing stable integration in the host genome. Excision of a monomeric genome, by either recombination or replication, results in extrachromosomal replicating copies of the genome. In experiments similar to those described for ACMV above, the CP orf from DNA A of TGMV was replaced by the *neo* gene, which encodes neomycin phosphotransferase II (NptII). Following agroinoculation with either monomeric or partially dimeric copies of the vector DNA it was established that only the partial dimer genome was able to 'escape' and give rise to extrachromosomal replicons. Levels of NptII activity were approximately ten-fold higher in plants transformed with the partial dimer of DNA A compared to plants transformed with the monomeric form, reflecting an increase in gene expression due to high-level accumulation of the extrachromosomal DNA (Hayes *et al.*, 1988). When the transforming binary vector carried multimeric DNA B, to provide movement functions, in addition to the partially dimeric DNA A, systemic movement of the vector virus was observed and NptII levels were further elevated to levels approximately two-fold higher than seen in the absence of DNA B (Hayes *et al.*, 1988). The highest levels of NptII activity obtained were seen when the partial dimer of DNA A-*neo* was agroinoculated to stably transformed plants carrying the dimeric DNA B. In this case the level of NptII activity was six-fold higher than in non-transformed plants agroinoculated with the same DNA A construct (Hayes *et al.*, 1988). In a subsequent study it was established that when recombinant DNA A molecules carrying the *UidA* gene, which encodes β-glucuronidase (GUS), were stably integrated into the host genome the 'escaped' extrachromosomal DNA replicated and maintained the *UidA* insert (Hayes *et al.*, 1989). In contrast when the same modified DNA A genome was introduced to DNA B transgenic plants by agroinoculation the insert sequence was rapidly lost. These results suggest that expression of genes over approximately 1 kb in length from a TGMV-based vector would require integrated 'master copies' of the genome in a stably transformed plant. Other studies, involving replacement of the CP gene in the DNA A component of the bipartite geminiviruses, have essentially given similar results; small gene insertions allow replication and movement of the viral genome through plants (in the presence of DNA B), large insertions move poorly and show genetic instability (for reviews see Palmer and Rybicki, 1997; Timmermans et al., 1994). In considering the use of geminiviruses as vectors it is also worth noting that the majority of geminiviruses are either completely or largely restricted to phloem and phloem-associated tissues and thus do not provide a good means for achieving whole-plant infections.

The first successful attempt to express a foreign gene from an RNA virus was the demonstration of chloramphenicol acetyl transferase (CAT) activity in protoplasts transfected with a modified brome mosaic virus (BMV) genome. This pioneering study used a modified RNA3 of BMV in which the CP ORF was replaced by the gene encoding CAT (French *et al.*, 1986). Although this vector produced CAT in infected plant protoplasts it was incapable of cell-to-cell movement, and therefore effective gene expression *in planta*, due to a requirement for CP for intercellular transport. The first RNA viral vectors that were capable of cell-to-cell movement in plants were based on infectious cDNA clones of TMV in which the CP orf was replaced with the gene encoding CAT. These TMV-CAT vectors showed local (cell-to-cell) movement on inoculated leaves and infected leaves were shown to accumulate CAT by *in vitro* enzyme activity assays (Dawson *et al.*, 1988, 1989; Takamatsu *et al.*, 1987). Determination of CAT activity by Takamatsu *et al.* (1987) led to an estimate of 1 μg CAT per gram of fresh leaf tissue, however, in these experiments no CAT activity was detected in non-inoculated leaves. The lack of systemic movement of the chimeric TMV genome was not unexpected as neither CP nor assembly-deficient forms of TMV enter the vascular system, suggesting that the TMV CP is a prerequisite for efficient long-distance movement (Dawson *et al.*, 1988). Since the initial reports describing plant virus-based gene replacement vectors, infectious clones of a wide range of different viruses have become available and many have been tested for their potential to serve as vectors (Boyer and Haenni, 1994). Most available gene replacement vectors do not move systemically, which limits their utility in biotechnological applications. In other applications, however, for example screening genes for function, this is not a drawback as the analysis can be limited to the inoculated leaf.

4.3.3 Gene insertion vectors

As stated previously most plant viruses have extremely compact genomes of 10 kb or less and with few exceptions gene replacement strategies were found to result in vectors that were compromised in accumulation and/or local and systemic movement. The next advance in the evolution of plant virus-based vectors was the move from gene replacement vectors to gene insertion vectors to avoid the adverse consequences associated with gene deletion. Despite the potential advantages of gene insertion vectors the issue of packaging constraints becomes more serious as any additional sequence increases the genome length compared to the wild-type progenitor virus. For this reason it was expected that viruses that encapsidate their genomes in rod-shaped particles (e.g. the tobamoviruses, potexviruses and potyviruses) would be more tolerant of larger gene insertions as there is no *a priori* limit on the size of genome due to packaging constraints.

The first gene insertion vectors based on a rod-shaped virus used TMV. In early experiments two gene insertion vectors were prepared; TMV-CAT-CP and TMV-CP-CAT, which differed in the relative position of the CAT and CP orfs. In both cases expression of the 3′-terminal gene in the viral genome was under the transcriptional control of a duplicated sequence spanning the CP subgenomic promoter. In the case of TMV-CAT-CP, in which the CP remains at the 3′-terminus of the viral genome, a 252-nucleotide duplication served as promoter for the CP. This modified virus moved efficiently cell to cell on inoculated leaves and CAT activity was detected (Dawson *et al.*, 1989). In contrast TMV-CP-CAT, in which transcription of the 3′-terminal CAT orf was driven by a 628-nucleotide duplication, accumulated poorly in inoculated leaves and only low levels of CAT activity were detected. Significantly, CAT activity was occasionally detected in upper non-inoculated leaves of plants infected with TMV-CAT-CP suggesting that the modified virus was competent for long-distance movement. Analysis of cDNA prepared from virus in systemically infected tissue indicated that eight of nine clones analysed showed evidence of homologous recombination to regenerate wild-type virus (Dawson *et al.*, 1989). Although the outcome of these experiments was disappointing, in that recombination led to the rapid elimination of foreign sequence from the viral population, the fact that systemic movement of modified virus was observed marks these experiments as a landmark in the development of plant virus-based vectors.

Subsequently, more stable TMV vectors have been developed by mixing sequences from distantly related tobamoviruses to limit the propensity for homologous recombination within the duplicated promoter region. A hybrid viral RNA was constructed containing functionally similar sequences encompassing the CP subgenomic promoter from two tobamoviruses (TMV-U1 and odontoglossum ringspot virus (ORSV)). The 203 nucleotides duplicated sequences upstream of the TMV-U1 and ORSV CP initiation codons exhibit only 45% homology at the nucleotide sequence level. This ORSV CP-promoter region along with downstream sequences spanning the ORSV CP and 3′ untranslated region was positioned 3′ to the native TMV-U1 CP-promoter, thus freeing up the upstream promoter to drive transcription of foreign gene sequences (Donson *et al.*, 1991). Expression of DHFR and NptII was analysed and in both cases systemic movement was observed, however, in these experiments no deletion of foreign gene inserts was observed. Moreover, the *dhfr* insert was stably maintained through 10 serial passages (reinoculation of non-infected plants with virus extracted from an infected plant), representing 170 days of propagation from inoculation time. The *neo* insert was less stable and loss of the inserted sequence was observed to occur after between 12 to 47 days of propagation (Donson *et al.*, 1991). This difference in stability between the two constructs can be explained by nucleotide composition of the foreign sequences themselves or a deleterious effect, due to a larger size of the *neo* gene (832 nucleotides instead of 238 nucleotides for *dhfr*), affecting virus fitness. These results represented an important advance in the development of systemic plant virus-based expression vectors and improved versions of the vector described above remain the expression system of choice for many applications (Kumagai *et al.*, 1993, 1995; Shivprasad, 1999; Turpen, 1999). Although the development of TMV-based gene insertion vectors represented a major breakthrough the levels of NptII and DHFR accumulating in infected plants were much lower than the level of CP accumulating in either a wild-type TMV infection or the heterologous ORSV CP during infection with the vector. In part this lower promoter activity is a consequence of the fact that the vector carries an extra promoter competing for transcription factors. It also reflects the increased distance between the promoter driving foreign gene expression and the 3′-terminus of the viral genome as 3′-proximal promoters tend to be more actively transcribed than more distal promoters (Culver *et al.*, 1993).

In order to maximize the gene expression potential of tobamovirus-based vectors a comprehensive study was conducted by screening different permutations of promoter, CP and 3′-untranslated sequences from tobamoviruses in a TMV U1-based vector backbone (Shivprasad *et al.*, 1999). The most effective vector in this series contained sequences encoding the subgenomic promoter, CP and 3′-untranslated region from the U5 strain of TMV. Green fluorescent protein (GFP) expressed from this vector (30B) accumulated at levels of up to 10% of total soluble protein in leaves (Shivprasad *et al.*, 1999).

In the case of TMV-based vectors the need to minimize the degree of homology between the additional promoter and the native CP promoter reflects the fact that the necessary promoter sequences for high-level gene expression are relatively long (>150 nucleotides). In the case of vectors based on PVX much shorter duplications, down to 41 nucleotides, can be used to drive efficient transcription of the CP subgenomic RNA (Baulcombe *et al.*, 1995; Chapman *et al.*, 1992). Thus PVX-based vectors use shorter duplications than required in TMV-based vectors resulting in a relatively low level of homologous recombination without the need to employ sequences from heterologous potexviruses. Like the tobamoviruses the potexviruses also encapsidate in rod-shaped particles allowing for large gene inserts without constraints imposed by packaging. Insertion of the 1.8 kb *UidA* gene into a PVX-based vector led to efficient GUS expression in both inoculated and systemically infected leaves (Chapman *et al.*, 1992). Although foreign gene expression levels are generally lower from PVX-based vectors compared to TMV, one advantage of PVX is its ability to tolerate relatively large gene insertions and still give systemic infections.

The potyvirus group provides another example of rod-shaped viruses that have been developed as systemic gene insertion vectors. Potyviruses, as members of the picornavirus superfamily, express their genomes as polyproteins that are proteolytically processed to generate the mature viral gene products. Thus, to engineer a potyvirus-based vector requires that genes to be expressed are inserted in frame with the viral polyprotein and carry appropriate protease target sequences at one or both ends of the insert to permit proteolytic release of the foreign gene product. The first potyvirus to be successfully modified as a vector was TEV, insertion of the GUS gene between the amino-terminal P1 proteinase and the helper component-proteinase (HC-Pro) led to the production of an active GUS-HC-Pro fusion protein (Dolja *et al.*, 1992). Because of the insertion strategy employed when engineering potyviral vectors the lack of duplicated viral sequence eliminates the problem of homologous recombination seen in vectors prepared from viruses in the alphavirus superfamily carrying duplicated promoter sequence. However, as with all plant virus-based vectors generated to date, recombinant forms of the vector arise through non-homologous recombination (Dolja *et al.*, 1992, 1993). Despite the problem of recombination the TEV-based vector has been used to produce recombinant proteins in plants at levels of up to 10 $\mu g\ g^{-1}$ of fresh leaf tissue (Dolja *et al.*, 1998).

4.3.4 Peptide and protein presentation systems

An alternative to free protein expression from viral vectors is the use of protein fusion vectors in which the foreign gene to be expressed is fused to a virus-encoded protein. In practice this approach has largely focused on fusions between short peptide sequences and the viral CP with the objective of retaining an assembly competent CP that displays the incorporated peptide at the surface of the viral particle.

This expression strategy avoids the problems associated with duplicated promoter sequences in the viral genome that can lead to RNA recombination. An in-frame translational fusion to a highly expressed viral gene, for example the CP gene, produces large amounts of the foreign protein or peptide sequence fused to the CP. Based on the knowledge of 3-dimensional viral structure or serological data, the sites of insertion are chosen so that the resulting peptides project outwards on the virus particles in order to minimize interference in particle assembly. This strategy is attractive since the unique physico-chemical characteristics of plant virus particles should allow easy purification of modified virus. The most extensively tested application of peptide presentation on plant virus particles has been aimed at producing virus-borne vaccines by incorporation of antigenic peptide epitopes derived from animal pathogens and this application of peptide fusion vectors is discussed more fully in Section 4.5.2.

The use of fusions to the CP for peptide presentation at the particle surface has been successfully employed for viruses with icosahedral, bacilliform and rod-shaped particle morphologies (see *Figure 4.3*). TMV has been widely used for peptide presentation and several approaches have been developed to generate assembly-competent CP subunits. The earliest example of a TMV CP-fusion attempted to place a pentapeptide corresponding to Leu-enkephalin to the carboxy-terminus of the viral protein, however, this modification prevented assembly of virus particles (Takamatsu *et al.*, 1990). In contrast a random heptapeptide, tested during the same series of experiments, did not block particle assembly thereby demonstrating the feasibility of this strategy. A later development of this approach was designed to produce mixed pools of both free CP subunits together with the peptide-CP fusion subunits. This was achieved through the use of a translational-readthrough sequence, present within the TMV RdRp gene, that was duplicated and inserted between the carboxy-terminus of the TMV CP gene and sequence encoding the 12-amino acid peptide, angiotensin-I-converting enzyme inhibitor (ACEI). Incorporation of the sequence that promotes leaky readthrough of the CP stop codon led to the production of both free TMV CP and 3′-terminal CP-fusion. The modified virus, expressing the CP-ACEI fusion protein, accumulated at up to 100 $\mu g\ g^{-1}$ fresh tissue, with a ratio of modified CP/unmodified CP of between 1/200–1/20 (Hamamoto *et al.*, 1993). A similar strategy was used to express a 15-amino acid epitope from the malarial parasite fused to the carboxy-terminus of the TMV CP. In this study the modified virus was recovered at yields up to 1.2 mg g^{-1} fresh weight in systemically infected tissue (Turpen *et al.*, 1995).

The TMV CP has long been a model for protein structural studies (Namba and Stubbs, 1986) and attempts to modify the TMV CP have benefitted greatly from the high-resolution data on the structure of the TMV CP and assembled virion. This information allows a targeted approach for the design of surface-modified CP subunits. The core of a single TMV-CP subunit contains a bundle of four antiparallel α-helices projecting away from the RNA binding site. Three copies of the AGDR epitope sequence from the *Plasmodium vivax* CS protein (a total of 12 residues) were inserted in the surface loop between two α-helices (Turpen *et al.*, 1995). The modified virion was purified to high yield (0.4 mg g^{-1} fresh tissue) and cross-reacted with a monoclonal antibody raised against the AGDR epitope in both Western blots and ELISA tests.

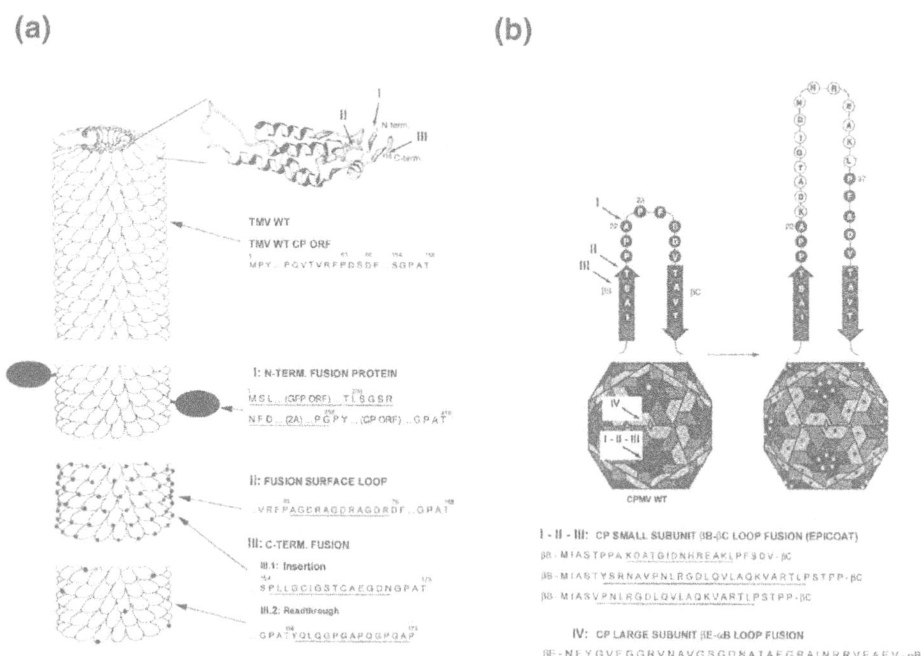

Figure 4.3. Illustration of epitope and protein presentation on the surface of virus particles. (a) Chimeric TMV. The upper part of the figure represents a wild-type TMV particle with its genomic RNA and CP subunits assembled to form a rod-shaped particle. On the upper right hand side a CP subunit is represented by a ribbon drawing and the CP residues used to generate chimeras either by fusion to the amino-terminus, surface loop or carboxy-terminus of the CP are indicated and numbered. Four representative examples of protein (I) or peptide (II and III) fusions are shown respectively as large black ellipses or small black dots. (b) Chimeric CPMV. The upper part of the figure illustrates residues in the βB-βC loops of CPMV and the insertion sites used for peptide presentation on the Small CP submit (I, II and III detailed below). The upper right hand side illustrates the insertion of the HRV-14 epitope (white circles indicate introduced amino acids). The lower part of the figure represents a CPMV virus particle decorated with foreign epitopes drawn as white or black dots when fused respectively to the Small (black) and Large (grey and light grey) CP subunits. Four representative examples of peptide presentation are indicated below (I, Porta *et al.*, 1994; II, Usha *et al.*, 1993; III, Lomonossoff and Johnson, 1996; IV, Brennan *et al.*, 1999b).

Other viruses with rod-shaped particles have also been tested for their utility as peptide presentation systems and in the case of PVX an approach was developed to allow relatively large fusions to the viral CP. Despite the lack of an X-ray-derived structure for PVX serological evidence indicated that the amino-terminus of each PVX CP subunit is exposed on the outer surface of the virus particle. As mentioned previously, PVX is encapsidated as a flexuous, rod-shaped particle which for the wild-type virus comprises approximately 1200 copies of CP per assembled virion. The flexible nature and relatively high helical pitch of the viral particle are probably both properties that contribute to the tolerance of PVX to incorporation of CP-fusion proteins into virus particles. The first successful application of a protein fusion vector strategy to PVX involved a fusion between the 27 kDa GFP and the 25 kDa viral CP (Santa Cruz *et al.*, 1996). In the case of this fusion it was found that assembly of virus particles would only occur in the presence of a pool of free (i.e. unfused) CP. Although this need for free CP could reflect an absolute requirement for unfused protein in nucleation of particle assembly it is more probable that steric constraints prevent the assembly of particles comprised solely from fusion protein. In order to provide both fused and free CP, a 16-amino-acid fragment of the 2A peptide from foot-and-mouth-disease virus (FMDV) was inserted between the GFP and CP components of the synthetic fusion protein. The short, 18-amino-acid, 2A region of apthoviruses is believed to be responsible for the hydrolytic cleavage of the viral polyprotein (Donelly *et al.*, 1997; Ryan *et al.*, 1991). Although the precise nature of the 2A-mediated cleavage is not fully understood the FMDV 2A sequence can cause co-translational cleavage of synthetic polyproteins both *in vitro* and *in vivo* (Ryan and Drew, 1994). By using only the carboxy-terminal 16 amino acids of the 2A peptide in the GFP-2A-CP fusion protein a cleavage efficiency of approximately 50% was achieved. Thus, in infected cells approximately half the CP was present as a fusion to GFP and the remainder was free CP (Santa Cruz *et al.*, 1996).

This approach has proven to be successful in the expression of larger fusion proteins than has been possible using other virus vectors and proteins of up to approximately 31 kDa have been expressed using this system (Santa Cruz *et al.*, 1996). Obviously peptide fusions are also possible using fusions to the PVX CP as recently demonstrated with a 38-amino-acid peptide fusion from the *Staphylococcus aureus* fibronectin-binding protein. PVX virions decorated with this peptide have been purified and used to raise an immune response against the fused peptide (Brennan *et al.*, 1999a). The strategy above, using the FMDV 2A peptide to create mixed pools of free and fusion CPs, has also been extended to another rod-shaped virus TMV (*Figure 4.3*) (C.L. and S.S.C. unpublished results). It is thus possible that other rod-shaped viruses will tolerate incorporation of large CP-fusion proteins into assembled virus particles.

As noted above, icosahedral viruses have also been exploited as peptide fusion vectors and considerable success has been obtained using CPMV as an epitope presentation system. As with TMV the ability to make viable CP-peptide fusions with CPMV was facilitated by the availability of high-resolution structural information on the CP subunit and assembled virion (Lomonossoff and Johnson, 1991). The icosahedral CPMV virion comprises 60 copies each of a large and a small CP subunit. Inserts between approximately 15–30 amino acids in the βB-βC loop in the small CP subunits (S-CP) become clustered around the 12 five-fold axes of symmetry on the outer surface of virus particles (*Figure 4.3*) (Usha *et al.*, 1993). Once it had been determined that all the S-CP amino acid sequence had to be retained and an optimal insertion site existed, it became possible to engineer genetically and structurally stable CPMV particles containing epitopes from FMDV, human rhinovirus 14, human immunodeficiency virus type 1 (HIV-1) and mink enteritis virus (MEV) (Dalsgaard *et al.*, 1997; Lomonossoff and Johnson, 1995; Porta *et al.*, 1994). Purified particles containing epitopes yielded around 1 mg virus g^{-1} leaf tissue and raised antibodies directed against the expressed peptide, when injected into rabbits or mice. Although an excellent vector for the presentation of short peptide sequences one limitation of the CPMV vector is the lack of assembly and intraplant movement in constructs engineered to carry peptides greater than about 30 amino acids (most epitopes tested ranged from 14–30 residues). Peptide display of antigen has also been developed using TBSV, another icosahedral virus, as vector. In this case a short, 13 amino acid, epitope from the HIV-1 glycoprotein 120 was shown to elicit an antibody response when purified virus was injected into mice. Interestingly this vector remained genetically stable over six plant-to-plant passages, suggesting that the TBSV CP is relatively tolerant of small insertions (Joelson *et al.*, 1997).

To overcome the limitation on the size of peptides that could be fused to viral CPs, Yusibov *et al.* (1997) took advantage of the relative genetic stability of the TMV insertion vector with duplicated CP promoters, to express the CP of alfalfa mosaic virus (AlMV) carrying amino-terminal fusions of foreign epitopes. The choice of AlMV CP as a carrier molecule was dictated by the fact that the AlMV genome is known to accommodate larger sequences by forming particles of different sizes and shapes. Moreover the amino-terminus of AlMV CP is located on the outer surface of virus particles and fused peptides do not appear to interfere with virus assembly. The autonomous expression of AlMV CP-peptide fusion from a TMV gene insertion vector required an origin-of-assembly sequence for AlMV CP that was provided by incorporating the 3′-untranslated region of AlMV RNA4 (the CP mRNA). The authors were able to express and purify two antigenic peptides fused to the AlMV virion. The inserted epitopes, of 40 and 47 amino acids, corresponded to the V3 loop of HIV-1 and a chimeric epitope comprising fragments from the rabies virus G5-24 glycoprotein and 31D nucleoprotein, respectively. Both modified viruses elicited specific virus-neutralizing antibodies against the target pathogen in immunized mice (Yusibov *et al.*, 1997).

4.3.5 Complementation systems

The strategies discussed above, for the insertion of foreign genetic material into viral genomes, will inevitably affect viral replication, are likely to slow down both local and long-distance movement and may also affect encapsidation. To overcome some of these problems helper-dependent systems have been explored in which foreign genes are inserted into defective (sub)viral components. Complementation of the defective viral function can be provided in *trans*, either by expression of the deleted viral gene in a transgenic plant or by co-infection with a helper virus to rescue the function deleted from the vector. In theory, complementation strategies have several potential advantages; the deletion of an essential viral gene provides a means to contain the vector while increasing the space made available for the insertion of foreign sequence. However, although there are many examples where *trans*-complementation has been demonstrated to occur, there are few instances where complementation has been exploited in the development of gene expression vectors. For example, complementation of a movement deficient TMV mutant by transgenically expressed MP was demonstrated as long ago as 1987 (Deom *et al.*, 1987; Holt and Beachy, 1991). Yet, despite the widespread use of TMV vectors and the constraints on insert size in currently available TMV-based vectors, MP *trans*-complementation strategies have not been explored as a possible means of enhancing the stability of TMV vectors.

Similarly, complementation strategies have been developed for geminiviruses but in general such complementation approaches have not been applied for practical purposes. As discussed in Section 4.3.2, the accumulation and gene expression level obtained using a TGMV CP-replacement vector were both enhanced when the vector was inoculated to plants carrying DNA B as a stably integrated transgene (Hayes *et al.*, 1988). In the case of TGMV, transgenic plants expressing under a single promoter either all the complementary sense orfs of TGMV (Hanley-Bowdoin *et al.*, 1989) or the TGMV Rep protein (Hanley-Bowdoin *et al.*, 1990) could complement TGMV replication-deficient mutants and the transgenic plants could support both replication and systemic movement.

An alternative to transgene-mediated complementation is to use a helper virus to provide *in trans* a function missing from a virus-based vector. Functional complementation by co-inoculation with a helper virus has been demonstrated for CaMV. Rescue was performed either by wild-type CaMV, or by a second defective virus bearing another deletion, to avoid any selective advantage in favour of wild-type helper virus. This strategy, however, is very inefficient, because the defective CaMV genomes readily recombine *in vivo*, restoring an infectious wild-type genotype (Mushiegan and Shepherd, 1995). To reduce the frequency of virus recombination, co-inoculation of a pair of CaMV mutants with long deletions was tested. In one genome, the deletion spanned most of the essential gene *I* and non-essential gene *II*, while in the other, most of gene *II* and the essential gene *III* were deleted (Hirochika and Hayashi, 1991). This overlap should prevent recombination events and result in forced genetic complementation, and co-existence of two defective CaMV mutants. However, this complementation approach has not been developed further, presumably due to the inherent genetic instability of modified caulimoviruses.

In the cases of monopartite geminiviruses, in which no genes are dispensable for infectivity, the only option left for generation of infectious gene vectors is to test *trans*-complementation of movement and/or replication. As exemplified by MSV, point mutations or small nucleotide insertion or deletions in the V1 and V2 orfs can be complemented in *trans* by wild-type virus or complementary mutants (Boulton *et al.*, 1989). However, the complementing mutants often recombine to generate wild-type virus. Another cross-complementation scheme has been described based on WDV. A WDV shuttle vector was created in which the *nptII* reporter gene and a prokaryotic origin of DNA replication replaced the virus CP gene (plasmid pWI-11; Ugaki *et al.*, 1991) with the result that NptII was expressed in plant protoplasts. As an assay to test whether the replication protein from WDV was *trans*-acting a replication-deficient vector carrying the GUS gene was coelectroporated in maize endosperm cells, together with pWI-11 providing the replication protein *in trans*. GUS activity was detected and the copy number of this WDV-based vector reach up to 30 000 copies per cell (Timmermans *et al.*, 1992).

Another possible approach to complementation is *via* the use of subviral agents 'parasitic RNAs' that depend on a helper virus for replication, encapsidation and movement. Satellite RNAs (sat RNAs) are small RNA molecules with only limited sequence homology to their helper virus. A sat RNA naturally associated with bamboo mosaic virus (BaMV) is a linear molecule of 836 nucleotides that is encapsidated by the BaMV CP into rod-shaped particles. In the first demonstration of a sat RNA-based vector for gene expression Lin *et al.* (1996) replaced a non-essential orf in sat BaMV with CAT to give a hybrid sat BaMVCAT.

Transcript inoculation of sat BaMVCAT, along with wild-type BaMV, to *Chenopodium quinoa* plants led to accumulation of sat BaMVCAT (0.5 mg g^{-1} of leaves) and high-level production of CAT enzyme (2 μg g^{-1} of leaves; Lin *et al.*, 1996). This report was also one of the first describing the construction of an RNA virus-based vector from a virus infecting monocotyledenous plants. Although only tested in barley protoplasts, rather than intact plants, this approach is potentially interesting since it could provide an alternative to geminiviruses as a vector for monocotyledonous plants.

Similarly, defective interfering (DI) RNA replicons may be employed as the helper-dependent foreign gene vector. DI RNAs are deletion mutants of viruses that replicate only in the presence of the parent (helper) viral genome. As with sat RNA, DI RNAs could theoretically be used to develop vectors that would be supported *in trans* by the helper virus. The only demonstration to date of a DI RNA engineered to express heterologous sequence involved cymbidium ringspot tombusvirus (CyRSV) modified to express the CP of the unrelated virus tomato aspermy virus (Burgyan *et al.*, 1994). CyRSV DI RNA has been shown to consist of a mosaic molecule made up entirely of non-contiguous regions of genomic RNA generated *de novo* during viral replication through progressive deletions of the parental genome (Rubino *et al.*, 1990). Gene insertions were tested at different recombination breakpoints within a cloned DI RNA of CyRSV and both DI RNA replication and accumulation of the foreign gene product were detected in inoculated leaves (Burgyan *et al.*, 1994).

4.4 Applications of plant virus-based vectors in basic research

The preceding section outlined a wide variety of potential strategies for engineering virus-based vectors using a range of different viruses. In the section that follows the various uses for which different virus-based vectors have been tested will be described in more detail.

With the exception of reporter gene studies (see below) it is notable that only a few of the many vectors that have been developed have been put to any practical use. Ultimately, this is perhaps the most stringent test of all with regard to the utility of any given viral vector. Only those vectors which live up to expectation (in terms of ease of genetic manipulation, foreign gene expression levels, fast cell-to-cell movement, genetic stability etc.) will become accepted and used as research tools.

4.4.1 Expression of reporter genes: studies of viral pathogenesis

To date one of the most frequent uses of plant virus-based vectors has been the development of reporter gene-tagged viral genomes for studying basic aspects of viral pathology. The use of reporter gene-tagged viruses depends on the ability to detect, either directly or indirectly, the product of the gene inserted into the viral genome allowing either quantification of gene expression and/or virus localization. A number of reporter genes that permit quantitation of viral gene expression have been described including luciferase, DHFR and CAT; however, although these genes were of use in developing and refining strategies for virus-based gene expression (see Section 4.3), their utility as reporter proteins is limited by the fact that assaying activity of the introduced gene is destructive and can only be determined *ex situ* in homogenized tissue extracts.

The first example of a reporter gene expressed from a plant viral genome that permitted easy *in situ* detection of protein activity was GUS. This enzyme was initially developed as a sensitive reporter of gene expression in stably transformed plants (Jefferson *et al.*, 1987). Both fluorimetric and luminescent substrates for GUS allow accurate and sensitive quantitation of enzyme activity, however, the real novelty of GUS was the availability of a chromogenic substrate that allowed detection of enzyme activity *in situ*.

The earliest examples of plant viruses engineered to express GUS were TEV and PVX. Both TEV.GUS and PVX.GUS gave systemic infections on *Nicotiana tabacum* that were accompanied by detectable GUS activity in the infected tissues (Chapman *et al.*, 1992; Dolja *et al.*, 1992). The use of GUS, as a reporter of virus gene expression, thus provided the first opportunity to map accurately the progress of local and systemic virus infections *in situ*. Furthermore, the availability of GUS as a reporter, coupled with the use of reverse genetics has allowed the effect of mutations in different viral genes to be correlated directly with effects on both local and systemic virus movement. Thus in the case of PVX deletion of the CP in PVX.GUS led to infection foci, determined by histochemical staining, that were restricted to single or very limited numbers of cells thereby demonstrating the key role of the CP in cell-to-cell movement of PVX (Chapman *et al.*, 1992). Similarly, a requirement for CP in local movement was also demonstrated using a CP deletion mutant of TEV.GUS (Dolja *et al.*, 1994). Using TEV.GUS-carrying mutations in the helper component protease it was also demonstrated that this gene is an important determinant of long-distance, systemic, movement of TEV (Cronin *et al.*, 1995). For PVX the role of the 25 kDa protein in regulating the plasmodesmatal pore size and viral cell-to-cell movement was established using PVX.GUS harbouring a mutation in the 25 kDa protein orf (Angell *et al.*, 1996).

Although for the reasons discussed below GUS has not been widely used as a reporter of virus movement, several viruses have been engineered successfully to express this reporter protein. Studies using TBSV expressing GUS demonstrated conclusively that the p22 protein of TBSV was required for viral cell-to-cell movement (Scholthof *et al.*, 1995). In the case of beet necrotic yellow vein virus, fusions between the RNA 3 orf N and GUS were used to demonstrate the very low level of expression of this gene *in vivo* (Jupin *et al.*, 1991). In a recent study a strategy based on translational fusions between GUS and virus-encoded proteins, was used to determine both the timing and level of expression of three different proteins encoded by beet yellows colosterovirus (BYV). Sampling of transfected plant protoplasts at various times after infection allowed the temporal regulation of individual BYV subgenomic RNA promoters to be compared (Hagiwara *et al.*, 1999).

The use of GUS as a reporter of virus infection represented a significant breakthrough compared to previously tested reporter proteins, however, a number of limitations have restricted the general application of GUS in plant virological studies. First, the size of the *UidA* gene, at 1.8 kb is likely to create problems in terms of packaging if introduced into the genome of viruses with icosahedral particles, thus effectively restricting the use of GUS to viruses either able to move independently of encapsidation or having rod-shaped particles. An additional problem associated with the size of *UidA* is that the gene is too large to be stably maintained in many virus-based vectors and several studies have been published in which the specific nature of the recombination events involved in elimination of *UidA* sequences have been investigated (Chapman *et al.,* 1992; Dolja *et al.,* 1993; Guo *et al.,* 1998). Thus, despite the early successes with TEV and PVX expressing GUS, few other viruses engineered to express GUS have been shown to give systemic infection. Further problems associated with the use of GUS as a reporter of viral infections include the destructive nature of the assay procedure, which involves vacuum infiltration of (usually) detached plant organs, and the need to remove plant pigments using organic solvents to clearly detect the coloured reaction product. Another significant problem is that a highly diffusible reaction intermediate is produced by GUS that then undergoes oxidative dimerization to form the coloured reaction product. Because the reaction intermediate is diffusible and can move between cells GUS cannot be considered a truly cell-autonomous reporter of gene expression (Guivarc'h *et al.*, 1996).

A major advance in reporter gene technology came with the cloning and subsequent expression in heterologous organisms of the jellyfish GFP (Chalfie *et al.*, 1994). The widespread application of GFP in cell biological studies of many different organisms has been reviewed extensively (Cubitt *et al.*, 1995; Haseloff and Amos, 1995). GFP, unlike all previously available reporter proteins, requires no substrate or co-factors and can be detected non-invasively simply by excitation using an appropriate lightsource. In addition the *gfp* gene is small (~700 nucleotides) and thus more likely to be maintained stably in the genome of a recombinant virus than the *UidA* gene. In terms of its application in plant virology GFP has proved to be an extremely useful tool that has revolutionized many areas of plant virus research. The ability to detect GFP in intact plant tissue permitted non-invasive analyses of the time course of both local and systemic virus movement in more detail than was previously possible (Roberts *et al.*, 1997; Szecsi *et al.*, 1999).

The first published study of a GFP-tagged plant virus used a PVX-based gene insertion vector to demonstrate that local and systemic movement of PVX.GFP could be monitored, both microscopically and macroscopically, using non-destructive techniques (Baulcombe *et al.,* 1995). This study extended the analysis of the PVX CP deletion mutant originally prepared in the PVX.GUS vector by showing conclusively that in the absence of CP no cell-to-cell movement of the virus occurs. In the study of PVX.GUS the localization of the CP deletion mutant to individual cells had not been possible due to the problem, discussed above, of diffusion of the reaction intermediate leading to non-cell autonomous labeling. The availability of reliable vectors also led to the early testing of TMV modified to express the GFP and, as with PVX, expression of GFP from a TMV-based gene insertion vector led to both local and systemic infection (Casper and

Holt, 1996). In the case of TMV GFP provided the first opportunity to map systemic virus movement *in situ* as TMV tagged with GUS, although giving rise to local infections, did not give GUS activity in systemic tissues.

Other examples of viruses engineered to express the GFP as a free protein include CPMV (Verver *et al.*, 1998), TEV (Kasschau and Carrington, 1998), zucchini yellow mosaic virus (Amit Gal-On, personal communication), pea seed borne mosaic virus (Elisabeth Johansen, personal communication), cucumber mosaic virus (Canto *et al.*, 1997), white clover mosaic virus (Tony Lough, personal communication), barley stripe mosaic virus (Andy Jackson, personal communication), tobacco rattle virus (Stuart MacFarlane, personal communication), bean dwarf mosaic virus (Sudarshana *et al.*, 1998) and potato leaf roll virus (Micha Taliansky, personal communication). Although all of these GFP-tagged viruses were infectious and directed GFP expression not all were capable of systemic movement and in some cases even cell-to-cell movement was not observed. Undoubtedly further improvements in vector design will overcome problems associated with restricted movement and further examples of infectious GFP-tagged viruses will be added to this list. Although the GFP-tagged viruses listed above have been prepared with specific research goals in mind, all of them in theory have the potential to provide insights into patterns of local and systemic virus movement, tissue tropism and induced host responses as well as allowing the analysis of virus gene function through reverse genetics. With a growing repertoire of GFP-tagged viral genomes comparative studies will allow general patterns of plant virus infections to be distinguished from features that are specific to particular viruses.

Expressing GFP as a free protein from a plant virus vector provides the opportunity to localize very accurately virus-infected tissue *in vivo* (Oparka *et al.*, 1996). In addition, the ability to tag a viral protein with GFP can provide even more detailed information by providing insight into the subcellular targeting of virus-encoded protein. The GFP has been shown to be extremely tolerant of modification at either its amino- or carboxy-terminus and many reports attest to the continued fluorescence of GFP in fusion proteins. However, although GFP tolerates modification it is not always true that the fusion partner will behave likewise. The caveat, in interpreting information regarding the subcellular localization of GFP-tagged proteins, is that the tagged protein must retain full biological function if meaningful conclusions are to be drawn from experiments exploiting fusions between GFP and virus-encoded proteins.

Two early studies of GFP-tagged viruses in plants expressed fusion proteins between the MP of TMV and GFP. These studies, in addition to confirming the known plasmodesmatal localization of the TMV MP, also pointed to a previously unsuspected association between the MP and the plant cytoskeleton (Heinlein *et al.*, 1995; McLean *et al.*, 1995). These studies demonstrated a clear association between the TMV MP and tubulin and they provided the first demonstration of an *in vivo* association between a virus-encoded MP and the endogenous intracellular transport apparatus of the plant cell. Numerous further studies of TMV expressing a GFP-tagged MP have since been published that provide a detailed *in vivo* analysis of various aspects of the cellular and molecular biology of the TMV movement process. These studies have addressed the subcellular distribution of the labeled TMV MP (Mas and Beachy, 1998; Padgett *et al.* 1996) cellular perturbations induced by TMV infection (Reichel and Beachy, 1998), molecular determinants of MP localization (Kahn *et al.* 1998) as well as the ontogeny of the TMV infection process (Szecsi *et al.*, 1999). Together the studies referred to above represent the first detailed cell biological analysis of a plant virus MP and have revealed the complex and changing interactions between a virus-encoded protein and the host cell.

Although GFP represents the reporter of choice for most studies of tagged viral genomes the development of viral vectors expressing fusions between viral proteins and GFP has been hampered due to both inherent constraints regarding the viability of modified genomes and in addition the consideration raised above concerning the retention of biological activity of fusion proteins. Apart from MP-GFP fusions the only other example of virus-based expression of a biologically active fusion protein between a viral protein and GFP is the GFP-CP fusion described for PVX (see Section 4.3.4). In this example the GFP-CP fusion protein expressed from the modified viral genome became incorporated into virus particles that could be imaged *in situ* by fluorescence microscopy (Santa Cruz *et al.*, 1996). As stated above, assembly of virus particles incorporating the GFP-CP fusion subunit required the availability of free CP in addition to the fusion protein and in the absence of free subunits viral infections were limited to individual epidermal cells. This result suggested that the absolute requirement for the PVX CP in intercellular movement, determined using the CP deletion mutant of PVX.GFP (Baulcombe *et al.*, 1995), reflected a requirement for assembled virus in the intercellular movement process rather than some additional, assembly-independent, role of the CP in transport (Santa Cruz *et al.*, 1998).

The examples above, of applications of GFP in basic studies of plant virus pathogenesis, illustrate the varied ways in which viral pathology can be studied using virus-based vectors expressing GFP. It is worth pointing out, however, that GFP as a reporter also provides a readily scoreable marker both for virus movement and gene expression levels thus making it an invaluable tool in studies aimed at improving virus-based vectors. Thus, in a recent study Shivprasad *et al.* (1999) compared a wide variety of vectors based on TMV and related tobamoviruses that were engineered to express GFP. The ability to score virus movement phenotypes rapidly by eye and even to perform crude estimates of gene expression based on the intensity of green fluorescence will greatly facilitate comparative assessments of vectors.

In addition to allowing the study of viral mutants affected in movement, viral vectors also have the potential to assist in investigating the genetics of host susceptibility and resistance. In a novel application of virus-based vectors TEV has been adapted to express either a herbicide resistance gene as selectable marker or a proherbicide sensitivity gene as a counterselectable marker. In order to identify plant mutants showing a gain of susceptibility to virus infection TEV was engineered to express the bacterial *bar* gene that confers resistance to the herbicide GA. When TEV-bar was inoculated to a fully susceptible ecotype of *Arabidopsis* infected plants were resistant to normally lethal doses of the herbicide. In contrast restrictive *Arabidopsis* ecotypes, in which long-distance movement of wild-type TEV is blocked, died following a similar treatment with GA. TEV-bar was then inoculated to in excess of 80 000 M2 plants from mutagenized populations of a restrictive *Arabidopsis* ecotype. Genetic analysis of progeny plants derived from mutagenized lines that showed susceptibility to systemic

TEV infection revealed two, independent, recessive loci conferring the susceptibility phenotype (Whitham *et al.*, 1999). As pointed out by the authors, the coupling of a highly selective screening procedure with high throughput analysis provides an extremely powerful tool for the identification of plant genes required for susceptibility to systemic virus transport. Both positive selection, as described above, and negative selection, using counter selectable markers such as $P450_{SU1}$, which confers sensitivity to the sulfonylurea proherbicide R7402, have the potential to uncover diverse classes of plant genes involved in both positive and negative control of plant virus infections. Furthermore, the successful application of such screens will not only provide invaluable insight into the dynamics of plant–virus genome interactions but may also allow the selection of plant genotypes showing enhanced susceptibility to virus movement and/or replication. Mutants with enhanced susceptibility to virus infection could prove useful in biotechnological applications of plant virus-based vectors where the objective is rapid, high-level, production of recombinant protein.

4.4.2 Viruses as tools to study induced plant resistance responses

One of the most frequent uses of the PVX-based vector has been for the analysis of phenotypes conferred by expression of genes from unrelated pathogens. In terms of dissecting the specific role of individual pathogen genes, virus-based expression has the advantage that mature, differentiated tissue can be infected leading to expression of the gene under study. This approach has been used to investigate pathogen avirulence gene products of viral, bacterial and fungal origin.

An investigation of the TBSV genes encoding p19 and p22 showed that both proteins were capable of eliciting a necrotic response when expressed from a PVX-based vector (Scholthof *et al.*, 1995). Interestingly PVX.p22 and PVX.p19 differed in the specific hosts on which a necrotic response was triggered, a result that highlights the ability of plants to recognize and respond to different viral products. In a follow-up study of TBSV p22 the PVX vector was used to analyse site-specific mutants allowing residues responsible for the necrotic phenotype to be discriminated from residues essential for biological activity of p22 in TBSV cell-to-cell movement (Chu *et al.*, 1999). In contrast to the work described above, where PVX was used to determine species-specific elicitors of necrosis, PVX has also been used to study cultivar-dependent elicitor activity of viral proteins. In a recent study expression of the pepper mild mottle virus CP from a PVX-based vector was shown to specifically trigger a hypersensitive (necrotic) response on pepper plants carrying the L^3 gene. On pepper plants lacking this resistance gene the chimeric PVX vector, expressing the PMMoV CP, did not induce a host response and gave rise to infections phenotypically similar to those obtained using wild-type PVX (Gilardi *et al.*, 1998).

PVX itself can elicit a hypersensitive response on potato cultivars carrying strain-specific resistance genes (Cockerham, 1970) and in the case of the resistance gene *Rx* it is known that the viral CP is the elicitor of the host response (Kavanagh *et al.*, 1992). In experiments performed using a TMV-based vector the TMV CP was replaced with the CP of PVX from either a strain that induces *Rx*-mediated resistance, to give TMV-TK, or a strain that fails to induce resistance, giving TMV-KR. When these hybrid viral genomes were introduced to protoplasts prepared from a potato cultivar carrying the *Rx* gene, TMV-TK, which carries the functional elicitor, failed to accumulate, whereas TMV-KR, which does not express an elicitor of the *Rx* response, was able to accumulate normally (Bendahmane *et al.*, 1995). In potato protoplasts of a cultivar lacking the *Rx* gene both TMV-KR and TMV-TK accumulated to similar levels, demonstrating that the PVX CP was sufficient to induce the specific host resistance response and that this response was able to suppress TMV accumulation (Bendahmane *et al.*, 1995).

In addition to the studies of species and cultivar specific viral elicitor proteins described above, PVX has also been used to identify and study elicitor peptides and proteins from other classes of plant pathogen. The small extracellular peptide AVR9, produced by the plant pathogenic fungus *Cladosporium fulvum*, acts as a specific elicitor of the tomato *Cf-9* resistance gene. Races of *C. fulvum* carrying *Avr9* induce localized cell death and fail to infect tomato plants carrying *Cf-9*. In order to investigate the death-inducing property of AVR9 the PVX vector was engineered to express a synthetic peptide comprising the 28-amino-acid AVR9 product fused to the signal peptide sequence of the *N. tabacum* PR1a gene. Infection of *Nicotiana clevelandii* with the modified virus, PVX:Avr9, resulted in attenuated symptoms compared to wild-type virus. When used to infect near isogenic tomato cultivars, either carrying or lacking *Cf-9*, a severe necrotic response frequently culminating in the death of the plant ensued only on the *Cf-9* carrying cultivar and not the *Cf-0* control (Hammond-Kosack *et al.*, 1995). In addition to confirming the ability of AVR9 to induce a *Cf-9* specific cell death response independently of any other *C. fulvum* gene products, the PVX-based expression system also allowed rapid structure function mapping of the AVR9 peptide. Using PVX to express AVR9 carrying specific amino acid substitutions, the residues responsible for the elicitor activity of the peptide were mapped and a hydrophobic β-loop was identified as a crucial determinant of AVR9 activity (Kooman-Gersmann *et al.*, 1997).

Similar studies to those described above have been performed using PVX to express the *Phytophthora infestans* INF1 elicitor peptide that induces necrotic cell death in tobacco (Kamoun *et al.*, 1999). Infection of tobacco with PVX::inf1 resulted in a necrotic response and the modified virus was restricted to the site of inoculation. In contrast, expression of a mutant form of *inf1* gave rise to a systemic necrotic response. Resistance to recombinant virus was also seen in tomato plants carrying the resistance gene *Pto* when they were infected with a PVX vector expressing the corresponding avirulence gene product from *Pseudomonas syringae* pv. *tomato*, AvrPto (Tobias *et al.*, 1999). The results obtained using PVX::inf1 and PVX.AvrPto are extremely significant in that they indicate that not only can bacterial elicitors trigger a resistance-related response but that the response induced is sufficient to contain a viral pathogen (Kamoun *et al.*, 1999; Tobias *et al.*, 1999). This demonstration underlines the commonality of plant disease resistance responses and indicates that similar mechanisms are utilized by plants in determining resistance against diverse classes of pathogen despite the highly specific nature of the elicitors involved in triggering the host response. In all of the examples discussed in the preceding section, virus-based expression of pathogen genes has allowed the analysis and dissection of phenotypes independently of the pathogen from which the genes were derived. Furthermore, as the genes in question are involved in triggering host responses that are frequently lethal, virus-based expression provides a means to investigate the phenotype of the proteins in question in mature tissue thereby avoiding the problems associated with regeneration of stably transformed plants expressing lethal genes.

In addition to identifying elicitors of host responses *via* expression of candidate genes from viral vectors, it has even proven possible to identify host receptors by function following expression of candidate genes from a viral vector. In experiments performed by Brian Staskawicz and colleagues, PVX was used to express the resistance gene *Pto* (see above) and its homologue *Fen* that confers sensitivity to the herbicide fenthion. When PVX expressing *Pto* was used to infect transgenic tomato plants carrying the *AvrPto* avirulence gene, a resistance response was not observed, suggesting that viral expression of Pto was not sufficient for accumulation and/or activity of the receptor protein (Rommens *et al.*, 1995). However, expression of *Fen* from PVX converted a fenthion insensitive tomato cultivar to fenthion sensitivity indicating that, at least in some cases, viruses can be used to express functional plant receptors (Rommens *et al.*, 1995).

4.4.3 Viral vectors in cell biological studies

The last few years have seen an increasing number of publications describing the use of plant virus-based vectors in cell biological studies. These studies all share the common feature that *in vivo* fluorescent labeling of specific cellular compartments has been achieved through the expression of fusion proteins between GFP and targeting sequences that confer organelle-specific labeling with the fluorescent protein thereby allowing real-time imaging of organelle dynamics.

The earliest example showing *in vivo* labeling of a subcellular compartment by expression of a GFP fusion protein expressed from a viral vector was provided by Boevink *et al.* (1996). In this study an amino-terminal signal peptide from the seed storage protein sporamin was used to direct GFP to the plant secretory pathway. This fusion protein was expected to accumulate in the apoplast (extracellular space) as this is the default destination for plant proteins entering the secretory pathway. However, expression of the sporamin signal peptide-GFP fusion resulted in barely detectable levels of extracellular fluorescent signal in tissue infected with the viral vector, suggesting that GFP was either unstable or incorrectly processed resulting in a non-fluorescent protein (Boevink *et al.*, 1996). In contrast, when the same fusion protein was further modified through the addition of a carboxy-terminal tetrapeptide KDEL motif, which directs retrieval of proteins from Golgi to the ER, bright fluorescence was observed specifically in the ER of cells infected with the viral vector. As well as demonstrating the utility of a virus-based vector for targeted labeling of a plant organelle with GFP this study also provided a tool for basic cellular analysis of plant ER dynamics. In a follow-up study Boevink *et al.* (1999) demonstrated the effects of both low temperature and treatment with the secretory inhibitor brefeldin-A on ER morphology and secretory capacity by direct *in vivo* analysis of the labeled ER in vector-infected tissues.

A similar approach, based on exploiting the PVX-based vector, has also been used to study the relationship between the ER and the Golgi in plant cells. In this study, fusions were prepared between GFP and either the signal anchor sequence of a rat sialyl transferase or the *Arabidopsis* H/KDEL receptor homologue ERD2. In the former case GFP labeling was restricted exclusively to the Golgi of cells infected with the viral vector, whereas in the latter case both ER and Golgi labeling was observed (Boevink *et al.*, 1998). This approach allowed the relationship between these organelles to be studied for the first time in living plant cells. *In vivo* labeling of the ER with GFP coupled with detection of F-actin using a fluorescent probe allowed the close association between the architecture of the ER and the actin cytoskeleton to be confirmed. Furthermore, the use of chemical inhibitors of actin-based motility demonstrated *in vivo* the dependency of Golgi movement on an actin-based transport system (Boevink *et al.*, 1998). Golgi labeling with GFP has also been reported using a TMV-based vector expressing a fusion between the cytoplasmic transmembrane stem of tobacco N-acetylglucosaminyltransferase I and GFP (Essl *et al.*, 1999). This study provided the first identification of a plant-derived Golgi targeting sequence and in addition showed that organellar targeting could be achieved using vectors other than PVX.

In theory, given appropriate targeting sequences, virally expressed proteins should be capable of being directed to any desired subcellular compartment or structure. For example actin depolymerizing factor from *Arabidopsis* has been expressed from a PVX-based vector as a carboxy-terminal fusion to GFP. In young infected tissue green fluorescence colocalized with the signal from the actin-specific marker rhodamine phalloidin. The labeling observed indicated that the labeled actin filaments in infected cells were shorter and more extensively bundled than filaments observed in non-infected cells. However, wild-type PVX infection causes rearrangements to the actin cytoskeleton thus preventing straightforward conclusions as to the phenotype of virally expressed actin depolymerizing factor (L. Simon Buela and S.S.C., unpublished results). It is thus important in any study involving virus-based gene expression that effects of the virus infection *per se* are taken into account when interpreting the phenotype(s) associated with the expression of a given gene. In cases such as the example above, where a standard and well accepted technique for imaging actin exists, a comparison can readily be made between phenotypes arising from expression of the foreign protein-GFP fusion and those directly resulting from the virus infection. In other circumstances, for example where no independent means is available for verifying possible effects of viral infection on organelle function and/or protein localization, any interpretation should be made with caution. Despite this reservation it still remains true that virus-based expression of GFP-fusion proteins provides a facile and rapid approach for obtaining preliminary information on the utility of fusion proteins for cell biological studies. Furthermore, as in other applications of virus-based gene expression, infection of mature, differentiated tissue bypasses potential problems associated with cytotoxic effects of long-term expression of stably integrated transgenes during plant regeneration.

4.4.4 Viruses as tools for investigating gene silencing

A relatively recent but extremely fast moving area of plant biology is the study of post-transcriptional gene silencing (PTGS). This class of gene silencing in plants was first uncovered in studies of genetically engineered plants where it was occasionally observed that transformation with endogenous genes, rather than boosting expression of the target gene product, resulted in a phenomenon referred to as co-suppression where both the transgene and the endogenous plant gene were no longer expressed (Napoli *et al.*, 1990, reviewed in Vaucheret *et al.*, 1998). While researchers investigating co-suppression were puzzling over possible mechanisms, virologists were encountering their own problems in interpreting experiments based on plants engineered to express virus-derived transgenes that were designed to confer resistance to subsequent infection from the virus from which the transgene was taken (Lindbo *et al.*, 1993). A confusing feature of the virus resistance observed in some transgenic lines was that the strength of resistance was inversely, rather than positively correlated with the accumulation of the transgene in question. This finding brought into question the previously accepted, and well documented, phenomenon of pathogen-derived resistance in which the constitutive presence of a pathogen-derived product in the host was able to interfere with subsequent pathogen infection (Beachy and Bendahmane, 1999). In a seminal study the first clue to the nature of the mechanism conferring resistance revealed that resistance in the transgenic lines was consistently associated with a high level of transgene transcription. However, this transcriptional activity did not result in either cytoplasmic mRNA or protein accumulation, implying that transgene mRNA was being degraded (Lindbo *et al.*, 1993). Furthermore the phenotype associated with this resistance mechanism appeared to be highly sequence specific, requiring a high degree of homology between the transgene sequence and the challenge virus (referred to as homology-dependent resistance).

In an early study investigating the link between PTGS and virus resistance, a PVX vector expressing all or part of the *UidA* gene (PVX.GUS) was inoculated to transgenic lines carrying the *UidA* gene. Inoculation of PVX.GUS to a transgenic line constitutively expressing GUS resulted in the accumulation of GUS activity in the viral infection foci that was readily detectable above the background of transgenically expressed GUS. In contrast, when PVX.GUS was inoculated to an independent transgenic line in which expression of GUS was severely downregulated due to PTGS the plants were effectively resistant to viral infection. The determinant of this resistance phenotype was demonstrated to be the homology between the silenced transgene and the *UidA* sequence in the PVX vector as the plants that were resistant to PVX.GUS were fully susceptible to infection with PVX.GFP (English *et al.*, 1996).

Subsequent studies have suggested that PTGS represents an RNA-mediated defence process that may have evolved specifically as a protection mechanism against viral and other rogue nucleic acids such as transposons (Kumpatala *et al.*, 1998; Ratcliff *et al.*, 1997). This model postulates that an endogenous mechanism exists in plants which permits surveillance of aberrant and/or excessive transcript accumulation that triggers a pathway leading to the specific degradation of the targeted RNA. Evidence for this hypothesis has been provided by studies of both RNA and DNA viruses that demonstrated that initial high-level accumulation of virus was subsequently followed by a phase of so-called recovery, when newly emerging leaves showed very low levels of pathogen accumulation (Ratcliff *et al.*, 1997). In an elegant study designed to investigate the nature of the recovery phenotype, David Baulcombe and colleagues utilized a PVX-based vector harboring sequence of an RNA nepovirus that was known to induce a recovery phenotype. This modified virus accumulated normally on healthy tobacco plants but failed to accumulate when inoculated on the upper, recovered leaves of plants previously infected with the nepovirus. In contrast, wild-type PVX was able to accumulate on both healthy tobacco as well as on plants showing recovery from a prior nepovirus infection (Ratcliff *et al.*, 1997). Thus, plant virus-based vectors have been used to demonstrate the commonalities between the phenomena of co-suppression and homology-dependent resistance (English *et al.*, 1996) and also between PTGS and a natural plant defence mechanism against viruses (Ratcliff *et al.*, 1997).

A further twist to the emerging story of gene silencing and its role in protection against viral infections was the finding that several viruses have specifically evolved counter-defensive strategies (reviewed in Carrington and Whitham, 1998). Two viral genes, which previous studies had identified as being involved in long-distance movement and which were therefore ascribed transport functions, have now been identified as negative regulators of transcriptional gene silencing. Thus, the potyvirus helper component protease has been shown to block the maintenance of PTGS in transgenic plants showing post-transcriptional silencing of either *UidA* or *gfp* (Anandalakshmi *et al.*, 1988; Brigneti *et al*, 1998; Kasschau and Carrington, 1998). In contrast, the CMV 2b protein cannot block maintenance of PTGS but it can prevent the initiation of gene silencing in the growing points of the plant (Brigneti *et al.*, 1998). More recently, Voinnet *et al.* (1999) surveyed a range of viruses for the ability to block either the maintenance or initiation of gene silencing and established that viral counter-defensive proteins are found in viruses from several different families. Curiously this study, which investigated the effects of virus infection on *N. benthamiana* plants exhibiting PTGS of *gfp*, found that two potexviruses, including PVX, could not block PTGS whereas three other potexviruses were efficient suppressors of silencing. Although this effect could simply reflect variation between the different potexviruses tested it is also plausible that the effectiveness of a 'counter-defensive' gene is host-dependent.

These recent discoveries, regarding the ability of some viruses to trigger an antiviral silencing mechanism and the fact that some viruses have evolved counter-defensive strategies, obviously has implications for virus vector design and application. Thus, overexpression of foreign proteins using virus-based vectors will benefit if the vector is not an inducer of gene silencing on the chosen host. Conversely, in applications of viral vectors that are aimed at exploiting gene silencing to knock out expression of endogenous genes (see Section 4.5.2) it is advantageous if the vector is a strong inducer of silencing and does not express a functional 'anti-silencing' protein. To date most studies of viral activation and suppression have relied on either *N. benthamiana* or other members of the genus *Nicotiana*, in terms of the practical application of plant virus vectors and particularly the development of vectors for new hosts, it will be important to understand the effects of PTGS in any given virus/host combination.

4.4.5 Applications of virus-based vectors in screening genes for function

At the present time there are relatively few examples of analyses of gene function outside the areas outlined above. An interesting early application of a viral vector for functional screening employed a PVX-based vector engineered to express a plant Myb homologue (Sablowski *et al.*, 1995). To investigate the molecular determinants for transcriptional activation of a gene encoding phenylalanine ammonia-lyase (PAL), transgenic plants were generated that carried transcriptional fusions between the PAL promoter sequence and the GUS orf. Transgenic plants infected with wild-type PVX and non-infected plants showed only background levels of GUS expression. In contrast, transgenic plants infected with a modified PVX vector, engineered to express a Myb homologue from snapdragon, showed substantially elevated levels of GUS activity (Sablowski *et al.*, 1995). This study, in addition to demonstrating the possibility of using viral vectors to study gene regulatory processes, also resulted in the production of sufficient quantities of the heterologous protein to permit direct biochemical analysis of the expressed protein. Another example of gene transactivation resulting from overexpression of a transcriptional regulatory protein was provided in a study of the ACMV AC2 protein that is required for transactivation of the viral CP promoter. Expression of AC2 from a PVX-based vector activated transcription from the ACMV CP promoter in transgenic plants engineered to carry a transcriptional fusion between the promoter and a plant ribosome inactivating protein (Hong *et al.*, 1997).

Screening for gene function in plants can also be extended to analyses of genes not originating from plants. In a recent example, the murine gene encoding the proapoptotic protein Bax was expressed from a TMV-based vector. Expression of Bax in either mammalian cell lines or yeast confers a lethal phenotype and extensive structure/function studies have allowed mapping of domains in the Bax protein required for lethality in either yeast or mammalian systems. When over-expressed in plants, Bax also induced cell death and functional mapping demonstrated that the requirements for Bax-mediated lethality in plants were similar to those in yeast. Interestingly many of the responses associated with Bax-induced cell death in plants were similar to the responses induced during the cell death programme that is triggered in tobacco carrying the *N* gene that confers resistance to TMV infection (Lacomme and Santa Cruz, 1999).

One of the most ambitious applications of a plant virus vector attempted to date is that described by Curtis Holt and colleagues (Karrer *et al.*, 1998) in which a TMV-based vector was used to screen a plant cDNA library for genes involved in plant-programmed cell death. In this study a cDNA library was prepared from leaves of an *N* gene-carrying tobacco cultivar undergoing the TMV-induced hypersensitive response (a localized form of programmed cell death occurring in response to some pathogens). Thus it was predicted that the library would contain cDNAs representing genes upregulated in plants undergoing a programmed cell death response to pathogen invasion. The cDNA library was cloned in a TMV-based vector and then 'plated' on leaves of a tobacco cultivar that lacked the *N* gene and were therefore fully susceptible (i.e. did not undergo the hypersensitive response) to wild-type TMV. Following expression from the TMV vector approximately 0.1% of the tested clones resulted in the induction of hypersensitive response-like symptoms and thus represented candidate cDNAs of genes involved in triggering programed cell death in plants. Of the subset of response-inducing cDNAs that were sequenced, several encoded genes either previously implicated in the hypersensitive response or which clearly encoded proteins with the potential to trigger cell death. Obviously the approach described above has great potential but is also subject to a number of pitfalls. Many genes, particularly when expressed ectopically to extremely high levels, may have cytotoxic phenotypes independent of any normal role in plant-programmed cell death. Ultimately the success or failure of the approach will be established with the further characterization of each of the independent cDNAs that conferred a death-inducing phenotype in order to establish the biological relevance of the identified cDNAs in plant-programed cell death.

An alternative strategy whereby viral vectors can be used for functional gene screening is through the exploitation of PTGS (see Section 4.4.4). In addition to inducing PTGS targeted against transgenes, plant virus-based vectors can be used to block expression of endogenous plant genes. Thus, expression of a fragment of phytoene desaturase, in either sense or antisense orientation from a TMV-based vector, resulted in the suppression of endogenous phytoene desaturase and resulted in leaves developing a bleached phenotype (Kumagai *et al.*, 1995). More recently PVX was used to express a 500-nucleotide fragment of ribulose-1,5-bisphosphate carboxylase oxygenase. Plants infected with this construct exhibited a stunted chlorotic phenotype due to down-regulation of the targeted enzyme (Jones *et al.*, 1999). These examples illustrate, at least in principle, the power of viral vectors to inhibit expression of endogenous plant genes and suggests the possibility of using such an approach to identify the function of unknown gene sequences. The exploitation of this approach is still in its infancy but if virus vectors can be used reliably to downregulate host gene expression, a powerful tool for gene identification will be available (reviewed by Baulcombe, 1999; and see Section 4.5.2).

4.5 Commercial applications of plant virus-based vectors

The unique properties of virus-based expression systems lend themselves to several applications that can impact commercial and medical goals. The first, most obvious, property of virus expression vectors is their ability to express large quantities of foreign proteins or RNAs in host plant cells. This expression potential can be exploited for the production of many proteins for vaccine and therapeutic applications. It is clear that to exploit these vectors for such medical uses, one must integrate their potential with fully functional agricultural and protein production capabilities. Virus expression vectors have proven to be potent tools in plants, but for these vectors to exert their production potential, one must be able to effectively transfer viruses to large numbers of host plants. This section aims to discuss the practical issues related to large-scale exploitation of virus-based vectors as well as summarizing the areas where plant viruses might be expected to have a realizable commercial potential in the near future.

4.5.1 Practical considerations in the commercial exploitation of viral vectors

Requirements for viable virus vectors. The first requirement for any commercial application of plant virus-based vectors is the development of vectors from virus backbones that can maintain foreign sequence insertions for multiple passages through plants. This necessity can be seen when one takes note of the processes that are necessary to move from concept to product. While fields or greenhouses of plants may be waiting as available 'bioreactors', one must generate a sufficient quantity of genetically stable virus vector inoculum to permit effective infection and transfer of the virus-expressed trait to all plants. Genetic stability has to be measured in two distinct ways: (i) lack of genetic drift in the foreign gene sequence held within the virus vector, and (ii) retention of the foreign gene insert for sufficient time and through different selective bottlenecks to allow economically relevant levels of protein production. Viruses evolve through standard population-based adaptation schemes and every virus population can be defined by a selection landscape. In this landscape, the predominant genotype may comprise more than 99.9% of all viral particles, but variants with differing sequences do exist. When viruses encounter selection, or encounter genetic bottlenecks, variants that were a minority of the total population can begin to exert dominance if they are suited to the new set of selective pressures. Early in the history of virus expression vectors, virologists, aware of the high mutational rates in virus populations, were dubious of the abilities of these vectors to maintain foreign genes without the accumulation of significant point mutations (van Vlotten-Dotting *et al.*, 1985). This possibility was directly tested by researchers and surprisingly, virtually no sequence drift occurs during multiple passages of virus vectors through plant hosts (Kearney *et al.*, 1993; see Section 4.5.3). Upon further reflection, it is clear that single-base mutations within a non-essential sequence would be predicted to have little positive effect on the adaptation of a virus vector population. In support of this view, such sequence drift has been rarely seen in cases of protein or peptide coding regions (G.P.P., unpublished results).

The most common rearrangements occurring in virus vectors are deletion events leading to either partial or complete loss of the inserted sequence. Clearly, virus vectors with smaller genetic insertions possess a selective advantage at the RNA replication and movement levels. This mutational propensity may seem unacceptable for the production of pharmaceutically relevant proteins, where there is no tolerance for deviation in protein coding sequence within a production run. However, the occurrence of deletion events in the foreign gene sequence will usually result in either the complete absence of the desired protein or greatly truncated protein products that can be eliminated by high-stringency protein purification procedures.

Plants as bioreactors – consideration of optimal host characteristics. Another essential element for applying virus vectors in production situations is the availability of plants that are compatible with modern agricultural procedures and still good hosts for protein production. Many standard cultivated plants are highly resistant to many viruses that are used to construct expression vectors. For example, most burley tobacco cultivars grown throughout the southern half of the United States contain the *N* gene, which confers resistance to TMV, and therefore cannot be infected by TMV-based vectors. In addition many prime hosts for virus vectors, for example *N. benthamiana*, are not adapted for field applications. For commercial applications, vectors must be able to establish vigorous systemic infections in host plants under field conditions. As those involved in agriculture know, 'every year is an exception', thus vector and host plant systems must be synchronized to be very robust. Temperature sensitivities, induced resistance, other plant disease and poor infectivity of virus vectors can all reduce the success of virus vector expression. To date most development of plant virus vector systems has exploited tobacco and other species of the genus *Nicotiana* as hosts. For applications relating to protein and peptide production, using tobacco as a host does not present any problems as advanced cultivation procedures are available for this plant. For some of the other possible uses of vectors that are discussed below, such as metabolic engineering and genomic screening, other host species may be necessary or desirable. One of the more important challenges for the future will be the development of robust virus vectors for other plant species and in particular for the many agronomically important monocotyledenous crop species.

Technical considerations in scaling up virus-based protein production. Before scaling up the production of recombinant viruses or virally expressed proteins, a number of factors, arising from the inherent differences between laboratory and production scale processes, need to be addressed. One important issue to consider is the ability of researchers to effectively inoculate literally acres of plants. Many issues come into play here, including the amount of virion that can be produced effectively from ‘packaging hosts’, the genetic stability of the virion preparation and the relative infectivity of the virions on field-hardened plants. Researchers have adapted tools to effectively deliver TMV-based vectors on to field tobacco by using a mixture of virions and abrasive materials applied to leaf surfaces using high pressure spray devices. This approach has proved very effective, resulting in >95% infectivity under field conditions (Large Scale Biology Corporation (LSBC), unpublished results).

Another consideration for field-related plant production hinges on methods to harvest and extract proteins from field-cultivated plants on a large scale. Again, robustness of the virus expression system and the plant host is essential for the success of a production system. Much development work has gone into the many agronomic issues that will allow the use of standard agricultural implements to harvest virus-infected plants in the field and LSBC (formally Biosource Technologies, Inc.) has successfully harvested acres of tobacco plants for vaccine and therapeutic protein production (Turpen, 1999). The first challenge for anyone purifying proteins from plants is the removal of the abundant plant proteins involved in photosynthesis. The presence of these proteins complicates most biochemical separations. Following the elimination of these host proteins as well as host cell membranes, one can apply standard biochemical purification methods to reduce the volume of plant extracts and selectively obtain the protein of interest.

4.5.2 Commercial opportunities for plant virus-based protein production

Ongoing development of existing virus vector systems will hopefully allow all the above mentioned challenges to be overcome. There are many opportunities for production targets using these vector systems and the inability of plant viruses to infect mammalian hosts makes plants not only a cheap but potentially safe alternative to standard cell-culture-based production systems.

Development of novel vaccines. Vaccine targets are among the most attractive options for virus expression systems and much research has been directed towards the production of effective vaccines using a variety of plant virus-based expression strategies. Integration of protective epitopes on virus CPs (Lomonossoff and Johnson, 1995) and the adjuvancy effects of virions provide a great potential for cheap and safe vaccine production. In addition, plant viruses have been used to direct the synthesis of larger antigenic proteins as free proteins rather than as fusions to the virus CP (McCormick *et al.*, 1999; Wigdorovitz *et al.*, 1999).

The first plant virus to be vigorously tested as an epitope presentation system was CPMV, a bipartite virus that forms icosahedral particles with structural similarity to the particles of animal-infecting picornavirus such as poliovirus. This presentation system has proven to be efficient in raising an immune response in several vaccination studies (*Table 4.1*). Not all tested epitopes have proven similarly efficient in eliciting an antibody response, however, significant progress has been made in developing veterinary vaccines (Brennan *et al.*, 1999b; Dalsgaard *et al.*, 1997; McLain *et al.*, 1995). Many viruses have been used as presentation systems with varying degrees of success (*Table 4.1*) and it is likely that the presentation of a candidate antigenic epitope will vary depending on the particular virus being used as carrier and the method by which it is carried on the particle surface (i.e. carboxy- or amino-terminal fusion, internal loop, addition of linker sequences etc.).

There is only one example of a comparative trial of different viruses expressing the same epitope that was carried out using either PVX or CPMV expressing respectively 38aa or 30aa of the D2 fibronectin binding peptide (FnBP) from *Staphylococcus aureus* (Brennan *et al.*, 1999a). Both viruses were shown to be efficient carriers of FnBP D2 peptide and as little as 30 ng of conjugated peptide was able to raise an immune response. Further comparative trials will be necessary before any empirical rules as to preferred viral carriers for particular types of epitope can be derived.

Ultimately the goal of vaccination studies is not simply to demonstrate the induction of an antibody response but to confer protective immunity on the vaccinated individual to subsequent challenge with the pathogen from which the antigen was derived. To this end it is extremely encouraging that several examples have now been provided where a high degree of protection has been obtained following immunization with either epitope decorated virus particles or virally expressed recombinant antigens (*Table 4.1*). Both CPMV and TMV constructs, engineered to express epitopes from MEV and murine hepatitis virus respectively, have been demonstrated in laboratory trials to confer protection to subsequent pathogen challenge (Dalsgaard *et al.*, 1997; Koo *et al.*, 1999). Similarly, AIMV pseudovirus particles, assembled from AIMV CP subunits tagged with a 40-amino-acid residue epitope from the rabies GP3 protein, which were expressed from a TMV-based vector, were shown to confer protection in 40% of vaccinated mice against a challenge infection with a lethal dose of rabies virus (Modelska *et al.*, 1998). In the case of recombinant FMDV VP1, synthesized as a free protein *via* a TMV-based vector, 100% protective immunity was observed in mice subsequently challenged with a lethal dose of FMDV (Wigdorovitz *et al.*, 1999). Interestingly, in the latter experiments, vaccinations were performed by intraperitoneal injection of unpurified plant leaf extracts, indicating that even extremely crude preparations of antigen can confer an

Table 4.1. Development of plant virus-based vectors as a potential source of novel vaccines

Vector	Nature of the antigen	Disease or biological function	Insert size	Immunogenicity *(serum dilution for significant epitope binding/end-point titres on ELISA tests)	Protection (%survival or asymptomatic individual)	Observations	References
CPMV	FMDV VP1 CP	foot-and-mouth disease	25 a.a.	n. t., only western pAb	n.t.	Insertion in the βB-βC loop of small CP subunit (EPICOAT). No systemic movement.	Usha *et al.*, 1993
	FMDV VP1 CP	foot-and-mouth disease	20 a.a.	not detected	n.t.	No accumulation. No particle formation.	Usha *et al.*, 1993
	FMDV VP1 CP	foot-and-mouth disease	19 a.a.	not detected	n.t.	No systemic spread. Failed to purify due to strong binding to cell membranes.	Porta *et al.*, 1994
	HIV-1 GP41	AIDS	22 a.a.	*1/25,800; 2 injections	n.t.	Neutralization of HIV-1 IIIB at 1/200 antiserum dilution.	McLain *et al.*, 1995
	HRV-14 VP1	rhinopharyngitis	14 a.a.	1/16,000 dil. serum in western	no	Non-neutralizing antisera. Post-assembly cleavage of the inserted (linear) epitope.	Porta *et al.*, 1994
	MEV VP2 CP	rhinopharyngitis	17 a.a.	not detected	100%	Dose-dependent protection. Almost undetectable antibody response.	Dalsgaard *et al.*, 1997
	S.aureus FnBP	nosocomial bacteraemia	30 a.a.	*1/143,000; 2 injections	n.t.	60 epitope copies/virion. Antibodies inhibited binding of FnBP to fibronectin.	Brennan *et al.*, 1999a
	*P.a.*OMF10+18	chronic pulmonary infections	34 a.a.	*1/1,500; 5 injections	n.t.	Elicited antibody response.	Brennan *et al.*, 1999b
	*P.a.*OMF10+18	chronic pulmonary infections	34 a.a.	*1/11,000; 5 injections	n.t.	Insertion in the large (L) CP subunit. Elicited antibody response.	Brennan *et al.*, 1999b
TBSV	HIV-1 V3 loop	AIDS	13 a.a.	*1/1,500–4,500; 1 injection	n.t.	CP carboxy-terminal fusion. 180 epitope copies/virion	Joelson *et al.*, 1997
PPV	CPV VP2 CP	dog myocarditis and enteritis	15 a.a./30 a.a.	*1/400,000; 2 injections	n.t.	Insertion on CP surface loop. Neutralizing antibodies against CPV.	Rosario-Fernandez *et al.*, 1998
PVX	*S.aureus* FnBP	nosocomial bacteraemia	38 a.a.	*1/108,000; 2 injections	n.t.	Amino-terminal CP fusion with 2A linker, antibodies inhibited binding of FnBP to fibronectin.	Brennan *et al.*, 1999
TMV	ZP3 protein	antibody-med. contraception	13 a.a.	*1/64–4,000; 1 injection	no	CP carboxy-terminal fusion. Virions insoluble and difficult to purify, no impact on fertility.	Fitchen *et al.*, 1995
	Influenza HA	flu	8 a.a./18 a.a.	n.t., only western mAb	n.t.	Readthrough CP carboxy-terminal fusion. Systemic movement.	Sugiyama *et al.*, 1995
	HIV-1 gp120	AIDS	13 a.a.	n.t., only western mAb	n.t.	Readthrough CP carboxy-terminal fusion. Systemic movement.	Sugiyama *et al.*, 1995
	P. vivax CS	malaria	12 a.a.	n. t., only western mAb	n.t.	Insert in surface loop-region of CP.	Turpen *et al.*, 1995
	P. yoelli CS	malaria	15 a.a.	n. t., only western mAb	n.t.	Readthrough CP carboxy-terminal fusion. Systemic movement.	Turpen *et al.*, 1995
	Rabies GP	rabies	40 a.a.	*1/160 ser.dil.; 3 injections	40%	Free AlMV CP fusion, milder clinical symptoms by food-mediated delivery.	Modelska *et al.*, 1998
	HIV-1 V3 loop	AIDS	47 a.a.	*1/66,000 ser.dil.; 7 injections	n.t.	Free AlMV CP fusion.	Yusibov *et al.*, 1997
	scFv	N-H Lymphoma vaccination	270 a.a.	~100μg/ml IgG1-2a, 3 injections	70–90%	Free protein, protection equivalent to IgM-mediated vaccination.	McCormick *et al.*, 1999
	FMDV VP1 CP	foot-and-mouth disease	210 a.a.	*1/100–32,000; 5 injections	100%	Free protein. No purification of VP1, 100% protection by injection of diluted plant extract.	Wigdorovitz *et al.*, 1999
	antibody H&L	colon cancer	230a.a /455 a.a	n. t., only western PAb	n.t.	Co-inoculation of TMV constructs expressing antibody Heavy or Light chain. Correct mAb assembly.	Verch *et al.*, 1998
	MHV S protein	encephalomyelitis	10a.a. /15 a.a.	* 1/16,400: 3 subcut. injections	80%	CP carboxy-terminal insertion. Protection from MHV by subcutaneous or intranasal administration.	Koo *et al.*, 1999

FMDV, foot-and-mouth disease virus; HIV, human immunodeficiency virus; gp, glycoprotein; HRV, human rhinovirus; MEV, mink enteritis virus; *S. aureus*, *Staphylococcus aureus*; FnBP, fibronectin binding-protein; ACEI, agiotensin-converting-enzyme-inhibitor; ZP, zona pellucida; *P. vivax*, *Plasmodium vivax*; *P. yoelli*, *Plasmodium yoelli*; scFv, single-chain antibody fragment; Pab, polyclonal antibody; mAb, monoclonal antibody; ELISA, enzyme-linked immunosorbent assay; Ig, immunoglobulin. AIDS, acquired immunodeficiency syndrom; a.a., amino acid; N-H, Non-Hodgkin's; PPV, plum pox potyvirus; CPV, canine parvovirus; P.a.OMF10+18: *Pseudomonas aeruginosa* outer-membrane protein F tandem synthetic peptide 10 and 18; H&L: heavy and light chain; HA: hemagglutinin; MHV S protein: murine hepatitis virus spike protein.

effective immune response (Wigdorovitz *et al.*, 1999). A summary of the various epitopes expressed using plant virus-based vectors and their efficacy in eliciting immune responses is provided in *Table 4.1*.

An extremely promising recent development described the production of a tumor-specific vaccine that conferred a high degree of protection in mice subsequently challenged with the tumor. The strategy relies on the fact that each malignant B cell clone expresses a unique cell surface immunoglobulin that serves as a tumor-specific marker. A single chain variable region antibody, comprising just the hypervariable region of the immunoglobulin unique to the 38C13 mouse B cell lymphoma, was cloned into a TMV-based vector as a fusion to an amino-terminal secretory peptide. The secreted antibody, which accumulated in the interstitial fluid, was isolated at levels representing up to 30 mg kg^{-1} fresh plant material and purified. This plant derived 'tumor vaccine' raised an anti-idiotype immune response that conferred an equal level of protection to tumor challenge (80% survival) as the native 38C13 immunoglobulin administered as a keyhole limpet hemocyanin conjugate (McCormick *et al.*, 1999). The speed with which appropriate single chain antibody clones could be identified, approximately 4 weeks after molecular cloning, raises the exciting possibility of developing patient-specific treatment of human lymphomas through effective immunotherapy (McCormick *et al.*, 1999). It is clear from the numerous results discussed above that effective vaccines can be produced *in planta*, either as epitope fusions to viral CPs or as free antigens. The challenge now is to improve the technology to enable efficient and cost-effective production in plants that can compete commercially with extant vaccines.

Production of therapeutic proteins in plants. Therapeutic proteins represent another attractive target for plant production systems (*Table 4.2*). The ability of viruses to express economically viable amounts of proteins to augment the mammalian immune response, effect angiogenesis and fight cancer provides many opportunities for these vectors to affect a range of different health conditions. An example of the success of these approaches has been the production of human α-galactosidase A (Gal-A). An inexpensive form of this enzyme would be extremely desirable for the treatment of patients suffering from Fabry disease, a lysosomal storage disorder which results from a Gal-A deficiency. This enzyme has been purified at levels of 30–50 mg kg^{-1} fresh weight of plant material and the resulting enzyme preparations have specific activities at least as high, if not greater than, native human Gal-A (Turpen, 1999; LSBC, unpublished results). Gal-A showing this level of activity has been purified under both pilot and full-production conditions. The high activity of the Gal-A produced *in planta* argues strongly for the ability of plant cells to fold and modify mammalian enzymes as well as native systems, even with stringent requirements for glycosylation and proteolytic processing. This opens the door for the production of many other products including cytokines and other bioactive compounds. Some products may require specific glycosylation patterns not produced *in planta* while other protein products may either tolerate non-native modifications or not require such modification. One issue that remains unresolved is the effects of plant-specific glycan structures on the effectiveness of these plant-derived protein products. All types of responses have been proposed,

Table 4.2. Development of plant virus-based vectors for production of pharmaceuticals

Vector	Protein or peptide expressed	Biological function	Insert/protein size	Expression strategy	Observations	References
CaMV	Human interferon α-D (IFNaD)	Antiviral-antiproliferative activities	400 bp, approx. 130 a.a.	Gene II replacement	Up to 2 μg IFNαD/g tissue. Biologically active IFN (up to 200 000 IU/ml) Reduction of cytopathic effect of vesicular stomatitis virus on MDBK cells.	De Zoeten *et al.*, 1989
TMV	Leu-enkephalin	Opiate-like activities	15 bp, 5 a.a.	CP-C.term. fusion	100% incorporation. No particle assembly. Not tested for activity	Takamatsu *et al.*, 1990
	Angiotensin-I-Converting Enzyme Inhibitor (ACEI)	Antihypertensive	36 bp, 12 a.a.	CP-C.term. fusion. Leaky readthrough	5% readthrough. Up to 10 μg/g in tomato fruit. Not tested for activity	Hamamoto *et al.*, 1993
	α-Trichosanthin	Ribosome-inactivating protein, antiviral activity	450 bp, approx. 250 a.a.	Gene insertion. Duplicated promoter	Reach 2% TSP. Protein synthesis inhibition activity comparable to native protein	Kumagai *et al.*, 1993
	α-Galactosidase	Fabry disease	1250 bp, approx. 420 a.a.	Gene insertion. Duplicated promoter	Purified to homogeneity. S.A. 10% higher than other reported sources	Turpen, 1999

a.a., amino acids; bp, base pairs; TSP, total soluble protein; CP-C.term, C.terminal extremity of the viral coat protein; S.A., specific activity.

from adverse reaction to tolerance. Ultimately any product and production process will be judged by the same criteria applied to any other source of recombinant protein with issues such as genetic stability, microbial contamination, product purity and product comparability all requiring consideration (Miele, 1997).

Metabolic engineering. An under-investigated application of plant virus-based vectors is their use as tools to modify pre-existant plant metabolic pathways in order to over-produce primary and secondary metabolites with pharmaceutical, nutritional or industrial applications. In theory there are a number of approaches that could be used to attain this objective: (i) expression of an enzyme that converts a plant metabolite to a desired end product; (ii) expression of an enzyme that redirects metabolic intermediates, either to give a novel product or to provide the substrate for further enzymatic modification; iii) inhibition of a plant-encoded enzyme, either to terminate a metabolic pathway or enhance flux through an alternative pathway. To date the only examples of directed metabolic engineering using virus-based vectors involve the modification of plant carotenoid biosynthetic pathways. In the first example TMV was used to express either phytoene synthase or a partial cDNA clone encoding phytoene desaturase. Over-expression of sense transcripts encoding phytoene synthase led to the development of a bright orange phenotype in infected leaves that was associated with a ten-fold increase in phytoene accumulation over non-infected control plants (Kumagai *et al.*, 1995). In contrast, expression of a fragment of the phytoene desaturase, in either sense or antisense orientation, inhibited carotenoid synthesis downstream of carotene and caused the development of a white, pigment-less, phenotype in systemically infected leaves (Kumagai *et al.*, 1995). In the latter case, the ability of both sense and antisense transcripts to induce a similar 'bleached' phenotype suggested that endogenous phytoene synthase may have been subject to homology dependent PTGS (see Section 4.4.4). Whatever the mechanism by which inhibition of phytoene desaturase was achieved the phenotype obtained demonstrated the possibility of downregulating the expression of an endogenous plant gene through the use of a virus-based vector. In a more recent report, again using TMV as a vector, the over-expression of capsanthin-capsorubin synthase was used to direct the synthesis of the non-native xanthophyll, capsanthin (Kumagai *et al.*, 1998). In this study up to 36% of the total carotenoid pool was redirected to capsanthin, which was recruited to the light-harvesting complexes of photosystem II (Kumagai *et al.*, 1998). The results discussed above, describing manipulation of the carotenoid biosynthetic pathway, clearly show the potential for virally expressed genes (or gene fragments) to modify endogenous plant metabolic pathways, either to provide a desired product or to enhance plant productivity. It is likely, given the wide range of plant primary and secondary metabolites that are derived from plants, that the future will see many more examples of targeted metabolic engineering using plant virus-based vectors.

Gene discovery. The dominant path to gene discovery in multicellular organisms focuses on structural approaches to genomics, gene mapping, isolation and sequencing. Despite, impressive technological progress in DNA sequencing and DNA database-driven bioinformatics, the actual contributions of genes to the biology of plants and other complex organisms are still primarily determined by more laborious, traditional molecular biological approaches. Application of T-DNA knockout libraries and transgenic plant construction provide insight into the functions of a proportion of genes currently under investigation. However, the occurrence of embryonic lethality, gene silencing and low expression associated with DNA-based expression systems obscure the determination of the functions of many genes. The structural genomics approach is further hampered by the inability of bioinformatics to ascertain the actual functions of many cloned genes or, when a particular function can be related to a given sequence, the pleiotropic functions of any gene on the biochemistry and physiology of the whole plant may not be apparent.

Genomics approaches in micro-organisms successfully relied on the power of mutagenesis and genetics to identify critical gene functionality. Only after interesting phenotypes were obtained did one map and clone novel genes. In order to apply the power of a more genetics-based gene discovery programme, virus vectors can be exploited as an approach to assign actual function to a gene or gene fragment, and the encoded protein, before the actual DNA sequence is determined. This is accomplished by using plant virus vectors to carry unknown sequences systemically in plants and mediate their expression in many plant cell types. These vectors can allow either the expression of full-length genes to assay for 'gain of function' phenotypes, or deliver a systemic gene 'knock-out' phenotype with partial gene fragments through anti-sense or sense-gene silencing approaches (for a recent review of silencing strategies see Baulcombe, 1999). In the expression approach, genes from virtually any organism can be inserted into the virus-based vector for activity testing in plants. The impact of such an introduced gene on augmenting innate plant enzymatic functions, generation of novel molecular architectures, resistance to herbicide chemistries as well as improved agronomic and host defence properties can all be assayed using various biochemical and physical selection approaches. The effectiveness of viruses in inducing silencing of endogenous plant genes is supported by several reports starting with silencing of the phytoene desaturase gene by the TMV vector (Kumagai *et al.*, 1993) and now extending to other virus vector systems (Jones *et al.*, 1999).

The use of virus expression vectors in a gene function discovery programme is complementary to other 'structural' (genome sequencing, protein family/homology assignments) and 'inferential' (mRNA profiling, proteomics, metabolic profiling) approaches. Genes that are defined by one of the above approaches can be directly tested for function using virus expression vectors. Conversely, genes with interesting functions as defined by virus expression vector research can be tested for relationships at the structural or biochemical level by standard genomics techniques.

4.5.3 Biological containment and safety

No discussion of recombinant protein production in plants would be complete without some consideration of the biological safety and containment issues raised. Indeed, at first glance, the use of viral vectors in the field might appear to provide a greater opportunity for horizontal gene transfer than do field-grown transgenic crops. Obviously it is important to evaluate potential environmental risks and hazards very closely before any deployment of viral vectors in the field. As stated previously plant viruses are completely non-pathogenic to man and animals and given the large quantities of plant virus proteins regularly consumed by all of us in food there is no suggestion that native viral proteins *per se* are in any way harmful to human health. Even so, the controversial principle of substantial equivalence need not be extended to plants infected with viral vectors, as there is currently no intention of using vector-infected crops for food products. In the case of vaccines and other therapeutic proteins that have been discussed in the preceding sections, the stringent regulations laid down by the Food and Drug Administration in the US, and equivalent bodies world-wide, provide a high degree of protection against the introduction of therapeutics with harmful side effects (Miele, 1997).

The most important issue to consider then is not the safety of the product it is intended to make but the environmental issues associated with virus dissemination to other hosts, either crop or weed, and the problems that this could cause. Only a limited amount of work has been done to address this issue and to date the only well studied virus in this regard is TMV. In data reported by LSBC, eight separate field trials have shown no evidence of transmission to adjacent non-inoculated tobacco or weed species of genetically modified TMV-based vectors. Furthermore, after two seasons of crop rotation tobacco was cultivated again on test sites with no carry-over of infectivity (Turpen, 1999). Thus, with good cultural practices the risk of inadvertent transmission of genetically modified TMV would appear to be low. In this regard TMV should be easier to manage than some virus-based vectors as it is not transmitted by insects, fungi or nematodes and moreover it is not transmitted *via* either seed or pollen. Viruses that do not meet these criteria regarding transmissibility would clearly require additional control and/or containment measures to prevent their dissemination in the environment. More recently LSBC have tested a TEV-based vector that had been genetically rendered non-transmissible by aphids. Prior to field release the non-transmissibility of the vector had to be demonstrated under containment conditions. Subsequently, a field trial of the TEV vector showed that the modified virus was contained within the inoculated plot and was not transmitted either to weeds or to non-inoculated plants (G.P.P., unpublished results). Thus, with an understanding of the genetic basis of interactions between viruses and their vectors, it is possible to modify viral genomes to generate non-transmissible variants.

However low the risk of transmission may be, it is nevertheless necessary to consider the consequences that could arise should escape occur. In discussing this point there are two important factors to bear in mind. Firstly, with the sole exception of the expression of *bar*, none of the genes expressed to date from plant virus-based vectors conferred any selective advantage on the modified virus. Even in the case of *bar*, which was used in a wholly experimental context to provide a positive screen for long-distance virus movement, selection for retention of the insert would only be maintained by continuous application of the herbicide GA. The second point to consider is the issue of genetic stability of plant virus-based vectors that has been raised constantly through this review. It is something of a paradox that for a reliable and robust vector it is essential that genetic stability of the recombinant virus be high, yet, as protection against the consequences of inadvertent escape, a degree of genetic instability is desirable

(Scholthof *et al.*, 1996). Ultimately any additional sequence placed in a viral genome, unless conferring a selective advantage, will place a genetic load on the virus that will render it less competitive than the wild-type progenitor virus. Furthermore the capacity of viruses to recombine and lose inserted sequences is an inherent feature of viral replication and therefore not something that can be easily modified. Thus, although virologists will continue to increase the genetic stability of recombinant viral vectors it is inevitable that any modified virus will be less fit than the parental virus and sooner or later recombination will occur. The selective advantage of deletion variants derived from viral vectors is most pronounced at genetic bottlenecks and plant-to-plant transfer provides the most stringent bottleneck of all. Although the risk of accidental release cannot be ignored, it must be evaluated on a virus-to-virus basis while taking into account the safeguards that can be implemented to reduce the risk of inadvertent release. Ultimately the environmental consequences of accidental escape are likely to be low given the relatively limited number of plant-to-plant transfers that can occur before the inserted sequence is lost.

4.6 Conclusions

From the many examples of plant virus-based vectors discussed in the preceding sections it is clear that no virus or foreign gene expression strategy is suited to all the many uses to which plant viral vectors could be applied. In fact, the number of virus-based vectors and novel gene expression strategies is likely to increase with time as new molecular tools and techniques become available. What is clear is that remarkable progress has been made in less than two decades and plant viral vectors have advanced from purely theoretical entities to becoming extremely efficient and robust gene expression systems that are likely to play an increasingly significant role in recombinant protein production in plants.

For the full potential of virus-based vectors to be realized, both as research tools and for the production of recombinant proteins, a number of issues will need to be addressed. These issues can be divided between factors that relate to current limits of virus vector technology and those factors requiring a better understanding of basic plant biological processes. In terms of virus biology an understanding of the factors limiting virus accumulation and movement will help identify those aspects of the replication and transport processes that could be targeted for selective improvement. Other features that would be desirable in future vectors include increased tolerance of large inserts, improved genetic stability, vectors for simultaneous expression of multiple genes and vectors for a wide range of crop species including monocotyledenous plants. In addition to improvements to virus vector technology, a further contribution to vector improvement will come from an increased understanding of the plant host. Obviously, the limits to virus accumulation and spread are not solely determined by the virus but reflect the dynamic interaction between pathogen and host phenotype. At the present time phloem-dependent movement is a relatively poorly understood process (Santa Cruz, 1999) and a better understanding of the basis of phloem loading and unloading of viruses is an important goal if viral vectors with fast systemic movement phenotypes are to be derived. Another area of plant biology with obvious relevance to virus vectors is gene silencing. Exploiting gene silencing will provide real opportunities for virus-based strategies for gene discovery; however, to successfully develop virus vectors as reliable inducers of gene silencing will require a better understanding of the triggers and executors of PTGS.

The factors outlined above are all the subject of current research and if the progress made over the last two decades is anything to go by then many of the goals stated above will be attained in the foreseeable future. Plant virus-based vectors are now well established tools in plant molecular biology and plant pathology and it is likely that their use as basic and applied research tools will continue to grow. For the commercial applications of virus-based vectors a wide range of factors will determine whether widespread field production of recombinant proteins becomes a reality. Clearly the single most important determinant will be economic. However, other issues, particularly those relating to public acceptability, will need to be addressed if virus-derived products are to reach the market. In the short-term virus-based gene discovery, epitope decorated virus-based vaccines and high-value speciality therapeutic proteins, are likely to provide the immediate focus for commercialization and these relatively low-volume processes can be accommodated under glass rather than in the field. Ultimately it is to be hoped that success in these areas will provide the incentive to develop plant virus-based vectors for a far wider range of applications, thereby reaping the benefits of the biosynthetic capacity of plants for the targeted production of valuable macromolecules.

References

Ahlquist P, French R, Janda M, Loesh-Fries LS. (1984) Multicomponent RNA plant virus infection derived from cloned viral cDNA. *Proc. Natl. Acad. Sci. USA* **81**: 7066–7070.

Anandalakshmi R, Pruss GJ, Ge X, Marathe R, Mallory AC, Smith TH, Vance VB. (1998) A viral suppressor of gene silencing in plants. *Proc. Natl. Acad. Sci. USA* **95**: 13079–13084.

Angell SM, Davies C, Baulcombe DC. (1996) Cell-to-cell movement of potato virus X is associated with a change in the size exclusion limit of plasmodesmata in trichome cells of *Nicotiana clevelandii*. *Virology* **215**: 197–201.

Baulcombe DC. (1999) Fast forward genetics based on virus-induced gene silencing. *Curr. Opin. Plant Biol.* **2**: 109–113.

Baulcombe DC, Chapman S, Santa Cruz S. (1995) Jellyfish green fluorescent protein as a reporter for virus infections. *Plant J.* **7**: 1045–1053.

Beachy RN, Bendahmane M. (1999) Structural and cellular basis of pathogen derived resistance to virus infection *FASEB J.* **13**: 1339.

Bendahmane A, Köhm BA, Dedi C, Baulcombe DC. (1995) The coat protein of potato virus X is a strain-specific elicitor of Rx1-mediated virus resistance in potato. *Plant J.* **8**: 933–941.

Birch RG. (1997) Plant transformation: problems and strategies for practical application. *Annu. Rev. Plant Physiol. Plant Mol. Biol.* **48**: 297–326.

Boevink P, Santa Cruz S, Hawes C, Harris N, Oparka KJ. (1996) Virus-mediated delivery of the green fluorescent protein to the endoplasmic reticulum of plant cells. *Plant J.* **10**: 935–941.

Boevink P, Oparka K, Santa Cruz S, Martin B, Betteridge A, Hawes C. (1998) Stacks on tracks: the plant Golgi apparatus traffics on an actin/ER network. *Plant J.* **15**: 441–447.

Boevink P, Martin B, Oparka KJ, Santa Cruz S, Hawes C. (1999) Transport of virally expressed green fluorescent protein through the secretory pathway in tobacco leaves is inhibited by cold shock and brefeldin A. *Planta* **208**: 392–400.

Boulton MI, Steinkellner H, Donson J, Markham PG, King DI, Davies JW. (1989) Mutational analysis of the virion-sense genes of maize streak virus. *J. Gen. Virol.* **70**: 2309–2323.

Boyer J-C, Haenni A-L. (1994) Infectious transcripts and cDNA clones of RNA viruses. *Virology* **198**: 415–426.

Brennan FR, Jones TD, Longstaff M *et al.* (1999a) Immunogenicity of peptides derived from a fibronectin-binding protein of *S. aureus* expressed on two different plant viruses. *Vaccine* **17**: 1846–1857.

Brennan FR, Jones TD, Gilleland LB *et al.* (1999b) *Pseudomonas aeruginosa* outer-membrane protein F epitopes are highly immunogenic in mice when expressed on a plant virus. *Microbiology* **145**: 211–220.

Brigneti G, Voinnet O, Li L, Ding S, Baulcombe DC. (1998) Viral pathogenicity determinants are suppressors of transgene silencing in *Nicotiana Benthamiana. EMBO J.* **17**: 6739–6746.

Brisson N, Paszkowski J, Penswick JR, Gronengorg B, Potrykus I, Hohn T. (1984) Expression of a bacterial gene in plants by using a viral vector. *Nature* **310**: 511–514.

Buck KW. (1996) Comparison of the replication of positive-stranded RNA viruses of plants and animals. *Adv. Vir. Res.* **47**: 159–251.

Buck KW. (1999) Replication of tobacco mosaic virus RNA. *Phil. Trans. R. Soc. London B.* **354**: 613–627.

Burgyan J, Salanki K, Dalmay T, Russo M. (1994) Expression of homologous and heterologous viral coat protein-encoding genes using recombinant DI RNA from cymbidium ringspot tombusvirus. *Gene* **138**: 159–163.

Canto T, Prior DAM, Hellwald KH, Oparka KJ, Palukaitis P. (1997) Characterization of cucumber mosaic virus. 4. Movement protein and coat protein are both essential for cell-to-cell movement of cucumber mosaic virus. *Virology* **237**: 237–248.

Carrington JC, Whitham SA (1998) Viral invasion and host defense: strategies and counter-strategies. *Curr. Opin. Plant Biol.* **1**: 336–341.

Carrington JC, Kasschau KD Mahajan SK, Schaad MC. (1996) Cell-to-cell and long-distance transport of viruses in plants. *Plant Cell.* **8**: 1669–1681.

Casper SJ, Holt CA. (1996) Expression of the green fluorescent protein-encoding gene from a tobacco mosaic virus-based vector. *Gene* **173**: 69–73.

Chalfie M, Tu Y, Euskirchen G, Ward WW, Prascher DC. (1994) Green fluorescent protein as a marker for gene expression. *Science* **263**: 802–805.

Chapman S, Kavanagh T, Baulcombe D. (1992) Potato virus X as a vector for gene expression in plants. *Plant J.* **2**: 549–557.

Chu M, Park JW, Scholthof HB. (1999) Separate regions on the tomato bushy stunt virus p22 protein mediate cell-to-cell movement versus elicitation of effective resistance responses. *Mol. Plant-Microbe Interact.* **12**: 285–292.

Cockerham G. (1970) Genetical studies on resistance to potato viruses X and Y. *Heredity* **25**: 309–348.

Covey S. (1991) Pathogenesis of a plant pararetrovirus: CaMV. *Sem Virol.* **2**: 151–159.

Cronin S, Verchot J, Haldeman-Cahill R, Schaad MC, Carrington JC. (1995) Long-distance movement factor: a transport function of the potyvirus helper component-protease. *Plant Cell* **7**: 549–559.

Cubitt AB, Heim R, Adams SR, Boyd AE, Gross LA, Tsien RY. (1995) Understanding, improving and using green fluorescent proteins. *Trends Biochem. Sci.* **40**: 448–455.

Culver JN, Lehto K, Close SM, Hilf ME, Dawson WO. (1993) Genomic position affects the expression of tobacco mosaic virus movement and coat protein genes. *Proc. Natl. Acad. Sci. USA* **90**: 2055–2059.

Dalsgaard K, Uttenthal A, Jones TD *et al.* (1997) Plant-derived vaccine protects target animals against a viral disease. *Nat. Biotechnol.* **15**: 248–252.

Dawson WO, Beck DL, Knorr DA, Grantham GL. (1986) cDNA cloning of the complete genome of tobacco mosaic virus and production of infectious transcripts. *Proc. Natl. Acad. Sci. USA* **83**: 1832–1836.

Dawson WO, Bubrick P, Grantham GL. (1988) Modifications of the tobacco mosaic virus coat protein gene affecting replication, movement and symptomatology. *Phytopathology* **78**: 783–789.

Dawson WO, Lewandowsky DJ, Hilf ME *et al.* (1989) A tobacco mosaic virus-hybrid expresses and loses an added gene. *Virology* **172**: 285–292.

Deom CM, Oliver MJ, Beachy RN. (1987) The 30-kiloDalton gene product of tobacco mosaic virus potentiates virus movement. *Science* **237**: 389–394.

De Zoeten GA, Penswick JR, Horisberger MA, Ahl P, Schultze M, Hohn T. (1989) The expression, localization, and effect of a human interferon in plants. *Virology* **172**: 213–222.

Ding B. (1998) Intercellular protein trafficking through plasmodesmata. *Plant Mol. Biol.* **38**: 279–310.

Dolja VV, McBride HJ, Carrington JC. (1992) Tagging of plant potyvirus replication and movement by insertion of β-glucuronidase into the viral polyprotein. *Proc. Natl. Acad. Sci. USA* **89**: 10208–10212.

Dolja VV, Herndon KL, Pirone TP, Carrington JC. (1993) Spontaneous mutagenesis of a plant potyvirus genome after insertion of a foreign gene. *J. Virol.* **67**: 5968–5975.

Dolja VV, Haldeman R, Robertson NL, Dougherty WG, Carrington JC. (1994) Distinct functions of capsid protein in assembly and movement of tobacco etch potyvirus in plants. *EMBO J.* **13**: 1482–1491.

Dolja VV, Peremyslov VV, Keller KE, Martin RR, Hong J. (1998) Isolation and stability of histidine-tagged proteins produced in plants via potyvirus gene vectors. *Virology* **252**: 269–274.

Donnelly MLL, Gani D, Flint M, Monaghan S, Ryan MD. (1997) The cleavage activities of aphtovirus and cardiovirus 2A proteins. *J. Gen. Virol.* **78**: 13–21.

Donson J, Kearney CM, Hilf ME, Dawson WO. (1991) Systemic expression of a bacterial gene by a tobacco mosaic virus-based vector. *Proc. Natl. Acad. Sci. USA* **88**: 7204–7208.

Drugeon G, Urcuquilnchima S, Milner M *et al.* (1999) The strategies of plant virus gene expression: models of economy. *Plant Sci.* **148**: 77–88.

English JJ, Mueller E, Baulcombe DC. (1996) Suppression of virus accumulation in transgenic plants exhibiting silencing of nuclear genes. *Plant Cell* **8**: 179–188.

Essl D, Dirnberger D, Gomord V, Strasser R, Faye L, Glössl J, Steinkeller H. (1999) The N-terminal 77 amino acids from tobacco *N*-acetylglucosaminyltransferase I are sufficient to retain a reporter protein in the Golgi apparatus of *Nicotiana benthamiana* cells. *FEBS Lett.* **453**: 169–173.

Fischer R, Drossard J, Commandeur U, Schillberg S, Emans N. (1999a) Towards molecular farming in the future: moving from diagnostic protein and antibody production in microbes to plants. *Biotech. Appl. Biochem.* **30**: 101–108.

Fischer R, VaqueroMartin C, Sack M, Drossard J, Emans N, Commandeur U. (1999b) Towards molecular farming in the future: transient protein expression in plants. *Biotech. Appl. Biochem.* **30**: 113–116.

Fitchen J, Beachy RN, Hein MB. (1995) Plant virus expressing hybrid coat protein with added murine epitope elicits autoantibody response. *Vaccine* **13**: 1051–1057.

French R, Janda M, Ahlquist P. (1986) Bacterial gene inserted in an engineered RNA virus: efficient expression in monocotyledonous plant cells. *Science* **231**: 1294–1297.

Ghoshroy S, Lartey R, Sheng J, Citovsky V. (1997) Transport of proteins and nucleic acids through plasmodesmata. *Annu. Rev. Plant Physiol. Plant Mol. Biol.* **48**: 27–49.

Gilardi P, Garcia-Luque I, Serra MT. (1998) Pepper mild mottle virus coat protein alone can elicit the *Capsicum* spp. L^3 gene-mediated resistance. *Mol. Plant-Microbe Interact.* **11**: 1253–1257.

Grill LK (1983) Utilizing RNA viruses for plant improvement. *Plant Mol. Biol. Rep.* **1**: 17–20.

Guivarc'h A, Caissard JC, Azmi A, Elmayan T, Chriqui D, Tepfer M. (1996) *In situ* detection of the *gus* reporter gene in transgenic plants: ten years of blue genes. *Transgenic Res.* **5**: 281–288.

Guo HS, Lopez-Moya JJ, Garcia JA. (1998) Susceptibility to recombination rearrangements of a chimeric plum pox potyvirus genome after insertion of a foreign gene. *Virus Res.* **57**: 183–195.

Hagiwara Y, Peremyslov VV, Dolja VV. (1999) Regulation of closterovirus gene expression examined by insertion of a self processing reporter and by northern hybridization. *J. Virol.* **73**: 7988–7993.

Hamamoto H, Sugiyama Y, Nakagawa N *et al.* (1993) A new tobacco mosaic vector and its use for the systemic production of angiotensin-I-converting enzyme inhibitor in transgenic tobacco and tomato. *Bio-Technology* **11**: 930–932.

Hammond-Kosack KE, Staskawicz BJ, Jones JDG, Baulcombe DC. (1995) Functional expression of a fungal avirulence gene from a modified potato virus X genome. *Mol. Plant-Microbe Interact.* **8**: 181–185.

Hanley-Bowdoin JS, Elmer JS, Rogers SG. (1989) Functional expression of the leftward open reading frames of the A component of tomato golden mosaic virus in transgenic tobacco plants.

Plant Cell **1**: 1057–1067.

Hanley-Bowdoin JS, Elmer JS, Rogers SG. (1990) Expression of functional replication protein from tomato golden mosaic virus in transgenic tobacco plants. *Proc. Natl. Acad. Sci. USA* **87**: 1446–1450.

Haseloff J, Amos B. (1995) GFP in plants. *Trends Genet.* **11**: 328–329.

Hayes RJ, Petty ITD, Coutts RHA, Buck KW. (1988) Gene amplification and expression in plants by a replicating geminivirus vector. *Nature* **334**: 179–182.

Hayes RJ, Coutts RHA, Buck KW. (1989) Stability and expression of bacterial genes in replicating geminivirus vectors in plants. *Nucl. Acids Res.* **17**: 2391–2403.

Heinlein M, Epel BL, Padgett HS, Beachy RN. (1995) Interaction of the tobacco mosaic virus movement proteins with the plant cytoskeleton. *Science* **270**: 1983–1985.

Hirochika H, Hayashi KI. (1991) A new strategy to improve a cauliflower mosaic virus vector. *Gene* **105**: 293–241.

Holt CA, Beachy RN. (1991) *In vivo* complementation of infectious transcripts from mutant tobacco mosaic virus cDNAs in transgenic plants. *Virology* **181**: 109–117.

Hong YG, Saunders K, Stanley J. (1997) Transactivation of dianthin transgene expression by African cassava mosaic virus AC2. *Virology* **228**: 383–387.

Howell SH. (1982) Plant molecular vehicles: potential vectors for introducing foreign DNA into plants. *Ann. Rev. Plant Physiol.* **33**: 609–650.

Jefferson RA, Kavanagh TA, Bevan MW. (1987) Gus fusions: β-glucuronidase as a sensitive and versatile gene marker in higher plants. *EMBO J.* **6**: 3901–3907.

Joelson T, Akerblom L, Oxefelt P, Strandberg B, Tomenius K, Morris TJ. (1997) Presentation of a foreign peptide on the surface of tomato bushy stunt virus. *J. Gen. Virol.* **78**: 1213–1217.

Jones L, Hamilton AJ, Voinnet O, Thomas CL, Maule AJ, Baulcombe DC. (1999) RNA–DNA interactions and DNA methylation in post-transcriptional gene silencing. *Plant Cell* **11**: 2291–2301.

Jupin I, Tamada T, Richards K. (1991) Pathogenesis of beet necrotic yellow vein virus. *Sem. Virol.* **2**: 121–129.

Kahn TW, Lapidot M, Heinlan M, Reichel C, Cooper B, Gafny R, Beachy RN. (1998) Domains of the TMV movement protein involved in subcellular localization. *Plant J.* **15**: 15–25.

Kamoun S, Honee G, Weide R *et al.* (1999) The fungal gene *Avr9* and the oomycete gene *infl* confer avirulence to potato virus X on tobacco. *Mol. Plant-Microbe Interact.* **12**: 459–462.

Karrer EE, Beachy RN, Holt CA. (1998) Cloning of tobacco genes that elicit hypersensitive response. *Plant Mol. Biol.* **36**: 681–690.

Kasschau KD, Carrington JC. (1998) A counterdefensive strategy of plant viruses: suppression of posttranscriptional gene silencing. *Cell* **95**: 461–470.

Kavanagh T, Goulden M, Santa Cruz S, Chapman S, Barker I, Baulcombe D. (1992) Molecular analysis of a resistance-breaking strain of potato virus X. *Virology* **189**: 609–617.

Kearney CM, Donson J, Jones GE, Dawson WO. (1993) Low-level of genetic drift in foreign sequences replicating in an RNA virus in plants. *Virology* **192**: 11–17.

Koo M, Bendahmane M, Lettierri GA *et al.* (1999) Protective immunity against murine hepatitis virus (MHV) induced by intranasal or subcutaneous administration of hybrids of tobacco mosaic virus that carries an MHV epitope. *Proc. Natl. Acad. Sci. USA* **96**: 7774–7779.

Kooman-Gersmann M, Vogelsang R, Hoogendijk ECM, de Wit PJGM. (1997) Assignment of amino acids residues of the AVR9 peptide of *Cladosporium fulvum* that determine elicitor activity. *Mol. Plant-Microbe Interact.* **10**: 821–829.

Kumagai MH, Turpen TH, Weinzettl N *et al.* (1993) Rapid, high-level expression of biologically active α-trichosanthin in transfected plants by an RNA viral vector. *Proc. Natl. Acad. Sci. USA* **90**: 427–430.

Kumagai MH, Donson J, Della-Ciopa G, Harvey D, Hanley K, Grill LK. (1995) Cytoplasmic inhibition of carotenoid biosynthesis with virus-derived RNA. *Proc. Natl. Acad. Sci. USA* **92**: 1679–1683.

Kumagai MH, Keller Y, Bouvier F, Clary D, Camara B. (1998) Functional integration of non-native carotenoids into chloroplasts by viral-derived expression of capsanthin-capsorubin synthase in *Nicotiana benthamiana*. *Plant J.* **14**: 305–315.

Kumpatala SP, Chandrasekharan MB, Iyer LM, Li G, Hall TC. (1998) Genome intruder scanning and modulation systems and transgene silencing. *Trends Plant Sci.* **3**: 97–104.

Lacomme C, Santa Cruz S. (1999) Bax-induced cell death in tobacco is similar to the hypersensitive response. *Proc. Natl. Acad. Sci. USA* **96**: 7956–7961.

Lacomme C, Smolenska L, Wilson TMA. (1998) Genetic engineering and the expression of foreign peptides or proteins with plant virus-based vectors. In: *Genetic Engineering Principles and Methods*. (ed. J.K. Setlow). **20**: 225–237.

Lin NS, Lee YS, Lin BY, Lee CW, Hsu YH. (1996) The open reading frame of bamboo mosaic potexvirus satellite RNA is not essential for its replication and can be replaced with a bacterial gene. *Proc. Natl. Acad. Sci. USA* **93**: 3138–3142.

Lindbo JA, Silva-Rosales L, Proebsting WM, Dougherty WG. (1993) Induction of highly specific antiviral state in transgenic plants: implications for regulation of gene expression and virus resistance. *Plant Cell* **5**: 1749–1759.

Lomonossoff G, Johnson JE. (1991) The synthesis and structure of comovirus capsids. *Prog. Biophys. Mol. Biol.* **55**: 107–137.

Lomonossoff G, Johnson JE. (1995) Eukaryotic viral expression systems for polypeptides. *Sem. Virol.* **6**: 257–267.

Maia IG, Séron K, Haenni A-L, Bernardi F. (1996) Gene expression from viral RNA genomes. *Plant Mol. Biol.* **32**: 367–391.

Mas P, Beachy RN. (1998) Distribution of TMV movement protein in single living protoplasts immobilized in agarose. *Plant J.* **15**: 835–8842.

McCormick AA, Kumagai MH, Hanley K *et al.* (1999) Rapid production of specific vaccines for lymphoma by expression of the tumor-derived single-chain epitopes in tobacco plants. *Proc. Natl. Acad. Sci. USA* **96**: 703–708.

McLain L, Porta C, Lomonossoff GP, Durrani Z, Dimmock NJ. (1995) Human immunodeficiency virus type 1-neutralyzing antibodies raised to a glycoprotein 41 peptide expressed on the surface of a plant virus. *AIDS Res. Hum. Retroviruses* **11**: 327–333.

McLean BG, Zupan J, Zambryski P. (1995) Tobacco mosaic virus movement protein associates with the cytoskeleton in tobacco cells. *Plant Cell* **7**:2101–2114.

Meshi T, Ishikawa M, Motoyoshi F, Semba K, Okada Y. (1986) *In vitro* transcription of infectious RNAs from full-length cDNA of tobacco mosaic virus. *Proc. Natl. Acad. Sci. USA* **83**: 5043–5047.

Miele L. (1997) Plants as bioreactors for biopharmaceuticals: regulatory considerations. *Trends Biotech.* **15**: 45–50.

Modelska A, Dietzschold B, Sleysh N *et al.* (1998) Immunization against rabies with plant-derived antigen. *Proc. Natl. Acad. Sci. USA* **95**: 2481–2485.

Mushiegan A, Shepherd RJ. (1995) Genetic elements of plant viruses as tools for genetic engineering. *Microbiol. Rev.* **59**: 548–578.

Namba K, Stubbs G. (1986) Structure of tobacco mosaic virus at 3.6 Å resolution: implications for assembly. *Science* **231**: 1401–1406.

Napoli C, Lemieux C, Jorgensen R. (1990) Introduction of a chimeric chalcone synthase gene into a petunia results in reversible co-suppression of a homologous gene in trans. *Plant Cell* **2**: 279–289.

Oparka KJ, Boevink P, Santa Cruz S. (1996) Studying the movement of plant viruses using green fluorescent protein. *Trends Plant Sci.***1**: 412–418.

Padgett HS, Epel BL, Khan TW, Heinlan M, Watanabe Y, Beachy RN. (1996) Distribution of tobamovirus movement protein in infected cells and implications for cell-to-cell spread of infection. *Plant J.* **10**: 1079–1088.

Palmer KE, Rybicki EP. (1997) The use of geminiviruses in biotechnology and plant molecular biology, with particular focus on mastreviruses. *Plant Sci.* **129**: 115–130.

Porta C, Lomonossoff GP. (1996) Use of viral replicons for the expression of genes in plants. *Molec. Biotechnol.* **5**: 209–221.

Porta C, Spall VE, Loveland J, Johnson JE, Barker PJ, Lomonossoff GP. (1994) Development of cowpea mosaic virus as a high-yielding system for the presentation of foreign peptides. *Virology* **202**: 949–955.

Ratcliff F, Harrison BD, Baulcombe DC. (1997) A similarity between viral defense and gene silencing in plants. *Science* **276**: 1558–1560.

Reichel C, Beachy RN. (1998) Tobacco mosaic virus infection induces severe morphological changes in the endoplasmic reticulum. *Proc. Natl. Acad. Sci. USA* **95**: 11169–11174.

Roberts AG, Santa Cruz S, Roberts IM, Prior DAM, Turgeon R, Oparka KJ. (1997) Phloem unloading in sink leaves of *Nicotiana benthamiana*: comparison of a fluorescent solute with a fluorescent virus. *Plant Cell* **9**: 1381–1396.

Rommens CMT, Salmeron JM, Baulcombe DC, Staskawicz BJ. (1995) Use of a gene expression system based on potato virus X to rapidly identify and characterize a tomato *Pto* homolog that controls fenthion sensitivity. *Plant Cell* **7**: 249–257.

Rosario Fernandez-Fernandez M, Martinez-Torrecuadrada JL, Ignacio Casal J, Antonio Garcia J. (1998) Development of an antigen presentation system based on plum pox potyvirus. *FEBS Lett.* **427**: 229–235.

Rubino L, Burgyan J, Grieco F, Russo M. (1990) Sequence analysis of cymbidium ringspot virus satellite and defective interfering RNAs. *J. Gen. Virol.* **71**: 1655–1660.

Ryan MD, Drew J. (1994) Foot-and-mouth disease virus 2A oligopeptide mediated cleavage of an artificial polyprotein. *EMBO J.* **13**: 928–933.

Ryan MD, King AMQ, Thomas GP. (1991) Cleavage of foot-and-mouth-disease virus polyprotein is mediated by residues located within a 19 amino acid sequence. *J. Gen. Virol.* **72**: 2727–2732.

Sablowski RWM, Baulcombe DC, Bevan M. (1995) Expression of a flower-specific Myb protein in leaf cells using a viral vector causes ectopic activation of a target promoter. *Proc. Natl. Acad. Sci. USA* **92**: 6901–6905.

Samuel G. (1934) The movement of tobacco mosaic virus within the plant. *Ann. Appl. Biol.* **21**: 90–111.

Santa Cruz S. (1999) Perspective: phloem transport of viruses and macromolecules – what goes in must come out. *Trends Microbiol.* **6**: 237–241.

Santa Cruz S, Chapman S, Roberts AG, Roberts IM, Prior DAM, Oparka K. (1996) Assembly and movement of a plant virus carrying a green fluorescent protein overcoat. *Proc. Natl. Acad. Sci. USA* **93**: 6286–6290.

Santa Cruz S, Roberts AG, Prior DAM, Chapman S, Oparka K. (1998) Cell-to-cell and phloem-mediated transport of potato virus X: the role of virions. *Plant Cell* **10**: 495–510.

Scholthof HB, Scholthof KB, Jackson AO. (1995) Identification of tomato bushy stunt virus host-specific symptom determinants by expression of individual genes from a potato virus X vector. *Plant Cell* **7**: 1157–1172.

Scholthof HB, Scholthof KB, Jackson AO. (1996) Plant virus gene vectors for transient expression of foreign proteins in plants. *Annu. Rev. Phytopathol.* **34**: 299–323.

Shivprasad S, Pogue GP, Lewandowsky DJ, Hidalgo J, Donson J, Grill LK, Dawson WO. (1999) Heterologous sequences greatly affect foreign gene expression in tobacco mosaic virus-based vectors. *Virology* **255**: 312–323.

Smolenska L, Roberts IM, Learmonth D, Porter AJ, Harris WJ, Wilson TMA, Santa Cruz S. (1998) Production of a functional single chain antibody attached to the surface of a plant virus. *FEBS Lett.* **441**: 379–382.

Sudarshana MR, Wang HL, Lucas WJ, Gilbertson RL. (1998) Dynamics of bean dwarf mosaic geminivirus cell-to-cell and long-distance movement in *Phaseolus vulgaris* revealed, using the green fluorescent protein. *Mol. Plant-Microbe Interact.* **11**: 277–291.

Sugiyama Y, Hamamoto H, Takemoto S, Watanabe Y, Okada Y. (1995) Systemic production of foreign peptides on the particle surface of tobacco mosaic virus. *FEBS Lett.* **359**: 247–250.

Szecsi J, Ding X, Lim CO, Bendahmane M, Cho MJ, Nelson RS, Beachy RN (1999) Development of tobacco mosaic virus infection sites in *Nicotiana benthamiana. Mol. Plant-Microbe Interact.* **12**: 143–152.

Takamatsu N, Ishikawa M, Meshi T, Okada Y. (1987) Expression of bacterial chloramphenicol acetyltransferase gene in tobacco plants mediated by TMV-RNA. *EMBO J.* **6**: 307–311.

Takamatsu N, Watanabe Y, Yanagi H, Meshi T, Shiba T, Okada Y. (1990) Production of enkephalin in tobacco protoplasts using tobacco mosaic virus RNA vector. *FEBS Lett.* **269**: 73–76.

Timmermans MCP, Das OP, Messing J. (1992) Trans replication and high copy numbers of wheat dwarf virus vectors in maize cells. *Nucl. Acids Res.* **20**: 4047–4054.

Timmermans MCP, Das OP, Messing J. (1994) Geminiviruses and their uses as extrachromosomal elements. *Annu. Rev. Plant Physiol. Plant Mol. Biol.* **45**: 79–112.

Tobias CM, Oldroyd GED, Chang JH, Staskawicz BJ. (1999) Plants expressing the *Pto* disease resistance gene confer resistance to recombinant PVX containing the avirulence gene *AvrPto*. *Plant J.* **17**: 41–50.

Turpen TH. (1999) Tobacco mosaic virus and the virescence of biotechnology. *Phil. Trans. R. Soc. Lond. B* **354**: 665–673.

Turpen TH, Reinl SJ, Charoenvit Y, Hoffman SL, Fallarme V, Grill LK. (1995) Malarial epitopes expressed on the surface of recombinant tobacco mosaic virus. *Bio-Technology* **13**: 53–57.

Ugaki M, Ueda T, Timmermans MCP, Viera J, Elliston KO, Messing J. (1991) Replication of a geminivirus derived shuttle vector in maize endopserm cells. *Nucl. Acids Res.* **19**: 371–377.

Usha R, Rohll JB, Spall VE, Shanks M, Johnson JE, Lomonossoff GP. (1993) Expression of an animal virus antigenic site on the surface of a plant virus particle. *Virology* **197**: 366–374.

Van Vloten-Doting L. (1983) Advantages of multipartite genomes of single-stranded RNA plant viruses in nature, for research, and for genetic engineering. *Plant Mol. Biol. Rep.* **1**: 55–60.

Van Vloten-Doting L, Bol J, Cornelissen B. (1985) Plant virus-based vectors for gene transfer will be of limited use because of the high error frequency during viral RNA synthesis. *Plant. Mol. Biol.* **4**: 323–326.

Vaucheret H, Béclin C, Elmayan T *et al.* (1998) Transgene-induced gene silencing in plants. *Plant J.* **16**: 651–659.

Verch T, Yusibov V, Koprowski H. (1998) Expression and assembly of a full-length monoclonal antibody in plant using a plant virus vector. *J. Immunol. Methods* **220**: 69–75.

Verver J, Wellink J, Van Lent J, Gopinath K, Van Kammen A. (1998) Studies on the movement of cowpea mosaic virus using the jellyfish green fluorescent protein. *Virology* **242**: 22–27.

Voinnet O, Pinto YM, Baulcombe DC. (1999) Suppression of gene silencing: a general strategy used by diverse DNA and RNA viruses of plants. *Proc. Natl Acad. Sci. USA* **96**: 14147–14152.

Ward A, Etessami P, Stanley J. (1988) Expression of a bacterial gene in plants mediated by infectious geminivirus DNA. *EMBO J.* **7**: 1583–1587.

Whitham SA, Yamamoto ML, Carrington JC. (1999) Selectable viruses and altered susceptibility mutants in *Arabidopsis thaliana*. *Proc. Natl Acad. Sci. USA* **96**: 772–777.

Wigdorowitz A, Perez Filgueira DM, Robertson N, Carillo C, Sadir AM, Morris TJ, Borca MV. (1999) Protection of mice against challenge with foot-and-mouth disease virus (FMDV) by immunization with foliar extracts from plants infected with recombinant tobacco mosaic virus expressing FMDV structural protein VP1. *Virology* **264**: 85–91.

Yusibov V, Modelska A, Steplewski K, Agadjanyan M, Weiner D, Hooper DC, Koprowsky H. (1997) Antigens produced in plants by infection with chimeric plant viruses immunize against rabies virus and HIV-1. *Proc. Natl Acad. Sci. USA* **94**: 5784–5788.

Chapter 5

Mammalian expression systems and vaccination

Miles W. Carroll, Gavin W.G. Wilkinson and Kenneth Lundstrom

5.1 Introduction

Viral vectors continue to provide the most efficient and reliable technology with which to deliver and promote expression of transgenes to eukaryotic cells both *in vitro* and *in vivo*. Viruses have adapted through evolution to become ultra-efficient agents for gene delivery and potentially any virus can provide the basis for a vector system. This chapter will review the development of some of the most powerful mammalian vector systems currently available derived from three virus families: the adenoviruses, alphaviruses and poxviruses. The three groups of viruses are unrelated but their associated vectors share a capacity to provide for high-level expression *in vitro* and particular emphasis will also be given to their application to vaccination and immunotherapy.

The vectors will be covered in historical order. The worldwide eradication of smallpox has been the paradigm for vaccine development uniquely providing over 200 years experience with a live vaccine. Since smallpox eradication vaccinia virus and other poxvirus-based vectors have actively developed to generate a popular range of vectors with enhanced safety and efficacy. Adenoviruses were first discovered nearly 50 years ago but already there has been nearly 40 years experience with a live vaccine. Adenovirus vector systems have come to prominence because their unrivalled capacity to promote efficient *in vivo* gene delivery has led to their extensive development and application in gene therapy (see Chapter 8). The application of adenovirus vectors in immunization protocols predated and facilitated their adoption by gene therapists. This section of the chapter will focus on the continued exploitation of adenovirus vectors to promote efficient antigen presentation and increasingly to modulate the nature of an induced immune response. Alphavirus vectors are relative newcomers in the last 10 years but are established as the most efficient RNA virus-based expression systems highly effective in both *in vitro* and *in vivo* applications.

Vectors continuously evolve to meet the requirements of specific applications. Each of these three virus families has spawned a wide variety of vectors, both replication-competent and replication-deficient, based both on human and animal viruses. For vaccination, there are many examples where a single immunization with a recombinant virus can elicit effective, long-lived protection. However, there are also problem

Genetically Engineered Viruses: Development and Applications, edited by C.J.A. Ring and E.D. Blair

situations in which to elicit effective protection multiple rounds of immunization have been required. In such circumstances it can be more effective to prime the immune response with one agent and boost with others. Such strategies can therefore combine a range of treatments including different virus vectors, cells transduced *ex vivo*, DNA immunization, peptide or purified antigens. Hopefully, this review will also assist researchers faced with the daunting task of selecting which vector system most suits their needs.

5.2 Recombinant poxvirus expression vectors

5.2.1 Background

This section provides an overview of the principles of using poxviruses as recombinant expression vectors and gives a brief account of their current applications. Other reviews for those interested in more in-depth studies on poxvirus molecular biology (Moss, 1996a), construction of recombinant poxviruses (Earl *et al.*, 1998a, 1998b) and their applications as vaccine vectors for infectious disease and cancer immunotherapy (Carroll and Restifo, 2000; Flexner and Moss, 1997; Moss, 1996b; Paoletti 1996) are essential reading.

Vaccinia virus (VV), is a member of the orthopoxvirus genus, within the poxvirus family (Fenner, 1996). Also within the Poxviridae is the avipox virus genus, of which fowlpox virus (FPV) (Boyle and Coupar, 1988) and canarypox virus (CPV) (Paoletti, 1996), have been developed as candidate human vaccine vectors. Though we will concentrate on aspects and applications of recombinant vaccinia virus (rVV), due to conservation of the expression machinery within the poxvirus family, many of the principles of rVV can be applied to other poxviruses.

Vaccinia virus was used in the successful eradication of smallpox which was completed in the late 1970s (Arita, 1979). The success of the programme was due in large part to several properties of VV, namely its simple and cheap manufacture process, its stability and antigenicity. After the eradication programme was completed it was logical to assume that research in VV would cease, however, in 1982 two reports were published illustrating the application of VV as a recombinant expression vector (Mackett *et al.*, 1982; Panicali and Paoletti, 1982).

Advantageous properties of the rVV system include: its ability to accommodate over 25 kilobasepairs (kbp) of foreign DNA (Smith and Moss, 1983), broad host range, cytoplasmic foreign gene expression, and authentic protein processing. Perhaps it is not surprising that this versatile vector has been used to express hundreds of recombinant genes for a variety of applications including protein function analysis, antigen processing and recombinant vaccine development. This point is more thoroughly illustrated by the fact that to date there are some 6000 publications that utilize the rVV system.

5.2.2 Vaccinia virus molecular biology and gene expression

Vaccinia virus contains a single copy of a double stranded DNA genome of approximately 200 kbp with hairpin loops at both ends. Vaccinia virus replication

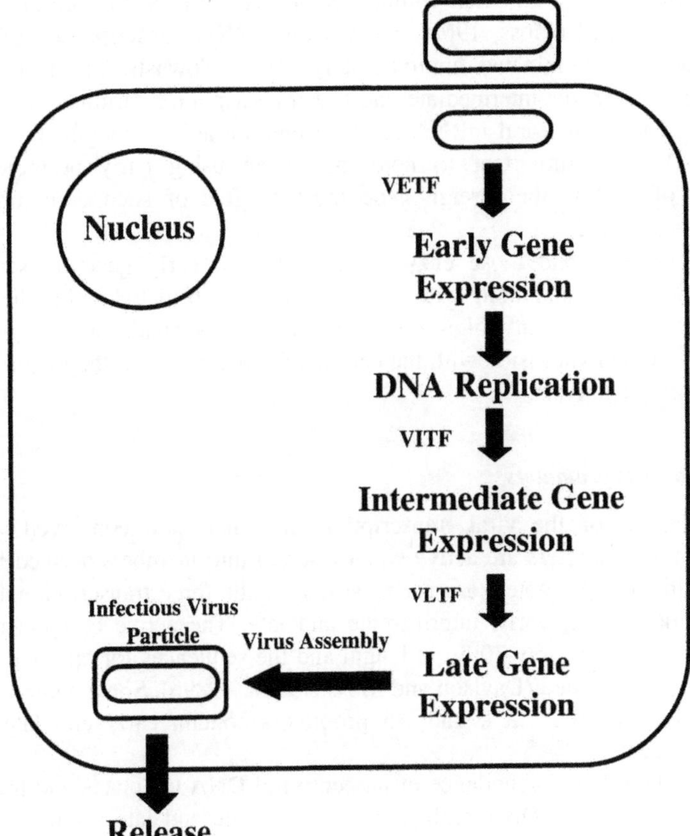

Figure 5.1. Poxvirus replication cycle.

occurs exclusively in the cytoplasm of the infected cell, as such it has evolved a number of genes that maintain a relative autonomy with the host cell nucleus. The virus encodes some 200 genes (for review see Moss, 1996a) including enzymes for the synthesis and processing of viral mRNAs and DNA (*Figure 5.1*). Vaccinia virus gene expression is a tightly regulated process (Moss, 1996a). After entry into the host cell via membrane fusion, early gene transcripts can be detected in as little as 15 minutes (Baldick and Moss, 1993) since the virion carries with it enzymes which are essential for gene expression. Early gene mRNAs are translated into a variety of proteins including immune modulators that dampen the hosts immune system (for a review see Smith *et al.*, 1997), viral DNA replication enzymes and intermediate gene transcription factors (VITF). Immediately after viral DNA replication, intermediate gene transcription is activated at which time viral late gene transcription factors (VLTF) are expressed that subsequently enable late gene transcription to initiate (Moss, 1996a). Several reports have suggested that VV transcription is not entirely independent from host cell nuclear factors. It is important to note that within several hours after VV infection host protein synthesis is abrogated (Moss, 1968). Early viral mRNA transcripts are of a defined length as they are terminated approximately 50 bp downstream of the sequence TTTTTNT. However, for intermediate and late transcription a termination sequence is not recognized and thus viral mRNA are long and variable in length (for review see Moss, 1996a). It is important to note that when using early promoters in the construction of a rVV, the foreign gene must be free of such early transcription termination sequences.

Structural proteins and those enzymes essential for early gene transcription are expressed at late times and along with viral DNA are encapsulated within the new virus particle. After wrapping in golgi-derived membranes a small percentage of virus particles are released via fusion with the cell membrane, however, the majority of virus is released during cell lysis.

5.2.3 Vaccinia virus promoters

As many features of the viral transcription machinery are conserved within the poxviridae, most promoters are active when inserted into members derived from other genera. Specific VV promoters are active at one of the three transcriptional stages of VV gene regulation i.e. early, intermediate and late. The active components of VV promoters are approximately 30 bp in length and the sequences for optimal expression levels have been identified (Davison and Moss, 1989a, 1989b). Some natural promoters e.g. the commonly used 7.5 k and H5 promoters contain early and late promoter elements in tandem.

Because of the relative abundance of nascent viral DNA templates and transcription factors present following DNA replication, intermediate and late promoters can give rise to over 10-fold higher protein levels than can early promoters (*Table 5.1*). However, due to several factors including: (i) prevention of late gene expression in some cell types e.g. dendritic cells (DC) and macrophages (Broder *et al.*, 1994; Bronte *et al.*, 1997; Drillien *et al.*, 2000), (ii) cytotoxic effects and (iii) shut off of host protein

Table 5.1. Kinetics and strength of vaccinia virus promoters

VV Strain	Promoter	Kinetics	β-galactosidase Expression (μg)	
			+Ara-C	−Ara-C
*MVA	$P_{7.5}$	E and L	0.2	0.5
*MVA	P_{H5}	E and L	1.1	2.6
WR	P_{11}	L	0.0	3.9
WR	P_{syn}	E and L	0.3	8.2

This table is a modification of the data presented by Wyatt *et al.* (Vaccine 1996) in which recombinant viruses containing the *E. coli* Lac Z gene, under transcriptional control of the various promoters, were used to infect monkey kidney BS-C-1 cells. To differentiate between early and late expression cells were incubated in the presence or absence of Ara C. Ara C inhibits viral DNA replication which prevents occurrence of late viral gene expression. Cells were harvested after 24 hours incubation and assessed for β-galactosidase expression.
* Average expression level from two independent recombinant viruses.

synthesis (Moss, 1968), the use of early promoters may be more advantageous depending on the final application of the rVV.

In early recombinant VV constructs only natural promoters were available. However, in-depth promoter mutagenesis studies (Davidson and Moss, 1989a, 1989b) led to the design of synthetic promoters (Chakrabarti *et al.*, 1997) which contain sequences that give optimal levels of gene expression, e.g. the commonly used synthetic early/late promoter sE/L. As described in *Table 5.1* there are a variety of promoters with a range of kinetics and strengths. The summary of the study illustrated in *Table 5.1* is unique in that a direct comparison was made between different promoters expressing the same reporter gene in the same cell type. This is not to say that these relative levels will be consistent if a different gene was expressed and a different cell line was used, as expression levels may be dictated by the toxicity or post-translational modifications required to express the recombinant protein. It is for this reason that high-level gene expression is not always desirable, e.g. some recombinant viruses containing foreign genes driven by the sE/L promoter may have a higher incidence of gene and/or promoter truncations.

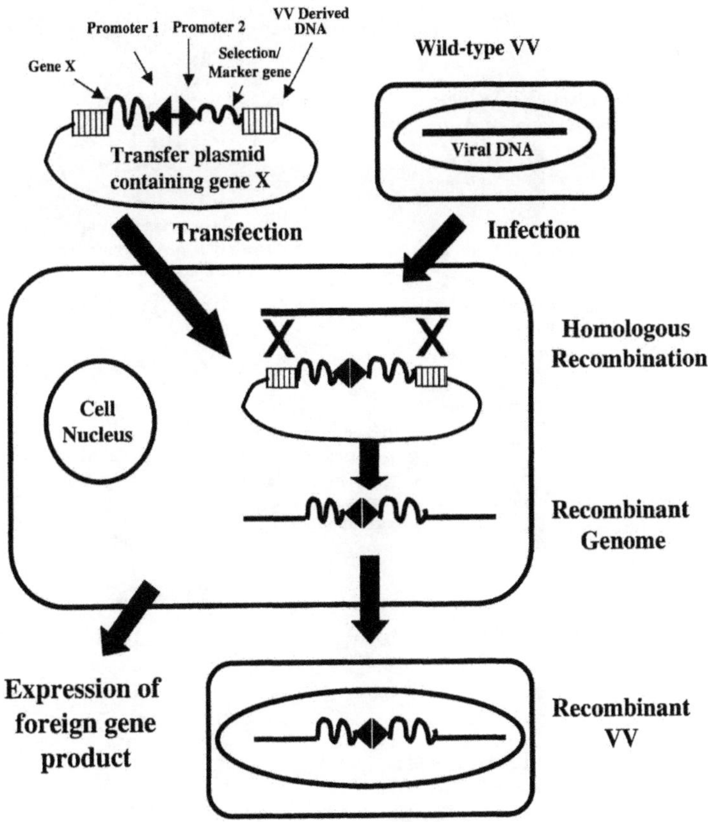

Figure 5.2. Vaccinia virus homologous recombination.

5.2.4 Construction of recombinant VV

The construction of rVV was initially carried out using homologous recombination (Mackett *et al.*, 1982; Panicalli and Paoletti, 1982) and to date this is still the most widely used technique. Cells are initially transfected with a transfer plasmid, which contains the foreign gene adjacent to a suitable VV promoter and flanked by viral derived DNA sequences (*Figure 5.2*). Cells are infected with wild-type VV and during replication, recombination occurs between the transfer plasmid and homologous regions within the virus genome at a frequency of approximately 1:1000. The recombinant genome is then packaged by wild-type virus and rVV is released from the cell. There have been many technical improvements since the first recombinant viruses were made that have greatly simplified the process for the construction of rVV. Most advances have been made by incorporating drug resistance or marker genes into the transfer vector. Initially, selection of the recombinant virus relied on the fact that recombination of the foreign gene was directed into the VV thymidine kinase coding sequence (tk), the subsequent tk negative (tk−) progeny could be selected, when grown on tk− cells, by the addition of bromodeoxyuridine (BrdU) into the cell culture media. Unfortunately, as BrdU is a mutagen there was a high incidence of spontaneous tk−

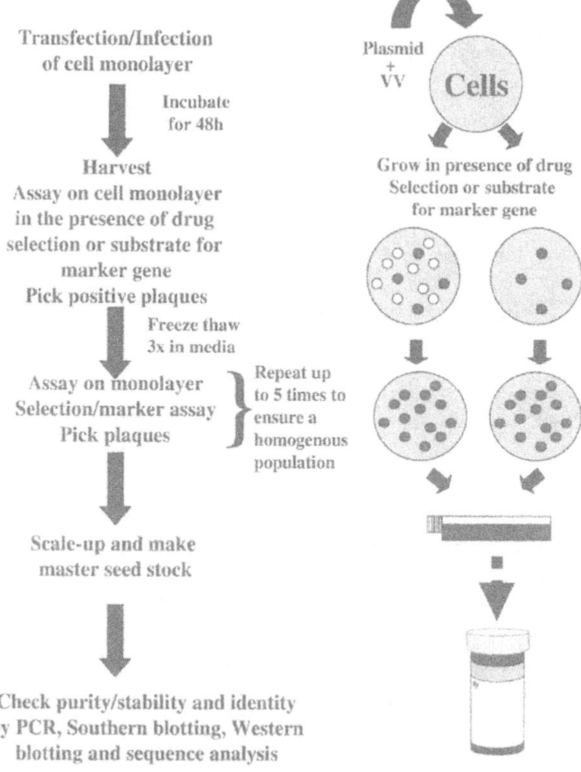

Figure 5.3. Production of recombinant VV.

virus which meant that a process of DNA hybridization screening using a radioactive probe was required to identify the rVV. To overcome this problem the *E. coli* Lac Z gene, coding for β-galactocidase (β-gal), was inserted adjacent to a second VV promoter, into the transfer vector such that after recombination it was transferred into the rVV genome (Chakrabarti *et al.*, 1985). True recombinants could thus be simultaneously selected and identified as blue plaques growing under an agar overlay media containing BrdU and the β-gal substrate, X-gal. Plaques are picked using a Pasteur pipette, plated onto fresh cell monolayers and subjected to several additional rounds of plaque purification under selection, to obtain a homogeneous population (*Figure 5.3*).

In the last decade a spectrum of genes has been added to transfer vectors to give further versatility to the rVV system. These include: antibiotic resistance markers e.g. *E. coli* neomycin gene (Franke *et al.*, 1985) and xanthine-guanine phosporibosyl transferase (*gpt*) gene (Falkner and Moss, 1988) and additional color markers to Lac Z e.g. green fluorescent protein (Wu *et al.*, 1995) and *E. coli* β-glucuronidase A gene (*gus*) (Carroll and Moss, 1995). The *gus* gene, is over 1 kb smaller than Lac Z and lacks commonly used restriction sites. Additionally, as the substrates for β-glucuronidase and β-gal are not cross-reactive both marker genes can be used simultaneously to make rVV with several inserts. Furthermore, transfer plasmids have been constructed using both *gus* as a marker and *neomycin* for drug selection (T. Shors and B. Moss, unpublished data). A further improvement of this marker-selection approach is the *gpt-gus* fusion gene driven by a single VV promoter (Cao and Upton, 1997). Other methods employ modified VV that has a plaque-deficient phenotype or host range restricted to a specific cell type. Genes that restore these deficiencies are included in the transfer vector and following successful recombination, enable replication in the host cell (Blasco and Moss, 1995; Perkus *et al.*, 1989).

In certain circumstances, e.g. when making a rVV for use in clinical trials or when inserting genes into several different regions of the genome, stable integration of selection or marker genes may not be desirable. For these reasons rVV transfer plasmids have been developed that after the initial recombination a selection gene is transiently inserted in the rVV genome. However, when the drug selection is reversed the virus genome will be prone to a second recombination event, which results in deletion of the marker or selection gene, but not the gene of interest, from the rVV genome (Falkner and Moss, 1990; Issacs *et al.*, 1990; Scheiflinger *et al.*, 1998; Spehner *et al.*, 1990). An alternative transient selection method relies on co-transfection of a plasmid, in addition to the transfer plasmid, that contains the *gpt* gene adjacent to a VV promoter but not flanked by VV derived DNA. Under drug selection, co-transfection of this plasmid transiently enriches for rVV (Kurilla, 1997). Alternatively, if the recombinant protein is expressed on the cell surface and an effective antibody is available, the recombinant virus can be identified by live immunostaining (Earl *et al.*, 1998b) which also obviates the requirement of additional marker/selection genes.

Due to the large size of the VV genome it was thought that *in vitro* manipulation would not be possible, however, it has since been shown that *in vitro* ligation is an alternative to homologous recombination (Merchlinsky and Moss, 1992; Scheiflinger *et al.*, 1992). The purified VV DNA can be digested with a restriction enzyme that cuts the genome at a unique site. The recombinant gene, along with a marker or selection gene, is ligated into the viral genome. As VV DNA is non-infectious the recombinant virus genome is transfected into cells that have previously been infected with a replication restricted helper poxvirus, e.g. a conditionally lethal mutant VV or fowlpox virus. The recombinant genome is packaged into infectious virus particles by the helper virus but the wild-type virus will not be able to replicate. Further improvements to these protocols have been made that allow easier identification of recombinant virus and more efficient foreign gene expression (Merchlinsky *et al.,* 1997; Pfleider *et al.,* 1995). *In vitro* ligation should allow efficient insertion of much larger regions of DNA compared to using transfer plasmids and homologous recombination. Furthermore, *in vitro* ligation may enable the construction of rVV carrying libraries of cDNA.

Initially in the construction rVV homologous recombination was targeted to the *tk* gene. There is now a spectrum of sites for insertion of foreign genes that allows the construction of complex recombinant viruses expressing multiple gene products (Carroll *et al.,* 1997; Tine *et al.,* 1996).

5.2.5 Chimeric VV-bacteriophage expression systems

High-level protein expression has been obtained using a hybrid expression system that utilizes VV promoters and bacteriophage RNA polymerases. The chimeric systems express the bactriophage T7 (Fuerst *et al.,* 1986), SP6 (Usdin *et al.,* 1993) or T3 (Rodriguez *et al.,* 1990) RNA polymerase gene under transcriptional control of a VV promoter. In the case of the T7 system, cells are infected with the rVV expressing the T7 RNA polymerase and transfected with a plasmid containing a recombinant gene under the regulation of the T7 promoter. Although transient gene expression was efficient, insertion of the encephalomyocarditis virus cap-independent ribosome binding site into the 5′-untranslated leader of foreign genes improved expression (Elroy-Stein *et al.,* 1989). A detailed description of the protocols and vectors required for the T7 system is available (Elroy-Stein and Moss, 1998). The T7 expression system has now been incorporated into two highly attenuated, host-restricted and extremely safe poxviruses, an attenuated VV (Sutter *et al.,* 1995; Wyatt *et al.,* 1995) and an avipoxvirus (Britton *et al.,* 1996). These systems offer a safe and efficacious method to produce authentic and potentially toxic recombinant protein, without the need to construct replication competent rVV.

For larger scale protein production, instead of transfecting a plasmid, the recombinant gene/T7 bacteriophage promoter cassette is stably integrated into a rVV (Barrett *et al.,* 1989; Elroy-Stein and Moss, 1997). Additional modifications to the T7 system utilize inducible mechanisms based on temperature or chemicals, and enable stringent transcriptional control of foreign gene expression (Ward *et al.,* 1995).

5.2.6 Replication-defective VV expression vectors

Vaccinia virus is infectious to humans and during its use in the smallpox eradication campaign some VV strains were associated with an unacceptably high incidence of complications (Lane *et al.,* 1969). VV is classified in the UK as a category II organism which requires the use of a class II microbial safety cabinet, in the USA the additional measure of pre-vaccination with VV is required. In Germany strict regulations on the use of replication-competent genetically engineered viruses make the use of VV that are replication-defective in mammalian cells a very attractive alternative. Two replication-defective VV, modified VV Ankara (MVA) and NYVAC have been developed and are now commonly used as expression vectors and experimental vaccines.

NYVAC is a genetically engineered VV strain that has 18 genes deleted that are associated with host range and pathogenesis (Tartaglia *et al.*, 1997). NYVAC replication is usually blocked at the stage after early gene expression in human cells but is able to replicate and produce infectious progeny in African green monkey kidney (VERO) and chick embryo fibroblast (CEF) cells (Tartaglia *et al.*, 1997).

MVA was derived from a replication-competent VV vaccine strain, Ankara, and was attenuated by passaging over 500 times in primary CEF cells (Mayr *et al.*, 1978). MVA was found to be non-pathogenic in a number of animal models and is replication-defective in mammalian cells (Carroll and Moss, 1997; Meyer *et al.*, 1991). Importantly MVA was used to vaccinate over 120 000 people, many of which were at risk of complications from vaccination with replication-competent strains, during the smallpox eradication programme (Mahnel and Mayr, 1994) without reports of complications.

MVA has subsequently been developed as an expression vector (Sutter and Moss, 1992) and there are now a variety of transfer plasmids that enable the construction of rMVA expressing multiple gene products under transcriptional control of a range of VV promoters (Earl *et al.*, 1998b).

Approximately 30 kbp of MVA genomic DNA has been deleted during >500 passages in CEF. It has been shown that its inability to replicate in mammalian cells is due to multiple gene defects (Wyatt *et al.*, 1998). The MVA replication deficiency is unusual since efficient early and late gene expression occur in most host range restricted cells (Carroll and Moss, 1997; Sutter and Moss, 1992). Immature circular virus particles are formed but the final step in virus particle maturation is defective.

Initially, it was reported that MVA replication was restricted to primary CEF, however, more recently it has been shown that baby hamster kidney cells (BHK-21) can support efficient replication (Carroll and Moss, 1997; Drexler *et al.*, 1998) and production of recombinant MVA (Carroll and Moss, 1997).

Furthermore, it has also been shown that transfer plasmids used in the construction of replication-competent VV, that direct recombination into the *tk* gene, can be used in the construction of rMVA (Carroll and Moss, 1997).

MVA has been used to induce Ab and CTL responses to a range of viral antigens including those derived from influenza virus (Sutter *et al.*, 1994), SIV (Hirsch *et al.*, 1996) and parainfluenza virus (Wyatt *et al.*, 1996). Additionally, it has been shown to be an effective vector for the induction of anti-tumor responses (Carroll *et al.*, 1997; Drexler *et al.*, 1999).

Moreover, a cell line-dependent VV system has been constructed which consists of a VV with a deleted uracil DNA glycosylase gene. The virus is strictly replication defective in all cells except in a helper cell line that expresses the VV uracil DNA glycosylase gene (Holzer and Falkner, 1997). This defective virus is an extremely safe expression vector and has shown much promise as a recombinant vaccine (Holzer *et al.*, 1999).

5.2.7 Non-vaccinia poxvirus vectors

The Poxviridae consists of a broad spectrum of members that primarily infect non-human hosts (Fenner, 1996; Moss, 1996a). Fortuitously the promoter sequences and transcription factors are generally conserved throughout the family so the techniques and concepts designed for construction of recombinant VV can be used in the construction of alternative poxvirus expression vectors (see *Table 5.2*).

Initially, members of the avian poxvirus group were primarily developed as vectors for potential recombinant vaccines for use in birds. However, two members of this genus, fowlpox (Boyle and Coupar, 1988) and canarypox (commercially known as ALVAC) (reviewed by Paoletti, 1996) have shown much promise as safe and effective vectors for use in animals and, more importantly, humans. Additionally, a large proportion of the population >30 years of age were immunized with VV during the smallpox eradication campaign. Experiments in murine models suggest such pre-existing immunity may have a detrimental effect on the induction of an immune response to recombinant proteins expressed by rVV. The avian poxviruses appear to be immunologically distinct from VV therefore pre-existing immunity should not have a detrimental effect (Wang *et al.*, 1995). Additionally, it should be noted that the effects of pre-existing immunity to VV have been overcome in murine models by delivering the rMVA via the intra-rectal mucosal route (Belyakov *et al.*, 1999).

Both ALVAC and FPV can be efficiently grown in CEFs. However, in most mammalian cells early gene expression is efficiently executed but subsequent DNA replication and late gene expression do not occur or occur extremely inefficiently (Somogyi *et al.*, 1993; Tartaglia *et al.*, 1997). More specifically in ALVAC-infected mammalian cells, cell death *via* dsRNA-mediated induction of apoptosis is readily observed. However, the precise timing varies between cell types. For instance, in human-derived cells ALVAC displays an abortive early phenotype, significantly shutting down host protein expression by 6 h post-infection (Tartaglia *et al.*, 1997; J.

Table 5.2. Properties of poxvirus expression vectors

Strain	Early expression	Late expression	Cell lysis within 72 hours	Inhibition of cell division
WR	Yes	Yes	Yes	Yes
MVA	Yes	Yes	Yes	Yes
NYVAC	Yes	No	Yes	Yes
RDVV	Yes	No	Yes	Yes
FPV	Yes	No	Yes	Yes
ALVAC	Yes	No	Yes	Yes
AmEPV	Yes	No	No	No

The above table describes the general properties of selected recombinant poxviruses based on infection of primate cells e.g. BS-C-1, CV-1 or HeLa. It is important to note that there are variations to these general properties in specific cell lines e.g. ALVAC and NYVAC will carry out early and late gene expression in VERO cells and NYVAC will actually replicate. Though generally all the above poxviruses with the exception of AmEPV will cause cell death/lysis within 72 hours, the lytic properties do vary considerably e.g. MVA has significantly less aggressive lytic effects on most mammalian cell lines compared to the replication competent VV strain Western Reserve (WR).

Tartaglia, personal communication). In contrast, in Vero cells, the virus displays an abortive late phenotype with clear production of immature particles in the cytoplasm of infected cells. In an attempt to enhance the efficacy of ALVAC as a vaccine vector, a second generation ALVAC has been constructed that contains VV derived genes, E3L and K3L, which significantly delays cell death by apoptosis in infected human cells and the virus appears to display an abortive late phenotype (Tartaglia *et al.*, 1997).

Recombinant poxviruses based on ectromelia (mouse pox) (Kochneva *et al.*, 1994) raccoonpox (Hu *et al.*, 1996), and members of the capripoxvirus (Romero *et al.*, 1994) and leporipox (Jackson *et al.*, 1996) genera have also been developed and may have applications in the study of poxvirus pathogenesis, and as wildlife and veterinary vaccines.

Amsacti moorei entomopox (AmEPV) a member of the insect poxviridae, has been developed as a recombinant expression system (Li *et al.*, 1997). Though AmEPV can be grown in insect cell lines it is strictly replication-defective in mammalian cells (Langridge, 1983). Li and colleagues have shown that recombinant AmEPV expressing the *Lac Z* gene can enter human liver cells and carry out effective early gene expression. A major problem for some applications of poxviruses is the fact that after infection the shut down of host protein synthesis rapidly occurs with eventual cell lysis, usually occurring within 48 hours. Importantly, AmEPV infected CV-1 cells have been shown to support early viral gene expression and still carry out cell division (Li *et al.*, 1997). Additionally, injection of AmEPV-*Lac Z* into the hind leg muscle of a mouse induced effective expression of β-gal (R. Moyer, personal communication). These properties indicate the potential application of entomopox as a gene therapy vector.

5.2.8 Recombinant poxviruses: laboratory applications

The recombinant poxvirus system has been used extensively for the analysis of protein–protein interactions, intracellular trafficking, post-translation modifications, antigen-processing and presentation and the identification of antigens and epitopes for the induction of humoral and cell-mediated immune responses (for a review see Mackett, 1994). It is perhaps not surprising that literally hundreds of foreign genes have been expressed in the rVV system when you consider that the foreign protein will be authentically expressed with respect to: carboxylation (De La Salle, 1985), phosporylation (Guy *et al.*, 1987), myristylation (Guy *et al.*, 1987), glycosylation (Mackett and Arrand, 1985) and proteolytic cleavage (Thomas *et al.*, 1986). Due to the toxicity of some gene products the rVV T7 transient expression system has been used widely in the study of a variety of proteins. Two major breakthroughs that employed the rVV T7 system include; the discovery of the HIV-1 T-cell coreceptor (Feng *et al.*, 1996) and the rescue of a negative-sense RNA virus entirely from a cDNA clone (Schnell *et al.*, 1994). Both these discoveries have allowed major progress in the fields of HIV and negative-stranded RNA virus research.

5.2.9 Poxviruses as vectors for recombinant vaccines and cancer immunotherapy

Recombinant vaccinia virus has been shown to induce both antibody (Mackett *et al.*, 1985) and CTL (Bennink *et al.*, 1984) responses to a broad spectrum of recombinant gene products, an in-depth description of such studies can be found in several detailed reviews (Flexner and Moss, 1997; Moss, 1996b).

Initially recombinant vectors based on replication-competent vaccinia virus strains were used in animal studies and showed great potential as vaccine vectors (Mackett *et al.,* 1985). A major success of this technology is the application of a rVV expressing the rabies glycoprotein as a wildlife vaccine that has been instrumental in the near eradication of rabies in the wildlife pool in northern Europe (Pastoret and Brochier, 1999). Recombinant vaccinia viruses have been assessed in Phase I/II clinical trials. A recombinant vaccinia virus, based on the smallpox vaccine strain Wyeth, expressing HIV-1 gp160 was injected into healthy volunteers and induced both HIV antibody and T cell responses (Cooney *et al.*, 1991).

However, due to environmental concerns and potential toxic side effects of replication-competent VV, especially in immunocompromised recipients, much progress has been made in the development and evaluation of vectors based on attenuated VV and poxvirus strains that do not replicate in human cells. More recently foreign proteins expressed in these alternative recombinant poxvirus systems have also illustrated efficacious immunogenic properties (Moss, 1996b; Paoletti, 1996; Tartaglia *et al.,* 1997).

Both recombinant NYVAC and MVA expressing a variety of recombinant proteins, have been assessed in numerous animals models and have both shown efficacious effects. Unfortunately these vectors have not been evaluated in parallel so it is impossible to determine which has the most effective properties. MVA expressing SIV gag-pol and env genes has been evaluated on several occasions in the Macaque model and has induced protection against challenge with SIV (Hirsch *et al.,* 1996; Ourmanov *et al.*, 2000). Furthermore, high levels of circulating gag-specific CD8 cells were identified by tetramer staining after inoculation with MVA-gag-pol (Seth *et al.,* 2000).

Both NYVAC and ALVAC have been evaluated in several clinical trials and have shown the ability to induce antibody and T cell responses to the recombinant proteins (Paoletti, 1996; Tartaglia *et al.,* 1997).

Due to their capacity to induce strong T cell responses, both rVV and ALVAC vectors, expressing tumor associated antigens (TAA) have been evaluated in cancer immunotherapy clinical trials (for review see Carroll and Restifo, 2000). Initial results have been very encouraging with both antibody and T cell responses induced against the tumor antigen expressed by the recombinant viruses.

5.2.10 Enhancing the efficacy of recombinant poxvirus vaccines

Poxviruses have been shown to be effective vectors for the induction of immune responses to recombinant proteins, however, there are several ways to further enhance their efficacy. Improvements to poxvirus efficiency fall into four general categories: (i) promoter kinetics, (ii) replication/lytic properties, (iii) co-expression of immune co-factors and (iv) inoculation regimes.

Vaccinia virus interference with host protein synthesis and the cellular antigen processing machinery, has meant that early promoters driving expression of the recombinant protein are beneficial for the induction of optimal CTL responses.

However, a study using a panel of synthetic early and late promoters driving expression of the model TAA β-galactosidase (β-gal) indicate that early and late promoters both induce CTL responses, however, early promoters are significantly more effective at inducing CTL-mediated tumor therapy in a β-gal murine model (Bronte *et al.*, 1997). Furthermore, it was illustrated that only DCs infected with a rVV expressing β-gal under an early promoter were able to prime murine β-gal specific CTL. This phenomenon is most likely due to the inability of VV to express proteins under late promoters in DCs (Bronte *et al.*, 1997).

A non-replicating strain of VV that is defective in the D4 gene, which is essential for viral DNA replication, has been developed (Holzer and Falkner, 1997). This vector and a replication competent VV, expressing tick-borne encephalitis (TBE) virus antigens, have been evaluated in an infectious disease model. It was found that 10-fold less of the replication-defective rVV was required to induce protection to a lethal challenge dose of TBE (Holzer *et al.*, 1999). As in the case of MVA the superior efficacy may be due to the less lytic activity of the vector favoring the induction of a more optimal immune response (Carroll *et al.*, 1997). Additionally, in an attempt to improve their efficacy, ALVAC vectors have been modified to express VV genes that delay apoptotic cell death of human cells and allow ALVAC late gene expression (Tartaglia *et al.*, 1997; Tartaglia, personal communication).

Poxviruses are able to induce effective immune responses to recombinant proteins, however, in some cases it may be vital to induce a more effective TH1 or TH2 response. The co-expression of immune co-factors that steer the immune system towards the different T helper responses may clearly be beneficial. Due to the large capacity of the VV genome for foreign DNA this is an option for recombinant poxviruses. IL-2 was the first immune co-factor to be expressed by vaccinia virus and was shown to have an attenuating effect in nude mice (Flexner *et al.*, 1987; Ramshaw *et al.*, 1987). Subsequently rVV expressing a spectrum of immune enhancers have been constructed with the ultimate aim of controlling the immune response by creating a localized immunologic micro environment in which the host immune system can interact with the recombinant protein. Many cytokines e.g. IL-2 and IL-12 have anti-viral and anti-tumor activity, however, their toxic side effects can be severe when delivered systemically. Reports show that co-expression of a model TAA and IL-12 by a rVV can obviate the requirement for systemic delivery of toxic levels of IL-12 (Carroll *et al.*, 1998). Additionally, rVV co-expression of IL-2 and a model TAA has also been shown to enhance the therapeutic activity of a rVV in a murine tumor model (Bronte *et al.*, 1997). Furthermore, co-stimulatory molecules e.g. B7.1, essential for effective induction of CD8 responses, co-expressed by a rVV have enhanced the tumor therapeutic activity of these viruses (Carroll *et al.*, 1998; Chamberlain *et al.*, 1996).

It is clear that the co-expression of immune co-factors offers a further dimension to the enhancement of poxvirus based therapeutic vaccines. Kauffman and Schlom describe the use of B7.1 within a recombinant poxvirus co-expressing CEA in a colon cancer clinical trial. Initial data indicate that there are no signs of auto-immune toxicity associated with such an approach (Kauffman *et al.*, unpublished data; Kauffman and Schlom, 2000). Additionally, co-expression of IL-2 and the tumor antigen MUC-1, within a rVV, appears to be non-toxic (Scholl *et al.*, 2000).

Various prime boost protocols have been designed to further enhance the efficacy of poxvirus inoculation regimes. Priming with a recombinant poxvirus expressing HIV gp160 followed by boosting with recombinant gp160 protein has been shown to enhance the level of neutralizing antibody (Hu *et al.,* 1991). Protocols which employ two immunologically non-cross-reacting poxvirus vectors e.g. FPV and MVA (Carroll *et al.,* 1997), and ALVAC and VV (Hodge *et al.,* 1997) have been evaluated and appear to improve survival and cell-mediated responses in animal models. Several reports show that the use of naked DNA as a priming vector and recombinant poxvirus to boost the response (Irvine *et al.*, 1997; Leong *et al.,* 1995; Schneider *et al.,* 1998) induces significantly more efficacious antibody and CTL responses to the recombinant protein. Interestingly the order of the regime is vital. It is likely that these vaccination regimes will be adapted to the clinic in the near future.

5.3 Recombinant adenovirus expression vectors

5.3.1 Background

In 1953 Rowe and co-workers described a progressive cytopathic effect in cells cultured from 33/53 adenoids removed from children (Rowe *et al.*, 1953). This cellular pathology was induced by virus infection and constituted the first report of adenovirus (Ad) isolation. The surprisingly high incidence of Ad recovery in these samples is thought to reflect the capacity of Ad to persist for long periods in lymphoid cells *in vivo* (Ginsberg *et al.,* 1987). About the same time Rowe and co-workers identified their adenoid degenerating virus, Hilleman and Weiner were in the process of isolating the causative agent responsible for acute respiratory disease (ARD). The viruses were clearly related and ARD was thereby identified as the first disease associated with Ad infection (Hilleman and Werner, 1954).

Although ARD is relatively rare in adults, severe outbreaks were routine during induction training of young military recruits in the US (and elsewhere) usually resulting in a short debilitating upper respiratory tract infection. A more severe disease in some individuals was associated with infection extending to the lower respiratory tract that occasionally could be fatal (Horwitz, 1996). This problem with ARD in the US armed forces led to the development of a vaccine specific for the relevant serotypes 4 and 7 (Rubin and Rocke, 1994). Preliminary studies using sub-unit vaccines were discouraging, hence a live Ad vaccine was produced. Ad isolates associated with respiratory tract infection were able to replicate in the gut without inducing disease. The vaccine developed for Ad4 and Ad7 was therefore based on tissue culture-adapted virus without a specific requirement for attenuation (Chanock *et al.*, 1966). The live vaccine is administered orally in a tablet consisting of lyophilized virus in the core and an enteric coat that protect the virus during passage through the stomach. The enteric coat also allows the tablet to be handled safely and prevents infection of the pharynx during administration. Virus may be excreted for 2–5 weeks following vaccination but recipients do not exhibit overt symptoms nor is there substantial transmission to non-immunized contacts (Mueller *et al.*, 1969; Rubin and Rocke, 1994; Stanley and Jackson, 1969). This inability to be re-transmitted is consistent with the vaccine strains having become attenuated through adaptation to growth *in vitro*; a contention also supported by the observation that the Ad7 vaccine strain has lost its capacity to induce tumors in rodents.

The live Ad vaccine has been administered to over a million US recruits since 1965 and has proved extremely effective in preventing ARD (Rubin and Rocke, 1994), although recent disruptions in vaccination for logistical reasons has led to a resurgence of disease (Barraza *et al.*, 1999; Hendrix *et al.*, 1999). The positive experience using the live Ad vaccines in US military personnel provided a foundation to further develop Ad for therapeutic purposes. It has already been demonstrated that the live vaccine programme could be extended to additional common Ad serotypes if required (Schwartz *et al.*, 1974).

A problem was however encountered in vaccine production. Simian cells do not normally support replication of human Ads and the growth of Ad vaccine virus in such cells was possible only because of a helper function provided by an SV40 contaminant. Replication in simian cells was further facilitated when recombination resulted in insertion of a segment of SV40 DNA into the Ad genome (Rapp *et al.*, 1964). Initially such Ad/SV40 hybrids were replication defective but viable viruses containing the SV40 helper function were subsequently generated (Grunhaus and Horwitz, 1992). Ad was thus able to capture a gene that provided a growth advantage.

5.3.2 Human and non-human adenoviruses

Adenoviruses constitute an extremely large family widespread in mammals (members of the genus Mastadenovirus) and birds (members of the genus Aviadenovirus). Some 47 distinct human serotypes have been defined and classified into subgroups based on biochemical, serological and biological properties (*Table 5.3*). This virus reservoir has been tapped to generate a diverse range of Ad vector systems. The well-characterized human Ad serotypes 2 and 5 are most widely utilized as vectors, although Ad 4- and Ad 7-based systems have also been developed to exploit the experience gained using the live oral vaccines (Abrahamsen *et al.*, 1997). Viruses derived from various animal species are increasingly being considered for both veterinary and human applications. Vectors have also been generated based on the avian CELO virus (Michou *et al.*, 1999), bovine (Reddy *et al.*, 1999b), canine (Kremer *et al.*, 2000), ovine (Hofmann *et al.*, 1999; Xu and Both, 1998; Xu *et al.*, 1997) and porcine (Reddy *et al.* 1999a) Ads.

Table 5.3. Classification of human Ad[1]

Subgenus	Serotype	Oncogenicity
A	12, 18, 31	High
B	3, 7, 11, 14, 16, 21, 34, 35	Weak
C	1, 2, 5, 6	No
D	8–10, 13, 15, 17, 19, 20, 22–30, 32, 33, 36–39, 42–47	No
E	4	No
F	40, 41	No

[1]Extracted from Sharp and Wadell (1995)

Vectors derived from the non-human Ads have advantages: they avoid potential problems with pre-existing immunity in man, rescue by and recombination with human viruses is less likely and biological safety can be enhanced by a capacity of the wild-type virus to infect, but not replicate, in human cells. Complete DNA sequence and transcriptional information is now available for a range of non-human Ads and this will facilitate further vector development (Chiocca *et al.*, 1996; Morrison *et al.*, 1997; Reddy *et al.*, 1998a, 1998b).

5.3.3 Adenovirus disease in man

One justification frequently cited for using Ad as a vector is that the virus is not a major human pathogen. The more common Ad serotypes are primarily associated with mild upper respiratory tract infections (e.g. Ad1, 2 and 5) but also pharyngoconjunctival fever (e.g. Ad3 and 7). Pneumonia and epidemic keratoconjunctivitis are occasionally observed as more severe consequences of infections (e.g. Ad19 and 37), whilst certain serotypes are associated with haemorrhagic cystitis (Ad11 and 21), infantile diarrhoea (Ad40 and 41) and even life-threatening infection in immunocompromised individuals that can present as pneumonia or hepatitis (Horwitz, 1996; Sharp and Wadell, 1995). Ad12 was identified as the first pathogenic human virus capable of inducing tumors in hamsters (Trentin *et al.*, 1962). Members of the subgenus A and subgenus B in particular are classified as tumorigenic (*Table 5.3*). Ads became a paradigm for molecular biologists following their designation as tumor viruses. An appreciation of the molecular mechanisms involved in virus infection and replication has been built up over years of intensive study, and included the discovery of RNA splicing. Despite extensive research, a clear association has yet to be established between Ad and any human cancer. Nevertheless, our knowledge of the molecular biology of this virus has proved invaluable for both vector development and application.

5.3.4 The Ad virion

A detailed understanding of virion structure has been built up using a range of techniques including biochemical and genetic investigations, conventional electron microscopy, cryo-electron microscopy and X-ray crystallography (Stewart and Burnett, 1995; Stewart *et al.*, 1993). The Ad particle consists of an internal DNA-containing core surrounded by a classical icosahedral capsid (80–110 nm in diameter). A total of 240 hexon capsomeres form the 20 triangular faces with additional penton capsomeres at each of the 12 vertices. Extending like 'antennae' from each vertex are the fibers (9–30 nm), each comprised of a homotrimeric protein complex (Russell, 1998; Shenk, 1996) (*Figure 5.4*). Interestingly, members of the subgroup F (Ad40 and Ad41), however, have two distinct fibers (although still one per vertex) while avian Ads can have two fibers attached at each vertex. The fiber protein itself can be divided into three elements: the N-terminal tail (~30 amino acids) that anchors the fiber to the penton base, a shaft of variable length consisting of a series of 15 amino acid inexact repeats and finally the globular head/knob domain.

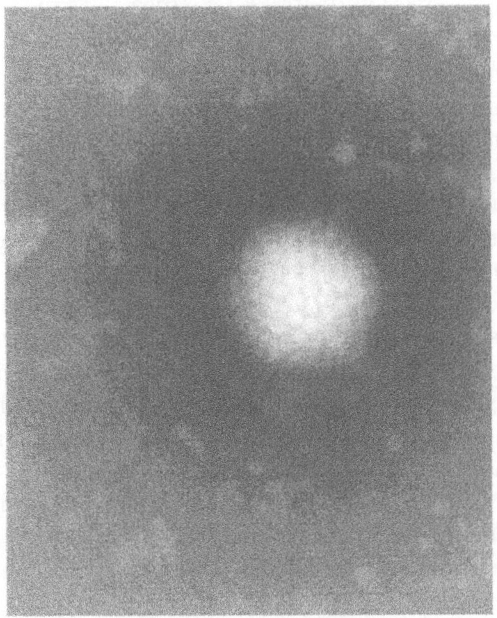

Figure 5.4. The Ad particle. An electron micrograph of an Ad particle. The fibers responsible for recognition of the primary receptor are just visible extending from each vertex (kindly provided by B. Dowsett, CAMR, Porton Down).

5.3.5 Ad infection of the cell

Ad infection is a sequential process initiated by fiber binding to the primary cellular receptor followed by a secondary interaction between the penton base protein and cellular integrins (Mathias *et al.*, 1994; Wickham *et al.*, 1993). The primary receptor for most human Ad serotypes (including Ad2 and Ad5) is a 43 kDa cell surface protein termed CAR for the Coxsackie and adenovirus receptor (Bergelson *et al.*, 1997; Tomko *et al.*, 1997). However, MHC-1 has also been proposed as an alternative receptor, the subgenus B serotypes Ad3 and Ad35 are known not to use CAR (DiGuilmi *et al.*, 1995; Stevenson *et al.*, 1995) and Ad37 has recently been shown to bind *via* conjugated sialic acid (Arnberg *et al.*, 2000). CAR is expressed on a wide range of cell types. Gene delivery by Ad5-based vectors is relatively inefficient in cells with low levels of CAR (e.g. lymphocytes) or where the receptor cannot readily be accessed (e.g. CAR is expressed only on the basolateral membrane of ciliated epithelium) (Boucher, 1999; Huang *et al.*, 1997; Walters *et al.*, 1999).

The tropism of an Ad vector can be modified by 'hijacking' the fiber from a different serotype (Segerman *et al.*, 2000) or by engineering a surface virion protein (fibre, penton or hexon) to re-direct binding to an alternative receptor. The need to modify the tropism of Ad vectors is governed not only by the need to re-direct to a different receptor but also to restrict binding to the natural receptors (Curiel *et al.*, 1991). For example, it may be necessary to avoid sequestration of the vector to cells expressing high levels of CAR (Smith *et al.*, 1993) or restrict delivery of a toxic therapeutic gene.

A secondary interaction between an RGD motif in the penton base protein and a cellular integrin ($\alpha_v\beta3$, $\alpha_M\beta2$ or $\alpha_v\beta5$) is commonly required to stimulate an intracellular signaling cascade that promotes virus endocytosis (reviewed by Nemerow and Stewart, 1999). Activation of phoshatidylinositol-3-OH kinase (PI3K) and Rho family members causes a re-organization of actin filaments and dynamin to promote rapid Ad uptake into clatherin-coated vesicles. The penton base proteins/α_v integrin interaction also promotes interleukin 8 production *via* stimulation of ERK1/ER2 MAP kinases (Bruder and Kovesdi, 1997). Following uptake into an endosomal vesicle, acidification triggers virus-induced disruption via an interaction with $\alpha_v\beta5$ and particle release into the cytosol (Wang *et al.*, 2000). Thus the act of virus infection in the absence of *de novo* gene expression has the potential to affect both cellular physiology and the cytokine balance. The particle is then transported to the nuclear pore and the genome, once released into the nucleus, is covalently attached *via* the Ad terminal protein to the nuclear matrix. This association with the nuclear matrix facilitates gene expression. Transcription and replication of the Ad genome is initiated within the nucleus adjacent to a defined nuclear body, known as PML-associated oncogenic domains (Doucas *et al.*, 1996).

5.3.6 The Ad genome

Ad5 has a linear ~36 kbp double-stranded DNA genome (*Figure 5.5*) and utilizes a cascade system of gene regulation in which first the E1a gene product activates early phase gene expression (E1b–E4) with late phase gene expression following virus DNA replication. The virus genome is transcribed in both directions making extensive use of RNA splicing and multiple polyadenylation signals. Multiple gene products are derived from each gene region although there is clustering of gene function (Wold *et al.*, 1999). The E1 and E4 regions are heavily involved with regulating transcription, RNA processing and controlling apoptosis (Leppard, 1997). While the E2 transcripts encode enzymes associated with replicating the viral genome (DNA polymerase, ssDNA-binding proteins, preterminal protein), the E3 gene region is dispensable for virus replication *in vitro* and thus often deleted from Ad vectors. The E3 region, however, encodes a range of functions implicated in suppressing the host immune response during infection, including down regulating MHC-1 antigen presentation (E3-gp19K), down-regulating cytokine receptors (the RID complex) and eliciting protection to both Fas and TNF (the RID complex and E3-14.7K) (Wold *et al.*, 1999). Thus, if expressed by a vector, the E3 gene region also has capacity to influence the immune response to Ad, expressed transgenes or tumor cell antigens. Genes encoding polypeptide IX (required for efficient virion assembly) and IVa2 are transcribed from additional promoters with intermediate kinetics. Two small virus-associated RNA polymerase III transcripts (VA RNA_I and VA RNA_{II}) produced in high abundance late in infection elicit protection against interferon and enhance expression of late genes (Liao *et al.*, 1998). The vast majority of late phase transcription is driven from a single late promoter and can be

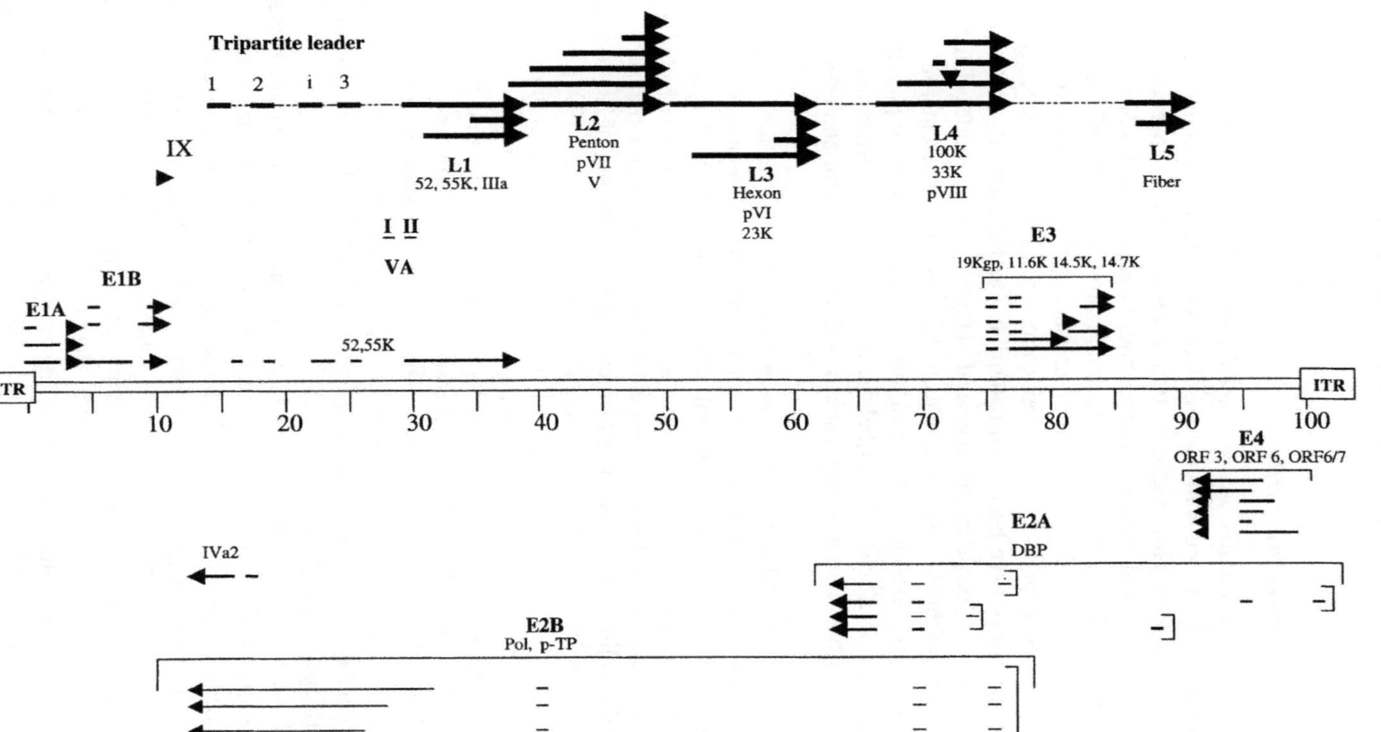

Figure 5.5. The Ad genome. Transcriptional map of the Ad genome showing the location of early (E1–E4) and late phase (L1–L5) transcription. The Ad5 genome (~ 35kb) is transcribed in both directions and makes extensive use of alternate RNA splicing. The E1 gene region (E1A and E1B) is essential for viral transcriptional regulation and promoting cell cycle progression. Replication-deficient Ad5 vectors are based on E1 deletion mutants (additional Ad genes may also be deleted in vectors) with the vector being propagated on a complementing cell line (e.g. 293 cells). The E3 region is often deleted from vectors because it is dispensible for growth *in vitro*; E3 encodes multiple genes contributing to immune evasion *in vivo*. Transgenes are commonly inserted in either E1 or E3 (Chapter 8).

conveniently divided into 5 blocks (L1–L5). Late phase genes mainly encode virion components.

5.3.7 Replication-competent adenovirus vectors

The successful application of the live Ad vaccines for more than 30 years provides a sound basis to go on and exploit replication-competent Ad vectors as vaccine carriers. With a requirement to maintain virus viability, transgenes are usually inserted in the E3 gene region with expression driven by an exogenous promoter, the E3 promoter or the major late promoter. Ad-encoded late proteins are produced in high abundance during lytic infection and similarly transgenes encoded by replication-competent vectors can be efficiently expressed. Both replication-competent and replication-deficient Ad (RDAd) vectors have been used extensively and successfully to induce immune responses to a wide range of defined viral antigens. The examples listed in *Table 5.4* use a variety of human and non-human Ads vectors, different routes of delivery, virus dosages, immunogens and animal models, thus it is seldom possible to make direct comparisons of efficacy. Nevertheless, much has been learned about the induction and evasion of host immune responses to both the Ad vector and expressed transgenes.

For a replication-competent Ad vector to act as an effective immunization agent, the recombinant virus must undergo a number of rounds of productive infection in a susceptible host (Lubeck *et al.*, 1997). Ad5 will infect a wide range of cell types, however, fully productive replication is restricted to relatively few species, including man, chimpanzee, the cotton rat and pigs (Torres *et al.*, 1996). The Ad genome can also be modified to enable replication in simian cells, thus facilitating the use of Ad vectors in SIV research (Buge *et al.*, 1997; Cheng *et al.*, 1992; Flanagan *et al.*, 1997). Most transgene insertions into replication-competent Ad vector are at E3, a region implicated in immune evasion (see above). Studies in chimpanzees indicate that Ad E3$^-$ constructs are at a growth disadvantage *in vivo*, are cleared rapidly *in vivo* and are thus not ideal immunization agents (Chengalvala *et al.*, 1997; Tacket *et al.*, 1992; Torres *et al.*, 1996). Consideration also has to be given to the overall effect of the transgene expression on virus viability. Ad recombinants expressing some transgenes have also been shown to have a growth disadvantage *in vivo*, resulting in a reduction in the level and duration of antigen expression (Buge *et al.*, 1997).

The strengths of Ad and its associated vectors as immunization agents are being increasingly acknowledged: they are inexpensive to produce, robust, withstand freeze-drying and can be readily administered orally. Replication-competent Ad recombinants have been exploited in a wide range of systems where they have been demonstrated to induce a comprehensive immune response, often following a single immunization, associated with protection against subsequent virus challenge (Both *et al.*, 1993; Breker-Klassen *et al.*, 1995; Callebaut *et al.*, 1996; Hammond *et al.*, 2000; Hsu *et al.*, 1992; McDermott *et al.*, 1989; Mittal *et al.*, 1996; Prevec *et al.*, 1989, 1990; Sanz-Parra *et al.*, 1999; Wesseling *et al.*, 1993). Oral vaccination, in addition to being logistically straightforward to administer, also induces comprehensive mucosal and cellular immune responses (Gallichan *et al.*, 1993; Gallichan and Rosenthal, 1995, 1996).

Table 5.4. Ad recombinants used as immunization agents for viral antigens

Virus	Antigen	Vector replication	Reference
HIV	env	Competent	Hung *et al.*, 1988, Dewar *et al.*, 1989, Chanda *et al.*, 1990, Natuk *et al.*, 1992, Lubeck *et al.*, 1997, Zolla-Pazner *et al.*, 1998
	gag		Prevec *et al.*, 1991, Robert-Guroff *et al.*, 1998
	gag+env		Natuk *et al.*, 1993, Lubeck *et al.*, 1994
	env	Deficient	Bruce *et al.*, 1999
SIV	Env	Competent	Buge *et al.*, 1997, Flanagan *et al.*, 1997, Buge *et al.*, 1999
Feline Immunodeficiency	Env	Deficient	Gonin *et al.*, 1995, Chenciner *et al.*, 1997
Hepatitis B	HbsAg	Deficient	Ballay *et al.*, 1985
	HbsAg	Competent	Morin *et al.*, 1987, Hung *et al.*, 1988, Lubeck *et al.*, 1989, Hung *et al.*, 1990, Mason *et al.*, 1990, Chengalvala *et al.*, 1991, Tacket *et al.*, 1992, Chengalvala *et al.*, 1994, Chengalvala *et al.*, 1997, Chengalvala *et al.*, 1999
	HBSAg+HbcAg+HbeAg	Competent	Ye *et al.*, 1991
	HbsAg	Both	Levrero *et al.*, 1991
Vesticular stomatitis	Glycoprotein	Competent	Prevec *et al.*, 1989
Rabies		Competent	Charlton *et al.*, 1992, Kalicharran *et al.*, 1992, Prevec *et al.*, 1990, Yarosh *et al.*, 1996
		Deficient	Tims *et al.*, 2000, Wang *et al.*, 1997, Xiang & Ertl, 1999, Xiang *et al.*, 1999, Xiang *et al.*, 1996
Measles	Nucleocapsid, Hemagluttinin, Fusion	Deficient	Fooks, 1995, Fooks *et al.*, 1996, Fooks *et al.*, 1998
RSV	Fusion, G	Competent	Hsu *et al.*, 1992, Connors *et al.*, 1992, Hsu *et al.*, 1994
Bovine parainfluenza 3	HN and F	Competent	Breker-Klassen *et al.*, 1995
Rotavirus	VP7sc	Competent	Both *et al.*, 1993
Mouse hepatitis	Spike, nucleocapsid, membrane	Competent	Wesseling *et al.*, 1993
Bovine coronavirus	HN-esterase	Competent	Baca-Estrada *et al.*, 1995
Porcine respiratory coronavirus	Spike	Competent	Callebaut *et al.*, 1996
Transmissible gastroenteritis coronavirus	Spike	Competent	Torres *et al.*, 1995, Torres *et al.*, 1996
Foot-and-mouth disease	Structural	Deficient	Mayr *et al.*, 1999
		Competent	Sanz-Parra *et al.*, 1999
Tick-borne encephalitis	NS1	Deficient	Jacobs *et al.*, 1992, Jacobs *et al.*, 1994, Timofeev *et al.*, 1997, Timofeev *et al.*, 1998
HCV	Core, E1	Deficient	Bruna-Romero *et al.*, 1997, Lasarte *et al.*, 1999
	Core-E1-E2	Competent	Seong *et al.*, 1998

Table 5.4 (cont)

Virus	Antigen	Vector replication	Reference
Bovine viral diarrhoea	Nucleocapsid	Deficient	Elahi *et al.*, 1999
Classical swine fever	gp55	Competent	Hammond *et al.*, 2000
Human papillomavirus	E6, E7	Deficient	He *et al.*, 2000
Herpes simplex	gB	Competent	McDermott *et al.*, 1989, Gallichan *et al.*, 1993, Gallichan & Rosenthal, 1995,Gallichan & Rosenthal, 1996
	gD, gD epitopes		Zheng *et al.*, 1993
Bovine herpesvirus 1	gD	Competent	Papp *et al.*, 1997, Zakhartchouk *et al.*, 1998, Papp *et al.*, 1999
		Both	Mittal *et al.*, 1996
Pseudorabies	gD	Deficient	Eloit *et al.*, 1990, Adam *et al.*, 1994, Ganne *et al.*, 1994, Endresz *et al.*, 1995,Gonin *et al.*, 1995, Gonin *et al.*, 1996, Le Potier *et al.*, 1997, Mittal *et al.*, 1996, Monteil *et al.*, 1997, Monteil *et al.*, 2000
Human cytomegalovirus	IE1, gB	Competent	Marshall *et al.*, 1990, Berencsi *et al.*, 1993, Berencsi *et al.*, 1996
Epstein Barr	gp340/220	Deficient	Ragot *et al.*, 1993
	EBNA3c	Deficient	Morgan *et al.*, 1996
	LMP2c	Deficient	Ranieri, 1999

Systematic studies have been conducted using replication-competent Ads encoding three relatively poor immunogens: HBV surface antigen, HIV-1 env and SIV env (*Table 5.4*).

Repeated immunization of rhesus macaques with the same Ad5-SIV env recombinant was effective in promoting mucosal and cell-mediated immunity that afforded partial protection to SIV challenge (Buge *et al.*, 1997, 1999). Thus, re-immunization in the presence of pre-existing immunity to an Ad vector can still elicit therapeutic benefit. Experience with various HIV env-based vaccines in chimpanzees indicated that multiple immunizations are required to induce only transient protection (Lubeck *et al.*, 1997). Protocols involving sequential delivery of Ad4, Ad5 and Ad7 recombinants encoding HIV env (gp160) combined with a final boost with purified HIV gp120 were therefore devised and tested. By exploiting multiple Ad serotypes, a comprehensive and long-lived immune response was generated capable of eliciting protection to challenge with both homologous and heterologous HIV-1 isolates (Lubeck *et al.*, 1997; Robert-Guroff *et al.*, 1998; Zolla-Pazner *et al.*, 1998).

Replication-competent Ad HBVsAg constructs proved effective at inducing a neutralizing antibody in hamsters (Morin *et al.*, 1987) and partial protection in chimpanzees (Lubeck *et al.*, 1989). Single dose oral adminstration of an Ad7 HBsAg to human volunteers, however, was unable to induce a detectable humoral response. The absence of a response was attributed to inefficient replication of the Ad7HBsAg recombinant in vaccinees, possibly due to a deletion in E3 (Tacket *et al.*, 1992).

5.3.8 Replication-deficient Ad vectors

The popularity of RDAd vectors is growing exponentially as the full utility of these vectors for research, therapeutic and biotechnology applications is appreciated. Extremely efficient Ad vector systems continue to be developed and with commercial RDAd vectors now being actively promoted (Microbix Biosystems Inc, Quantum Scientific and Clontech), Ad vectors are no longer the province of specialized laboratories but generally accessible to the research community. Procedures employed in the construction and application of such recombinants are provided by commercial suppliers and have been reviewed in detail elsewhere (Graham and Prevec, 1995; Precious and Russell, 1985; Scarpini *et al.*, 1999; see Chapter 8).

RDAd vectors are conceptually distinct from their replication-competent counterparts as vectors and immunization agents. Since effectively only the transgene is expressed in target cells, the immune response is totally focused on the relevant antigen. The RDAd recombinant does not require *in vivo* cell-to-cell spread, a process that can be inhibited by pre-existing specific and non-specific immune responses. Furthermore, this intrinsic inability of the vector to replicate is an important safety feature, particularly as vaccines inevitably get administered to individuals with immune defects or deficiencies.

The levels of transgene expression that can be achieved with a non-replicating vector are impressive. Using a truncated version of the HCMV major IE promoter up to 25% of total cellular protein can be comprised of the expressed transgene (Jacobs *et al.*, 1992; Wilkinson and Akrigg, 1992). A truncated HCMV promoter, which lacks repressor elements (Liu *et al.*, 1994), is also associated with prolonged transgene expression *in vivo* and thus sustained antigenic stimulation (Armentano *et al.*, 1999; Wilkinson and Akrigg, 1992). Recent reports indicate that the murine CMV is even stronger than the human CMV major IE promoter in some cell types (Addison *et al.*, 1997). Expression levels are dependent on the promoter associated with the transgene, but will also vary substantially with species, strains, cell type and the environment, most notably inflammation.

RDAd recombinants provide efficient *in vivo* gene delivery to a wide range of cell types. In gene therapy studies, extreme amounts of recombinant viruses are tested to maximize gene delivery. Inevitably, escalating concentrations of input virus eventually leads to the induction of both humoral and cell-mediated immune response against elements of the vector that can limit transgene expression and induce inflammatory disease (Yang *et al.*, 1994). The inflammatory response associated with the delivery of RDAd vectors can be alleviated or eliminated by reducing the amount of virus used, co-administering immunosuppressive agents or by utilizing the more disabled second generation and gutless RDAd vectors (see Chapter 8). RDAd recombinants are uniquely capable of promoting high level *in vivo* expression of a transgene for a period of a few days to many months supplying a sustained antigenic stimulus. A single inoculation with a RDAd recombinant is able to generate efficient, long-lived humoral and cell-mediated immune responses restricted to the expressed transgene (Juillard *et al.*, 1995) that can elicit protection against a subsequent virus challenge (Eloit *et al.*, 1990; Fooks, 1995; Jacobs *et al.*, 1992; Mayr *et al.*, 1999; Ragot *et al.*, 1993; Xiang *et al.*, 1996). In direct comparisons replication-deficient Ad recombinants have generated levels of protection superior to direct DNA immunization (Fooks *et al.*, 1996) and similar with comparable replication-competent Ad (Mittal *et al.*, 1996) or vaccinia virus recombinants (Fooks, 1995; Gonin *et al.*, 1996; Xiang *et al.*, 1996). Immunization with RDAd vectors administered parenterally or to a range of mucosal surfaces can be therapeutically effective (Fooks *et al.*, 1998; Xiang and Ertl, 1999; Xiang *et al.*, 1999). Indeed, a RDAd recombinant has recently been demonstrated to elicit an immune response when applied directly to skin (Shi *et al.*, 1999).

Immunizations are commonly performed in early childhood in order to provide protection before an individual becomes exposed naturally to the pathogen. Timing can be problematical as passive immunity during the first year of life often provides protection not only against the infectious agent but also against a live vaccine. A safe, non-replicating agent capable of inducing long-lived immunity in the face of passive protection has clear benefits. In this context a single immunization of piglets, born to immune mothers, with a RDAd recombinant encoding pseudorabies virus gD induced protection to subsequent challenge (Monteil *et al.*, 1997, 2000). Similarly, immunization in the presence of passive immunity has also been demonstrated with a RDAd rabies virus glycoprotein recombinant (Wang *et al.*, 1997). Maternal immunization with a recombinant Ad can also be exploited to induce passive protection in the newborn (Both *et al.*, 1993).

Many of the protocols described in *Table 5.4* are directed at veterinary problems. This is an important field that is being supported by the development of non-human Ad vectors. Significant success has been achieved in the control of rabies amongst wild animal populations both in Europe and America using vaccinia virus recombinants. To complement this approach, rabies vaccines have also been successfully developed based on both replication-competent and replication-deficient Ad vectors (Charlton *et al.*, 1992; Paolazzi *et al.*, 1999; Prevec *et al.*, 1990).

5.3.9 Ad tumor immunotherapy

Ad is currently the second most popular viral vector deployed in gene therapy and consequently considerable experience has been accruing from preclinical and, increasingly, clinical studies. Only a small proportion of gene therapy protocols (13.9%) involve the treatment of monogenic disease. Currently, 63.9% of all protocols and 69% of all patients in gene therapy clinical trials are participating in cancer treatments (see Chapter 8). Protocols designed to target tumor cells by means of genetic prodrug therapy, tumor suppressors or anti-oncogenes inevitably also inflict cellular damage that can have an indirect beneficial effect of promoting specific and non-specific immune responses to the tumors. However, cancers are increasingly being tackled directly by immunotherapies.

5.3.9.1 Ad immunization and tumor antigens. Approximately 20% of all cancers worldwide are virus-associated (e.g. human papilloma virus, Epstein Barr virus, hepatitis B and C) and are prime candidates for prophylactic vaccination or immunotherapeutic intervention during disease. The list of characterized host-encoded 'tumor antigens' also continues to expand (e.g. MART-1, MAGE-1, tyrosinase, gp100), providing an ever stronger foundation on which to design anti-cancer immunotherapies (Rosenberg, 1999). When RDAd are used to express a foreign gene *in vivo* an efficient, comprehensive immune response to the transgenes is generated that can limit the longevity of transgene expression (Juillard *et al.*, 1995; Morral *et al.*, 1997). Immunization with RDAd recombinants can also overcome tolerance to elicit responses to host-encoded protein. Pre-clinical studies in rodent models have demonstrated that RDAds encoding MART, gp100, gp75 and other tumor antigens were able to elicit appropriate anti-tumor immune responses and protection against tumor cell challenge (Hirschowitz *et al.*, 1998; Zhai *et al.*, 1996). This approach is being pursued in clinical trials. A RDAdLacZ construct was able to generate efficient humoral and cell-mediated transgene-specific immune responses following intra-tumoral delivery of four patients with lung cancer (Gahery-Segard *et al.*, 1997). A phase I trial has also been undertaken in 54 patients with metastatic melanoma who received escalating doses of RDAds encoding either MART or gp100. Although the treatment was well tolerated, indications of therapeutic benefit were limited (Rosenberg *et al.*, 1998).

5.3.9.2 Ads encoding co-stimulatory molecules and cytokine/chemokine therapy. As the immune system is dissected key cellular and molecular level interactions continue to be identified that can be manipulated to optimize the induction of an appropriate response. Co-stimulatory molecules, cytokines and chemokines are now being extensively used in RDAd vectors to enhance appropriate responses to antigen. The interaction between B7-1 (CD80) on antigen-presenting cells (APCs) and CD28 on T cells provides an important co-stimulatory signal for NK and $CD8^+$ T cells. Similarly, the CD40 ligand (CD154)–CD40 interaction plays a crucial role in T cell activation. Delivery of RDAd-B7-1 and RDAd-CD154 recombinants to tumors *in vivo* has been effective in promoting tumor regression in murine models (Emtage *et al.*, 1998; Kikuchi and Crystal, 1999; Putzer *et al.*, 1997). Interestingly, a RDAd-CD154 virus also enhanced immune responsiveness to respiratory syncytial virus antigens in a murine model (Tripp *et al.*, 2000).

Cytokines and chemokines have been used extensively in tumor immunotherapy protocols, reviewed in greater detail elsewhere (Powell and Wilkinson, 2000). RDAd vectors encoding transgenes that promote a Th1 response (e.g. IL-2, IL-7, IL12, GM-CSF, MIP-3α and γ-interferon) have been particularly successful in promoting protection against tumors in preclinical studies using characterized model systems. Cytokines can be provided either independently or in combination, relying on endogenous antigen expression from the tumor, however, synergistic responses are generated by co-delivery of cytokines and tumor antigens (Hirschowitz and Crystal, 1999). The development of this methodology so that is can be successfully applied in the clinic is a major challenge. Bowman and co-workers immunized 10 children with advanced neuroblastoma with autologous tumor cells transduced with a RDAd-IL-2 construct and observed a 'potent increase in antitumor activity' that also correlated with an encouraging clinical response (Bowman *et al.*, 1998). Similarly, Schreiber and co-workers used RAd-IL2-transduced, irradiated autologous, metastatic melanoma cells in a vaccine that was well tolerated and, although not curative, was accompanied by significant clinical benefit (Schreiber *et al.*, 1999). In a further phase I clinical trial, escalating doses of a RDAd IL2 recombinant were successfully administered by direct intratumoral injection to a group of metastatic breast cancer and melanoma patients without significant adverse reaction. Some clinical benefit was demonstrated by partial tumor regression at the site of injection in 24% of patients (Stewart *et al.*, 1999).

5.3.9.3 Ad and dendritic cells. A majority of cells in the body can present antigen in the context of MHC-1, however, a subset capable of presenting antigen in association with MHC-II are considered to be 'professional' APCs (e.g. macrophages, B cells and dendritic cells). Research in cancer immunotherapy has focused on dendritic cells (DCs) particularly because of their unique capacity to stimulate naive T cells, overcoming neonatal tolerance and immunological unresponsiveness, thus to drive CTL generation against tumor antigens (Calaco, 1999). DCs can be readily generated then infected *ex vivo* with a RDAd recombinant. Immature DCs are extremely efficient at taking up and processing antigens that will subsequently be presented following DC maturation. RDAd vectors are able to transduce and express transgenes in immature DCs while simultaneously facilitating their maturation (Rea *et al.*, 1999; Zhong *et al.*, 1999). Expression from a RDAd human gp100 construct could be detected in DCs for up to 7 days and the infected DCs were efficient at stimulating growth of specific CTL *in vitro* (Linette *et al.*, 2000). Standard Ad5 vectors infect DCs relatively inefficiently but this deficiency can be remedied by modifications to the fiber (see above) that redirect the vector to a DC-specific receptor (Tillman *et al.*, 1999).

Although much has yet to be done to optimize immunization with RDAd-infected DCs, already the therapeutic potential of this approach has been established in a succession of preclinical studies (Gong *et al.*, 1997; Kaplan *et al.*, 1999; Ribas *et al.*, 1999; Song *et al.*, 1997). Most encouragingly, RDAd-transduced DCs have proved significantly more efficient at eliciting tumor protection compared with direct delivery of the RDAd recombinants (Song *et al.*, 1997; Wan *et al.*, 1999) and thus constitutes the most potent approach currently available for inducing anti-tumor immunity.

5.3.10 Oncolytic adenovirus

A study in 1956 reported on the direct injection of wild-type Ads as a treatment for late stage, cervical carcinoma in which a transient beneficial effect was observed in 65% of tumors (Smith, 1956). The application of oncolytic viruses has recently been revisited using a series of Ad mutants designed to replicate preferentially in transformed cells. During virus replication E1A sequesters the tumor suppressor protein Rb while E1B-55kD inactivates p53 to promote cell cycle progression. Cells lacking functional p53 preferentially support replication of the Ad $E1B^-$ deletion *dl*1520 (renamed ONYX-015 and more recently CI-1402) (Barker and Berk, 1987; Bischoff *et al.*, 1996; Rogulski *et al.*, 2000). Since ~50% of tumors exhibit overt defects in p53 expression, the CI-1402 virus has the potential to be of therapeutic value against a wide range of cancers. Preclinical studies demonstrated the oncolytic Ad to be effective against a range of tumors (Heise *et al.*, 1999). Phase I and II trials, involving direct intra-tumoral injection of CI-1402 (ONYX-015) into recurrent head and neck carcinomas, were highly encouraging (Kirn, 2000) and so studies have been expanded to other cancers (pancreatic, ovarian, colorectal, lung and oral dysplasia) with over 200 patients having been treated (Heise and Kirn, 2000).

Oncolytic Ads are being further developed to enhance specific killing of tumor cells (Fueyo *et al.*, 2000; Rodriguez *et al.*, 1997). Although oncolytic viruses act primarily by directly inducing cytotoxicity in tumor cells, they also enhance an anti-tumoral immune response and sensitivity to chemotherapy (Heise and Kirn, 2000). Clearly there is the potential to incorporate therapeutic genes directly into the genome of the oncolytic Ads ('armed therapeutic viruses') (Hermiston, 2000). An oncolytic Ad over-expressing the 'Ad death protein' had enhanced anti-tumor activity in a mouse tumor model (Doronin *et al.*, 2000). Interestingly, an oncolytic Ad when delivered along with a RDAd recombinant will complement the deletion resulting in replication of both viruses. Co-infection of RDAd-IL2 or RDAd-IL12 recombinants with an oncolytic Ad enhanced both transgene expression (>100 fold) and tumor clearance (Motoi *et al.*, 2000).

5.4. Recombinant alphavirus expression vectors

5.4.1 Background

Alphaviruses, members of the *Togaviridae* family, have a positive (+) single stranded (ss) RNA genome packaged into an icosahedral nucleocapsid structure. The viral envelope consists of a lipid bilayer, in which two or three membrane glycoproteins are embedded. The biology of alphaviruses has been thoroughly reviewed by Strauss and Strauss (1994). Efficient expression vectors for Semliki Forest virus (SFV) (Liljeström and Garoff, 1991), Sindbis virus (SIN) (Xiong *et al.*, 1989) virus and Venezuelan equine encephalitis virus (VEE) (Davis *et al.*, 1989) have been constructed. There are basically two types of vector systems: (i) Replication-competent vectors based on the full-length genome plus a second subgenomic promoter and the foreign gene of interest. (ii) Replication-deficient vectors containing the replicase genes and a transgene. The structural genes are deleted and simultaneous transfection of a helper vector with the structural genes in *trans* is required to allow packaging of recombinant virus particles. In contrast to the replication-competent particles these suicidal SFV replicons are capable of only one round of infection with no further virus production occurring. However, the infected cells will produce extensive amounts of recombinant protein.

The following properties make alphavirus vectors highly efficient. (i) The host range is extremely broad including mammalian and insect cell lines as well as primary cell cultures (Liljeström and Garoff, 1991; Lundstrom, 1999) (*Table 5.5*). In the rare cases of low infection efficiency significant improvement can be achieved by addition of polyethylene glycol (PEG) to the cell culture medium (Arudchandran *et al.*, 1999). (ii) Release of positive (+) ssRNA directly into the cytoplasm excludes a requirement for RNA transport from the nucleus. (iii) The RNA molecules contain the viral nonstructural genes coding for the viral replicase complex (responsible for massive production of new +ssRNA copies). It is estimated that more than 100 000 copies of ssRNA molecules per cell are generated leading to extremely high levels of transgene expression. (iv) Several genes can be expressed in the same cell either by introduction of the genes in the same vector in tandem (Zhang *et al.*, 1997) (behind a second

Table 5.5. Cell lines and primary cell cultures infected by alphavirus vectors

Cell type	Reference
Acinar cells*	Arudchandran *et al.*, 1999
BAEC, bovine endothelial cell*	Lundstrom (unpublished results)
BHK-21, baby hamster kidney	Liljeström and Garoff, 1991
CEF, chicken embryo fibroblast	Xiong *et al.*, 1989
CHO-K1, Chinese hamster ovary	Lundstrom *et al.*, 1994
C6 rat glioma	Lundstrom (unpublished results)
C6/36, *Aedes albopictus* (insect)	Olson *et al.*, 1994
COS M6, Green Monkey kidney	Blasey *et al.*, 2000
COS-7, Green Monkey kidney	Liljeström and Garoff, 1991
DRG, dorsal root ganglion neurons	Lundstrom (unpublished results)
HeLa, human epithelial carcinoma	Liljeström and Garoff, 1991
HEK293, human embryonic kidney	Schlaeger and Lundstrom, 1998
HMC-1, human mast cell*	Arudchandran *et al.*, 1999
HPF, human primary fibroblast	Huckriede *et al.*, 1996
HOS, human osteogenic sarcoma	Liljeström and Garoff, 1991
Hybridoma 179	Blasey *et al.*, 2000
Jurkat*	Arudchandran *et al.*, 1999
KU 812, human basophil-like cell line*	Arudchandran *et al.*, 1999
Mast cells, rat peritoneal*	Arudchandran *et al.*, 1999
MDCK, canine kidney	Olkkonen *et al.*, 1993
MME, mouse mammary epithelial	Olkkonen *et al.*, 1993
MOLT-4, lymphoblastic leukemia	Paul *et al.*, 1993
Myoblast, human*	Arudchandran *et al.*, 1999
NIE 115, mouse neuroblastoma	Blasey *et al.*, 2000
NIH 3T3, mouse embryo*	Liljeström & Garoff, 1991
NK cells*	Arudchandran *et al.*, 1999
NT2N, rat neurons	Cook *et al.*, 1996
OLF, mouse olfactory epithelium	Monastyrskaia *et al.*, 1999
PCN, primary rat cortical neurons	Lundstrom (unpublished results)
Raji, human Burkitt lymphoma	Blasey *et al.*, 2000
RBL-2H3, mast cell line*	Arudchandran *et al.*, 1999
RHN, rat hippocampal neurons	Olkkonen *et al.*, 1993
RIN	Lundstrom *et al.*, 1997
RPMI 8226, human myeloma	Blasey *et al.*, 2000
SCG, superior cervical ganglion	Ulmanen *et al.*, 1997
SL3, *Drosophila melanogaster* (insect)	Lundstrom (unpublished results)
Swiss 3T3*	Arudchandran *et al.*, 1999
TK6, human lymphoblast	Blasey *et al.*, 2000
V79, hamster lung fibroblast	Lundstrom (unpublished results)

* Significant increase in infection rate obtained with PEG-treatment.

subgenomic promoter or an IRES sequence) or by multiple infections with several virus stocks (Scheer *et al.*, 1999).

5.4.2 Gene expression **in vitro**

5.4.2.1 Recombinant protein expression. Since the first reports on recombinant gene expression from alphavirus vectors (Davis *et al.*, 1989; Liljeström and Garoff, 1991;

Xiong *et al.*, 1989) many different genes have been expressed at high levels in various host cell lines. Among the recombinant proteins produced are enzymes, receptors, channels, viral proteins, transcription factors, etc. (Blasey *et al.*, 2000). Topologically different proteins, including nuclear, cytoplasmic, membrane-associated, transmembrane and secreted proteins, have been successfully expressed from alphavirus vectors.

G protein-coupled receptors (GPCRs) and ligand-gated ion channels, known to be particularly difficult to express at high levels, could be efficiently expressed from SFV vectors (Lundstrom, 1997, 1999). These receptors have important neurotransmitter and signal transduction functions and are therefore highly significant targets for drug development. The high receptor density achieved has allowed efficient binding studies in different cell lines (Lundstrom *et al.*, 1994), although the inhibition of the host cell protein synthesis results in relatively quick death of the infected cells. The cell viability varies from one host cell line to another, but is, interestingly, also dependent on the individual recombinant receptor expressed. For instance, cells expressing the human neurokinin-1 receptor showed massive cell death at 24 h post-infection (Lundstrom *et al.*, 1994), whereas α_{1b}-adrenergic receptor expressing cells were still viable at 60 h post-injection (Scheer *et al.*, 1999). The receptor densities achieved are extremely high (up to 8×10^6 receptors cell^{-1}) and the B_{max} values (up to 150 pmol receptor mg^{-1} protein), higher than for any other expression system reported so far.

5.4.2.2 Functional assays. Despite the dramatic shut down of the host cell protein synthesis after SFV-infections many studies have confirmed the feasibility of functional assays for receptors (Lundstrom *et al.*, 1994; Scheer *et al.*, 1999) and other recombinant proteins (Blasey *et al.*, 1997). Efficient coupling to G proteins could be demonstrated by intracellular Ca^{2+}-release, cAMP stimulation, inositol phosphate accumulation and GTPγS binding assays for various GPCRs. However, the inhibition of the endogenous expression of G-protein subunits led to a decrease in functional activity with time. This could be compensated by co-expression of both receptor and G-proteins from SFV vectors (Scheer *et al.*, 1999).

Expression of ligand-gated ion channels from SFV vectors yielded high functional activity (measured by intracellular Ca^{2+}-release) as well as high amplitude electrophysiological responses in whole-cell patch clamp recordings (Evans *et al.*, 1996). The possibility to infect primary cultures, like primary neurons (70–95% infection rate), has facilitated studies in an environment closely resembling the native tissue (Olkkonen *et al.*, 1993; Ulmanen *et al.*, 1997).

5.4.2.3 Safety aspects and scale-up. Conditionally infectious SFV particles were generated by the introduction of three point mutations at the cleavage region of E2 and E3 genes in the p62 precursor (Berglund *et al.*, 1993). This second-generation helper vector ensured that no replication-competent SFV particles were produced and guaranteed high safety standards for large-scale virus production. Blasey and co-workers (1997) managed to adapt BHK and CHO cells to suspension cultures and successful SFV-infection of spinner flask cultures on a litre scale resulted in high levels of human cyclooxygenase-2 (hCOX-2) activity. Further scale-up in 15 l bioreactors allowed production of large amounts of mouse serotonin 5-HT_3 receptor. Yields of

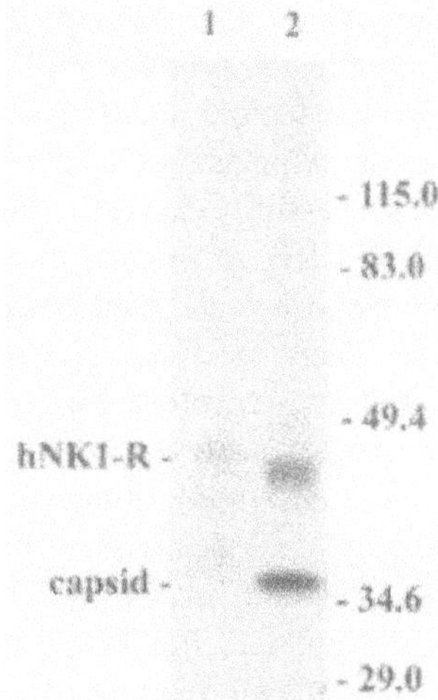

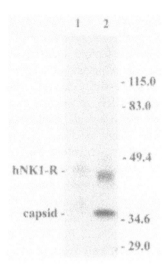

Figure 5.6. Expression of human neurokinin-1 receptor. BHK cells infected with SFV-hNK1-R (lane 1) and SFV-CAP hNK1R-HisFLAG (lane 2), respectively, were labelled with [^{35}S]-methionine at 16 h post-infection. The expression was visualized by 10% SDS-PAGE followed by autoradiography.

10–20 mg receptor protein could be reproducibly recovered from 11.5 l cultures of BHK cells (Lundstrom *et al.*, 1997). Purification and solubilization led to preliminary characterization of the 5-HT_3 receptor (Hovius *et al.*, 1998) and 2-D crystallography and cryo-electron microscopy are now in progress. A further improvement to increase protein yields was the discovery of a translation enhancement signal in the SFV capsid sequence. Fusion constructs containing the first 102 nucleotides of the viral capsid gene sequence increased the level of recombinant protein expression 5- to 10-fold (Sjöberg *et al.*, 1994). During the life-cycle of SFV the capsid protein is efficiently autocatalytically cleaved from the nascent polyprotein consisting of capsid-p62-E1 (Melancon and Garoff, 1987). Sjöberg and co-workers (1994) demonstrated that heterologous genes fused to the full-length capsid gene also resulted in efficient co-translational release of the recombinant protein by capsid protein-mediated autocatalytic cleavage. Although variations in the first few amino acids after the cleavage site were tolerated, a construct with a proline residue at position +2 after the fusion junction resulted in inefficient, less than 5% cleavage (Sjöberg *et al.*, 1994). We have recently obtained significantly increased expression levels of functionally active human neurokinin-1 receptor when fused to the capsid gene and the viral protein was efficiently cleaved off the recombinant receptor as shown in *Figure 5.6* (Lundstrom *et al.*, 1999b).

5.4.3 In vivo *expression*

5.4.3.1 Neuron-specific expression in hippocampal slice cultures. The highly efficient infection of primary neurons (Olkkonen *et al.*, 1993; Ulmanen *et al.*, 1997) inspired further investigations to use alphavirus vectors for *in vivo* applications. Rat

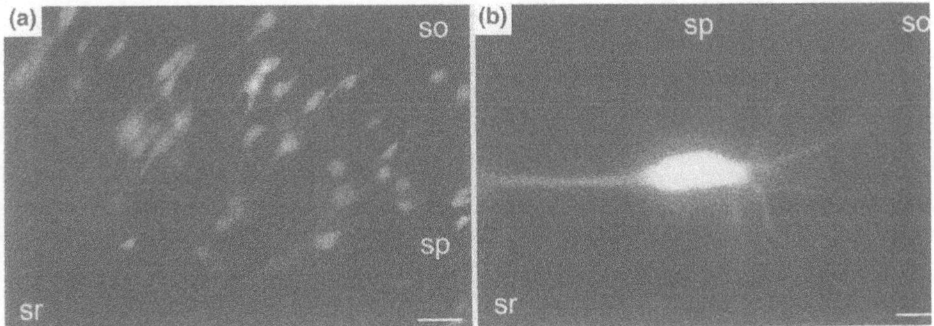

Figure 5.7. Expression of GFP in rat hippocampal *in vivo* slice cultures. Recombinant SFV-mediated expression of GFP in neurons of hippocampal slice cultures prepared from 6-day-old rats. (a) Fluorescence illumination of the CA1 region from a living slice at 14 days in culture and 5 days after injection of ~10^4 infectious particles into the pyramidal cell layer. (b) GFP-expressing CA3 pyramidal cell at 2 days post-infection; GFP-positive dendritic spines are visible. Abbreviations: so, stratum oriens; sp, stratum pyrimidale; sr, stratum radiatum. Bars 65 μm (a), 20 μm (b).

hippocampal slice cultures were infected with SFV and SIN vectors with either LacZ or GFP (green fluorescence protein) as reporter genes. Low concentration infections (200–1000 infectious particles per injection) revealed expression of β-galactosidase and GFP in pyramidal cells, interneurons and granule cells (Ehrengruber *et al.*, 1999). Over 90% of the GFP-positive cells in living slices were identified as neurons (*Figure 5.7*). SIN vectors expressing GFP were used to characterize the dendritic morphogenesis in response to synaptic activity in CA1 hippocampal neurons by two-photon imaging (Maletic-Savatic *et al.*, 1999). In another approach the GFP was fused to the N-terminus of the AMPA receptor (GluR1) and the delivery to CA1 hippocampal dendrites and the requirement of synaptic activation of NMDA receptors were demonstrated (Shi *et al.*, 1999).

5.4.3.2 Local injection into rat brain. Recently, high local delivery and expression of β-galactosidase into mouse nucleus caudata/putamen and nucleus accumbens septi were obtained with replication-deficient SIN vectors (Altman-Hamandzic *et al.*, 1997). In another study, high local reporter gene expression was demonstrated in rat brain after injection of recombinant SFV-LacZ particles into striatum and amygdala (Lundstrom *et al.,* 1999a). The SFV-injected rats showed no change in body weight and body temperature compared to control animals or differences when tested for exploratory locomotor behaviour and forced motor performance. Histological studies revealed strong local β-galactosidase expression (*Figure 5.8*). The highest expression levels were obtained at 1–2 days post-injection followed by an accelerated decline in staining with time. Surprisingly, some staining was still visible at 28 days post-injection. However, this was most likely due to the high stability of β-galactosidase. *In situ* hybridization data confirmed the transient nature of expression and showed dramatic decrease in LacZ mRNA levels at 4 days post-injection. High preference for neuronal infections was also observed *in vivo* (*Figure 5.8b*).

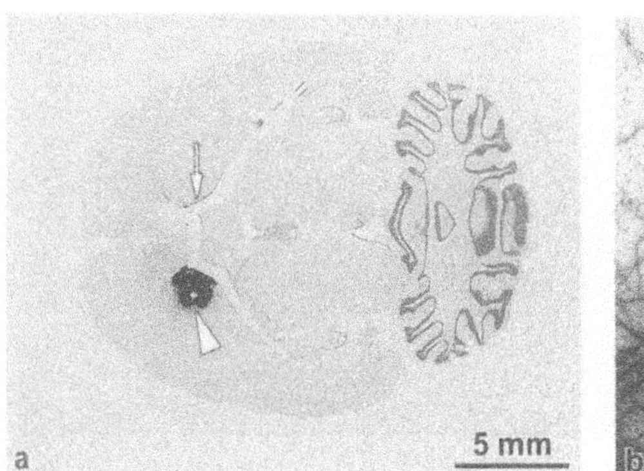

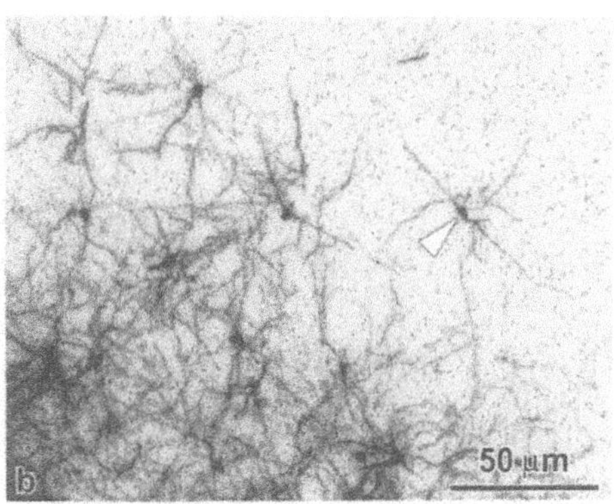

Figure 5.8. Expression of β-galactosidase in rat brain. SFV-LacZ (10^5 particles/injection) was injected into rat brain (striatum and amygdala) and X-gal staining performed at 1 day post-injection. (a) Whole brain section. Arrow indicates injection site, arrowhead spread of virus into the ventricles. The counter-staining seen in cerebellum is nonspecific. (b) Higher magnification of the edge of the injection site, showing expression of β-galactosidase in neuronal processes.

5.4.3.3 Cell/tissue-specific targeting. Due to the broad host range of alphaviruses it would be of great advantage to develop vectors with more specific cell/tissue recognition in order to be able to more precisely target the infection. Successfully targeted alphavirus vectors would be important tools for investigating many neurobiological functions and a prerequisite for future gene therapy applications. The surface structure of SIN virus was modified by inserting IgG-binding domains of protein A into the E2 membrane protein to produce chimeric envelope structures (Ohno *et al.*, 1997). This change in the envelope resulted in significantly reduced infectivity of BHK cells and human cell lines (approximately 10^5-fold decrease). However, chimeric viruses used in conjunction with monoclonal antibodies reacting with cell-surface antigens allowed efficient infection of human cell lines. Further development demonstrated that SIN virus with chimeric envelopes containing the α- and β-hCG genes had a dramatically changed host range. These SIN-hCG particles were no longer capable of infecting BHK cells or human cancer cells lacking LH/CG receptors (Sawai and Meruelo, 1998). In contrast, a reporter gene was efficiently delivered to choriocarcinoma cells. In another approach SIN vectors were used to *in vitro* transcribe biotinylated and self-replicating SIN genomic RNA with streptavidin-protein A fusion protein and monoclonal antibodies (mAbs). In the presence of cationic liposomes, to prevent RNA degradation, it was possible to obtain reporter gene transfection of specific cancer cells in a mAb dose-dependent manner (Sawai *et al.*, 1998).

5.4.3.4 Gene therapy applications. SFV vectors have been applied to produce infectious recombinant Moloney murine leukaemia particles in BHK cells (Li and Garoff, 1996). Insertion of the *gag-pol* and *env* genes as well as a recombinant retrovirus genome (LTR-ψ^+-*neo*-LTR) into individual SFV vectors generated extracellular virus-like particles that possessed reverse-transcriptase activity. Titres in the range of 3×10^7 transduction-competent particles could be obtained from 4×10^6 transfected cells.

Recently, it was demonstrated that retrovirus particles containing intron sequences were efficiently packaged by the SFV expression system (Li and Garoff, 1998). This packaging method should be very useful for gene therapy applications for several reasons. The virus production is easy and rapid, no helper virus was detected in the generated virus stocks and retroviruses with amphotropic envelope could be as efficiently produced as particles with ecotropic envelope.

In addition to the high efficiency of gene delivery obtained *in vivo* (Altman-Hamandzic *et al.*, 1997; Lundstrom *et al.*, 1999a), the strong shut-down of host cell protein synthesis and the apoptosis induced by alphavirus infections (Lundstrom *et al.*, 1997) can be advantageous in cancer gene therapy applications. SFV-LacZ virus was shown to infect and kill human prostate cancer cell lines and prostatic duct epithelial cells *ex vivo* (Hardy *et al.*, 2000). To increase the anti-tumor effect alphavirus vectors expressing cytokine genes were engineered and tested in a murine B16 tumor model (Asselin-Paturel *et al.*, 1999). The p35 and p40 subunits of interleukin-12 were expressed from the same SFV vectors in tandem using two subgenomic promoters (Zhang *et al.*, 1997). Injection of replication-deficient SFV-IL12 into murine B16 tumors resulted in significant tumor regression and inhibition of tumor blood vessel formation. No toxicity was detected in SFV-IL12 treated mice. Repeated injections enhanced the anti-tumor effect and, most encouragingly, no antiviral response to SFV could be observed, which has often been the case for some other viral vectors.

Recently, it was demonstrated that pre-immunization with self-replicating SFV-LacZ RNA could protect mice from tumor challenge (Ying *et al.*, 1999). Furthermore, therapeutic immunization prolonged the survival of BALB/c mice with established tumors.

5.4.4 Vector development

The obvious areas where further improvements of alphavirus vectors can be achieved are the reduction of the cytotoxicity associated with alphavirus infections and the elimination of the host cell protein synthesis inhibition. These factors have restricted to some extent the use of alphavirus vectors because of the relatively early onset of apoptosis. Another obstacle recently addressed is the enormous over-expression of recombinant proteins in infected cells. More moderate expression levels would more likely reflect the native physiological protein levels and would also facilitate kinetic studies on recombinant gene expression.

5.4.4.1 Promoter mutants. One approach to down-regulate the expression levels has been mutagenesis of the subgenomic alphavirus promoter. A 3-nucleotide insertion in the SIN subgenomic promoter resulted in a 100-fold lower promoter activity (Raju and Huang, 1991). Very recently, a series of mutants, including amino acid substitutions and insertions throughout the SFV 26S subgenomic promoter, were introduced upstream of reporter genes. Expression studies revealed novel SFV vectors with significantly lower expression levels, 30%, 3% and 1%, respectively, compared to genes expressed from the wild type promoter (Lundstrom, unpublished results). Despite the lower expression levels obtained most mutants still showed a cytopathogenic phenotype which resulted in the typically strong inhibition of host cell protein synthesis.

5.4.4.2 Non-cytopathogenic vectors. Replication-competent SIN particles expressing the selectable marker gene puromycin *N*-acetyltransferase were generated and virus particles capable of persistent, non-cytopathogenic infection of BHK cells were isolated (Agapov *et al.*, 1998). Genomic studies of this modified vector revealed that a single amino acid substitution in the nsP2 gene (Pro726Ser) was responsible for the change in phenotype. The recombinant protein expression levels were reasonably high but the non-cytopathogenic phenotype seemed to be restricted to mainly BHK cells. Similar vectors have now been developed for SFV containing some other point mutations in the nsP2 gene. These non-cytopathogenic vectors showed an increase in recombinant protein expression levels up to 10-fold and the survival of many mammalian cell lines (BHK, CHO, HEK293) and cultured primary rat hippocampal neurons was substantially prolonged (Lundstrom *et al.*, 1999b).

5.4.4.3 Temperature-sensitive mutants. Inducible expression systems are useful for many applications because it makes it possible to either turn on or off genes when required. Combination of the non-cytopathogenic SIN vector with the Pro726Ser mutation in the nsP2 gene and another mutation in the nsP4 gene at position 153 resulted in a temperature-sensitive (ts) non-cytopathogenic phenotype (Boorsma *et al.*, 2000). The SIN replicon was inactive at +37°C, but induction occurred when the temperature was below +35°C. This phenomenon could be observed for transfected RNA or RNA introduced by viral infections, as well as when a CMV promoter was introduced in front of the nonstructural genes to create a layered DNA vector (see below) for direct plasmid transfections. Introduction of ts-mutants in DNA vectors has opened up new possibilities to generate cell lines stably expressing heterologous genes from alphaviruses.

5.4.4.4 Packaging cell line. Although the generation of recombinant alphavirus particles is relatively simple and rapid, the construction of a packaging cell line (Polo *et al.*, 1991) will especially facilitate large-scale recombinant protein expression, where large volumes of viruses are required. The SIN structural protein genes were expressed stably in BHK cells and it was demonstrated that transfection of RNA or infection with alphavirus particles resulted in reasonably efficient (10^7 infectious units ml^{-1}) packaging of recombinant SIN particles. In one construct the capsid gene and membrane protein genes were separated on distinct vectors to reduce the possibility of homologous recombination and by this procedure eliminate the production of replication-competent SIN particles.

5.4.5 Vaccine production

The typically strong alphavirus RNA replication in the cytoplasm and the high expression levels of recombinant proteins have made alphavirus vectors potential candidates for vaccine production. Replication-deficient recombinant alphavirus particles can be inoculated intravenously into animals. Other approaches involve the use of *in vitro* transcribed RNA or plasmid DNA, directly injected into muscle tissue.

5.4.5.1 Recombinant virus particles. Replication-competent VEE was used to express the influenza virus HA gene (Davis *et al.*, 1996) and the human immunodeficiency virus type 1 (HIV-1) matrix/capsid (MA/CA) (Caley *et al.*, 1997) by subcutaneous inoculation in the footpads of mice. The advantages of using VEE are that most human and animal populations show no pre-existing immunity to VEE and that the virus replicates efficiently in lymphoid tissue to yield strong immune responses. The immunizations with VEE-influenza HA resulted in complete protection of mice against virulent influenza virus strains. Likewise, the VEE-HIV-1 MA/CA vector stimulated a comprehensive humoral and cellular immune response. Replication-deficient SFV vectors expressing the nucleoprotein of influenza virus were injected intravenously into BALB/c mice and humoral responses with high antibody titres were obtained (Zhou *et al.*, 1994). As few as 100 infectious SFV particles induced strong cytotoxic T-cell (CTL) responses and after one booster injection CTL memory was generated for more than 40 days (Zhou *et al.*, 1995). Immunization with SFV expressing simian immunodeficiency virus (SIV) gp160 protected from lethal disease in pigtail macaques (Mossman *et al.*, 1996).

5.4.5.2 RNA and DNA vectors. The delivery of naked RNA into muscle tissue resulted in high-level reporter gene expression (Johanning *et al.*, 1995) as well as in significant immune response against influenza nucleoprotein (Zhou *et al.*, 1994). The development of layered plasmid DNA alphavirus vectors has made it possible to directly use DNA for immunization. In principle, these vectors utilize an RNA polymerase II expression casette to drive the transcription of a self-amplifying RNA (replicon) vector. These types of vectors have been constructed for both SFV (Berglund *et al.*, 1998) and SIN (Dubensky *et al.*, 1996; Herweijer *et al.*, 1995). The *in vivo* efficacy of SIN DNA vectors expressing glycoprotein B of herpes simplex virus type 1 was demonstrated. One single intramuscular injection protected BALB/c mice from lethal virus challenge (Hariharan *et al.*, 1998). Dosing studies revealed that 100- to 1000-fold lower doses of SIN plasmid were needed to induce the same response as from conventional DNA vectors. Likewise SFV DNA vectors showed higher levels of humoral and cellular immune responses than conventional DNA plasmids and immunized animals acquired protective immunity against lethal challenges with influenza virus (Berglund *et al.*, 1998). Recently, injection of SFV DNA expressing the p80 (NS3) protein of bovine viral diarrhoea virus (BVDV) into quadriceps muscle of BALB/c mice resulted in high CTL activity and cell-mediated immune response against BVDV (Reddy *et al.*, 1999).

5.5 Conclusions

Alphaviruses have shown great potential in many different application areas. The rapid production of high-titre virus combined with the broad host range has made these vectors specifically attractive. The efficient RNA replication in the cytoplasm and the development of DNA-based vectors are further improvements. Small and large-scale protein expression, *in vivo* expression studies, potential gene therapy applications and vectors for vaccine production are all methods of today. Further vector and system development will make these viruses attractive tools for additional applications in the field of molecular biology and medicine in the new millennium.

Acknowledgements

The authors would like to thank Dr Gerald Kovacs, Dr Richard Harrop and Dr M. Powell for useful comments on this manuscript. Dr B. Dowsett kindly provided the electron micrograph of adenovirus (*Figure 5.4*). Drs Markus Ehrengruber (Brain Research Institute, Zurich, Switzerland) and Grayson Richards (Roche, Basel) are acknowledged for supplying figures of the rat hippocampal slice cultures (*Figure 5.7*) and the *in vivo* injection into rat brain (*Figure 5.8*).

References

Abrahamsen K, Kong HL, Mastrangeli A, Brough D, Lizonova A, Crystal RG, Falck-Pedersen E. (1997) Construction of an adenovirus type 7a E1A- vector. *J. Virol.* **71**: 8946–8951.

Adam M, Lepottier MF, Eloit M. (1994) Vaccination of pigs with replication-defective adenovirus vectored vaccines: the example of pseudorabies [published erratum appears in Vet Microbiol 1995 Oct;46(4):445]. *Vet. Microbiol.* **42**: 205–215.

Addison CL, Hitt M, Kunsken D, Graham FL. (1997) Comparison of the human versus murine cytomegalovirus immediate early gene promoters for transgene expression by adenoviral vectors. *J. Gen. Virol.* **78**: 1653–1661.

Agapov EV, Frolov I, Lindenbach BD, Pragai BM, Schlesinger S and Rice CM. (1998) Noncytopathogenic Sindbis virus RNA vectors for heterologous gene expression. *Proc. Natl. Acad. Sci. USA* 95: 12989–12994.

Altman-Hamandzic S, Groceclose C, Ma J-X *et al.* (1997) Expression of β-galactosidase in mouse brain: utilization of a novel nonreplicative Sindbis virus vector as a neuronal gene delivery system. *Gene Ther.* **4**: 815–822.

Arita I. (1979) Virological evidence for the success of the smallpox eradication programme. *Nature* 279(5711): 293–298.

Armentano D, Smith MP, Sookdeo CC *et al.* (1999) E4ORF3 Requirement for Achieving Long-Term Transgene Expression from the Cytomegalovirus Promoter in Adenovirus Vectors. *J. Virol.* **73**: 7031–7034.

Arnberg N, Edlund K, Kidd AH, Wadell G. (2000) Adenovirus type 37 uses sialic acid as a cellular receptor. *J. Virol.* **74**: 42–48.

Arudchandran R, Brown MJ, Song JS, Wank SA, Haleem-Smith H, Rivera J. (1999) Polyethylene glycol-mediated infection of non-permissive mammalian cells with Semliki Forest virus: application to signal transduction studies. *J. Immunol. Meth.* **222**: 197–208.

Asselin-Paturel C, Lassau N, Guinebretiere J-M *et al.* (1999) Transfer of the murine interleukin-12 gene *in vivo* by a Semliki Forest virus vector induces B16 tumor regression through inhibition of tumor blood vessel formation monitored by Doppler ultrasonography. *Gene Ther.* **6**: 606–615.

Baca-Estrada ME, Liang X, Babiuk LA, Yoo D. (1995) Induction of mucosal immunity in cotton rats to haemagglutinin-esterase glycoprotein of bovine coronavirus by recombinant adenovirus. *Immunology* **86**: 134–140.

Baldick CJ Jr, Moss B. (1993) Characterization and temporal regulation of mRNAs encoded by vaccinia virus intermediate-stage genes. *J. Virol.* **67**(6): 3515–3527.

Ballay A, Levrero M, Buendia MA, Tiollais P, Perricaudet M. (1985) In vitro and in vivo synthesis of the hepatitis B virus surface antigen and of the receptor for polymerized human serum albumin from recombinant human adenoviruses. *Embo. J.* **4**: 3861–3865.

Barker DD, Berk AJ. (1987) Adenovirus proteins from both E1B reading frames are required for transformation of rodent cells by viral infection and DNA transfection. *Virology* **156**: 107–121.

Barraza EM, Ludwig SL, Gaydos JC, Brundage JF. (1999) Reemergence of adenovirus type 4 acute respiratory disease in military trainees: report of an outbreak during a lapse in vaccination. *J. Infect. Dis.* **179**: 1531–1533.

Barrett N, Mitterer A, Mundt W *et al.* (1989) Large-scale production and purification of a vaccinia recombinant-derived HIV-1 gp160 and analysis of its immunogenicity. *AIDS Res. Hum. Retroviruses* **5**: 159–171.

Belyakov IM, Moss B, Strober W, Berzofsky JA. (1999) Mucosal vaccination overcomes the barrier to recombinant vaccinia immunization caused by pre-existing poxvirus immunity. *Proc. Natl Acad. Sci. USA* **96**(8): 4512–4517.

Bennink JR, Yewdell JW, Smith GL, Moller C, Moss B. (1984) Recombinant vaccinia virus primes and stimulates influenza haemagglutinin-specific cytotoxic T cells. *Nature* **311**(5986): 578–579.

Berencsi K, Rando RF, deTaisne C, Paoletti E, Plotkin SA, Gonczol E. (1993) Murine cytotoxic T cell response specific for human cytomegalovirus glycoprotein B (gB) induced by adenovirus and vaccinia virus recombinants expressing gB. *J. Gen. Virol.* **74**: 2507–2512.

Berencsi K, Gonczol E, Endresz V *et al.* (1996) The N-terminal 303 amino acids of the human cytomegalovirus envelope glycoprotein B (UL55) and the exon 4 region of the major immediate early protein 1 (UL123) induce a cytotoxic T-cell response. *Vaccine* **14**: 369–374.

Bergelson JM, Cunningham JA, Droguett G *et al.* (1997) Isolation of a common receptor for Coxsackie B viruses and adenoviruses 2 and 5. *Science* **275**: 1320–1323.

Berglund P, Sjöberg M, Garoff H, Atkins GJ, Sheahan BJ, Liljeström P. (1993) Semliki Forest virus expression system: Production of conditionally infectious recombinant particles. *Bio/Technology* **11**: 916–920.

Berglund P, Smerdou C, Fleeton MN, Tubulekas I, Liljeström P. (1998) Enhancing immune responses using suicidal DNA vaccines. *Nature Biotechnol.* **16**: 562–565.

Bischoff JR, Kirn DH, Williams A *et al.* (1996) An adenovirus mutant that replicates selectively in p53-deficient human tumor cells [see comments]. *Science* **274**: 373–376.

Blasco R, Moss B. (1995) Selection of recombinant vaccinia viruses on the basis of plaque formation. *Gene*; **158**(2): 157–162.

Blasey HD, Lundstrom K, Tate S, Bernard AR. (1997) Recombinant protein production using Semliki Forest virus expression system. *Cytotechnology* **24**: 65–72.

Blasey HD, Brethon B, Hovius R *et al.* (2000) Large scale transient serotonin 5-HT3 receptor production with the Semliki Forest virus expression system. *Cytotechnology* **32**: 199–208.

Boorsma M, Nieba L, Koller D, Bachmann MF, Bailey JE, Renner WA. (2000) A temperature-regulated replicon-based DNA expression system. *Nat. Biotechnol.* **18**: 429–432.

Both GW, Lockett LJ, Janardhana V *et al.* (1993) Protective immunity to rotavirus-induced diarrhoea is passively transferred to newborn mice from naive dams vaccinated with a single dose of a recombinant adenovirus expressing rotavirus VP7sc. *Virology* **193**: 940–950.

Boucher RC. (1999) Status of gene therapy for cystic fibrosis lung disease. *J. Clin. Invest.* **103**: 441–445.

Bowman L, Grossmann M, Rill D *et al.* (1998) IL-2 adenovector-transduced autologous tumor cells induce antitumor immune responses in patients with neuroblastoma. *Blood* **92**: 1941–1949.

Boyle DB, Coupar BE. (1988) Construction of recombinant fowlpox viruses as vectors for poultry vaccines. *Virus Res.* **10**(4): 343–356.

Breker-Klassen MM, Yoo D, Mittal SK, Sorden SD, Haines DM, Babiuk LA. (1995) Recombinant type 5 adenoviruses expressing bovine parainfluenza virus type 3 glycoproteins protect Sigmodon hispidus cotton rats from bovine parainfluenza virus type 3 infection. *J. Virol.* **69**: 4308–4315.

Britton P, Green P, Kottier S, Mawditt KL, Penzes Z, Cavanagh D, Skinner MA. (1996) Expression of bacteriophage T7 RNA polymerase in avian and mammalian cells by a recombinant fowlpox virus. *J. Gen. Virol.* **77** (Pt 5): 963–967.

Broder CC, Kennedy PE, Michaels F, Berger EA. (1994) Expression of foreign genes in cultured human primary macrophages using recombinant vaccinia virus vectors. *Gene* **142**: 167–174.

Bronte V, Carroll MW, Goletz TJ *et al.* (1997) Antigen expression by dendritic cells correlates with the therapeutic effectiveness of a model recombinant poxvirus tumor vaccine. *Proc. Natl Acad. Sci. USA* **94**: 3183–3188.

Bruce CB, Akrigg A, Sharpe SA, Hanke T, Wilkinson GW, Cranage MP. (1999) Replication-deficient recombinant adenoviruses expressing the human immunodeficiency virus Env antigen can induce both humoral and CTL immune responses in mice. *J. Gen. Virol.* **80**: 2621–2628.

Bruder JT, Kovesdi I. (1997) Adenovirus infection stimulates the Raf/MAPK signaling pathway and induces interleukin-8 expression. *J. Virol.* **71**: 398–404.

Bruna-Romero O, Lasarte JJ, Wilkinson G, Grace K, Clarke B, Borras-Cuesta F, Prieto J. (1997) Induction of cytotoxic T-cell response against hepatitis C virus structural antigens using a defective recombinant adenovirus. *Hepatology* **25**: 470–477.

Buge SL, Richardson E, Alipanah S *et al.* (1997) An adenovirus-simian immunodeficiency virus env vaccine elicits humoral, cellular, and mucosal immune responses in rhesus macaques and decreases viral burden following vaginal challenge. *J. Virol.* **71**: 8531–8541.

Buge SL, Murty L, Arora K *et al.* (1999) Factors associated with slow disease progression in macaques immunized with an adenovirus-simian immunodeficiency virus (SIV) envelope priming-gp120 boosting regimen and challenged vaginally with SIVmac251. *J. Virol.* **73**: 7430–7440.

Calaco A. (1999) Why are dendritic cells central to cancer immunotherapy. *Molecular Medicine Today* **January**, 14–17.

Caley IJ, Betts MR, Irlbeck DM, Davis NL, Swanstrom R, Frelinger JA, Johnston RE. (1997) Humoral, mucosal and cellular immunity in response to a human immunodeficiency virus type 1 immunogen expressed by a Venezuelan Equine Encephalitis virus vaccine vector. *J. Virol.* **71**: 3031–3038.

Callebaut P, Enjuanes L, Pensaert M. (1996) An adenovirus recombinant expressing the spike glycoprotein of porcine respiratory coronavirus is immunogenic in swine. *J. Gen. Virol.* **77**: 309–313.

Cao JX, Upton C. (1997) gpt-gus fusion gene for selection and marker in recombinant poxviruses. *Biotechniques* **22**: 276–278.

Carroll MW, Moss B. (1997) Host range and cytopathogenicity of the highly attenuated MVA strain of vaccinia virus: propagation and generation of recombinant viruses in a nonhuman mammalian cell line. *Virology* **238**: 198–211.

Carroll MW, Moss B. (1995) *E. coli* beta-glucuronidase (GUS) as a marker for recombinant vaccinia viruses. *Biotechniques* **19**: 352–354, 356.

Carroll MW, Restifo NP. (2000) Poxviruses as vectors for cancer immunotherapy. In *Cancer Vaccines and Immunotherapy* (Eds PL Stern, PLC Beverley, MW Carroll). Cambridge University Press, UK. In Press.

Carroll MW, Overwijk WW, Chamberlain RS, Rosenberg SA, Moss B, Restifo NP. (1997) Highly attenuated modified vaccinia virus Ankara (MVA) as an effective recombinant vector: a murine tumor model. *Vaccine* **15**: 387–394.

Carroll MW, Overwijk WW, Surman DR, Tsung K Moss B, Restifo NP. (1998) Construction and characterization of a triple-recombinant vaccinia virus encoding B7–1, interleukin 12, and a model tumor antigen. *J. Natl Cancer Inst.* **90:** 1881–1887.

Chakrabarti S, Brechling K, Moss B. (1985) Vaccinia virus expression vector: coexpression of beta-galactosidase provides visual screening of recombinant virus plaques. *Mol. Cell. Biol.* **5**: 3403–3409.

Chakrabarti S, Sisler JR, Moss B. (1997) Compact, synthetic, vaccinia virus early/late promoter for protein expression. *Biotechniques* **23**: 1094–1097.

Chanda PK, Natuk RJ, Mason BB *et al.* (1990) High level expression of the envelope glycoproteins of the human immunodeficiency virus type I in presence of rev gene using helper-independent adenovirus type 7 recombinants. *Virology* **175**: 535–547.

Chanock R, Ludwig W, Heubner RJ. (1966) Immunization by selective infection with type 4 adenovirus grown on human diploid tissue cultures: I. Safety and lack of oncogenicity and tests for potency in volunteers. *JAMA* **195**: 45–452.

Charlton KM, Artois M, Prevec L, Campbell JB, Casey GA, Wandeler AI, Armstrong J. (1992) Oral rabies vaccination of skunks and foxes with a recombinant human adenovirus vaccine. *Arch. Virol.* **123**: 169–179.

Chenciner N, Randrianarison-Jewtoukoff V, Delpeyroux F *et al.* (1997) Enhancement of humoral immunity to SIVenv following simultaneous inoculation of mice by three recombinant adenoviruses encoding SIVenv/poliovirus chimeras, Tat and Rev. *AIDS Res. Hum. Retroviruses* **13**: 801–806.

Cheng SM, Lee SG, Ronchetti-Blume M *et al.* (1992) Coexpression of the simian immunodeficiency virus Env and Rev proteins by a recombinant human adenovirus host range mutant. *J. Virol.* **66**: 6721–6727.

Chengalvala M, Lubeck MD, Davis AR, Mizutani S, Molnar-Kimber K, Morin J, Hung PP. (1991) Evaluation of adenovirus type 4 and type 7 recombinant hepatitis B vaccines in dogs. *Vaccine* **9**: 485–490.

Chengalvala MV, Bhat BM, Bhat R, Lubeck MD, Mizutani S, Davis AR, Hung PP. (1994) Immunogenicity of high expression adenovirus-hepatitis B virus recombinant vaccines in dogs. *J. Gen. Virol.* **75**: 125–131.

Chengalvala MV, Bhat BM, Bhat RA, Dheer SK, Lubeck MD, Purcell RH, Murthy KK. (1997) Replication and immunogenicity of Ad7-, Ad4-, and Ad5-hepatitis B virus surface antigen recombinants, with or without a portion of E3 region, in chimpanzees. *Vaccine* **15**: 335–339.

Chengalvala MV, Bhat RA, Bhat BM, Vernon SK, Lubeck MD. (1999) Enhanced immunogenicity of hepatitis B surface antigen by insertion of a helper T cell epitope from tetanus toxoid. *Vaccine* **17**: 1035–1041.

Chiocca S, Kurzbauer R, Schaffner G, Baker A, Mautner V, Cotten M. (1996) The complete DNA sequence and genomic organization of the avian adenovirus CELO. *J. Virol.* **70**: 2939–2949.

Connors M, Collins PL, Firestone CY *et al.* (1992) Cotton rats previously immunized with a chimeric RSV FG glycoprotein develop enhanced pulmonary pathology when infected with RSV, a phenomenon not encountered following immunization with vaccinia – RSV recombinants or RSV. *Vaccine* **10**: 475–484.

Cook DG, Sung J, Golde T *et al.* (1996) Expression and analysis of presenilin 1 in a human neuronal system: Localization in cell bodies and dendrites. *Proc. Natl Acad. Sci. USA* **93**: 9223–9228.

Cooney EL, Collier AC, Greenberg PD *et al.* (1991) Safety of and immunological response to a recombinant vaccinia virus vaccine expressing HIV envelope glycoprotein. *Lancet* **9**: 567–572.

Curiel DT, Agarwal S, Wagner E, Cotten M. (1991) Adenovirus enhancement of transferrin-polylysine-mediated gene delivery. *Proc. Natl Acad. Sci. USA* **88**: 8850–8854.

Davis NL, Willis LW, Smith JF, Johnston RE. (1989) *In vitro* synthesis of infectious Venezuelan equine encephalitis virus RNA from a cDNA clone: analysis of a viable deletion mutant. *Virology* **171**: 189–204.

Davis NL, Brown KW, Johnston RE. (1996) A viral vaccine vector that expresses foreign genes in lymph nodes and protects against mucosal challenge. *J. Virol.* **70**: 3781–3787.

Davison AJ, Moss B. (1989a) Structure of vaccinia virus early promoters. *J Mol. Biol.* **210**: 749–769.

Davison AJ, Moss B. (1989b) Structure of vaccinia virus late promoters. *J Mol. Biol.* **210**: 771–784.

de la Salle H, Altenburger W, Elkaim R *et al.* (1985) Active gamma-carboxylated human factor IX expressed using recombinant DNA techniques. *Nature* **316:** 268–270.

Dewar RL, Natarajan V, Vasudevachari MB, Salzman NP. (1989) Synthesis and processing of human immunodeficiency virus type 1 envelope proteins encoded by a recombinant human adenovirus. *J. Virol* **63**: 129–136.

DiGuilmi CM, Barge A, Kitts P, Gout E, Chorobocsek J. (1995) Human adenovirus serotype 3 (Ad3) and the Ad3 fiber protein bind to a 130-kDa membrane protein on HeLa cells. *Virus Research* **38**: 71–81.

Doronin K, Toth K, Kuppuswamy M, Ward P, Tollefson AE, Wold WS. (2000) Tumor-specific, replication-competent adenovirus vectors overexpressing the adenovirus death protein [In Process Citation]. *J. Virol.* **74**: 6147–6155.

Doucas V, Ishov AM, Romo A, Juguilon H, Weitzman MD, Evans RM, Maul GG. (1996) Adenovirus replication is coupled with the dynamic properties of the PML nuclear structure. *Genes Dev.* **10**: 196–207.

Drexler I, Heller K, Wahren B, Erfle V, Sutter G. (1998) Highly attenuated modified vaccinia virus Ankara replicates in baby hamster kidney cells, a potential host for virus propagation, but not in various human transformed and primary cells. *J. Gen. Virol.* **79**: 347–352.

Drexler I, Antunes E, Schmitz M *et al.* (1999) Modified vaccinia virus Ankara for delivery of human tyrosinase as melanoma-associated antigen: induction of tyrosinase- and melanoma-specific human leukocyte antigen A*0201-restricted cytotoxic T cells in vitro and in vivo.

Cancer Res.; **59:** 4955–4963.

Drillien R, Spehner D, Bohbot A, Hanau D. (2000) Vaccinia virus-related events and phenotypic changes after infection of dendritic cells derived from human monocytes. *Virology* **268**: 471–481.

Dubensky TWJr, Driver DA, Polo JM *et al.* (1996) Sindbis virus DNA-based expression vectors: Utility for *in vitro* and *in vivo* gene transfer. *J. Virol.* **70**: 508–519.

Earl P, Wyatt LS, Moss B, Carroll MW. (1998a) Generation of Vaccinia Virus Recombinant Viruses. *Current Protocols in Molecular Biology.* Supplement 43 Unit 16.17.1–16.17.19. John Wiley & Sons, Inc.

Earl P, Wyatt LS, Cooper N, Moss B, Carroll MW. (1998b) Preparations of Cell Cultures and Vaccinia Virus Stocks. *Current Protocols in Molecular Biology* Supplement 43 Unit 16.16.1–16.16.13. John Wiley & Sons, Inc.

Ehrengruber MU, Lundstrom K, Schweitzer C, Heuss C, Schlesinger S, Gähwiler BH. (1999) Recombinant Semliki Forest virus and Sindbis virus efficiently infect neurons in hippocampal slice cultures. *Proc. Natl Acad. Sci. USA* **96**: 7041–7046.

Elahi SM, Shen SH, Talbot BG, Massie B, Harpin S, Elazhary Y. (1999) Recombinant adenoviruses expressing the E2 protein of bovine viral diarrhea virus induce humoral and cellular immune responses. *FEMS Microbiol. Lett.* **177**: 159–166.

Eloit M, Gilardi-Hebenstreit P, Toma B, Perricaudet M. (1990) Construction of a defective adenovirus vector expressing the pseudorabies virus glycoprotein gp50 and its use as a live vaccine. *J. Gen. Virol.* **71**: 2425–2431.

Elroy-Stein O, Moss B. (1998) Generation of Vaccinia Virus Recombinant Viruses. *Current Protocols in Molecular Biology*. Supplement 43 Unit 16.19.1–16.19.11. John Wiley & Sons, Inc.

Elroy-Stein O, Fuerst TR, Moss B. (1989) Cap-independent translation of mRNA conferred by encephalomyocarditis virus 5' sequence improves the performance of the vaccinia virus/bacteriophage T7 hybrid expression system. *Proc. Natl Acad. Sci. USA* **86:** 6126–6130.

Emtage PC, Wan Y, Bramson JL, Graham FL, Gauldie J. (1998) A double recombinant adenovirus expressing the costimulatory molecule B7-1 (murine) and human IL-2 induces complete tumor regression in a murine breast adenocarcinoma model. *J. Immunol.* **160**: 2531–2538.

Endresz V, Berencsi K, Gonczol E. (1995) An adenovirus-herpes simplex virus glycoprotein B recombinant (Ad- HSV.gB) protects mice against a vaccinia HSV.gB and HSV challenge. *Acta Microbiol. Immunol. Hung.* **42**: 247–254.

Evans RJ, Lewis C, Virginio C, Lundstrom K, Buell G, Surprenant A, North RA. (1996) Ionic permeability of, and divalent cation effects on, two ATP-gated cation channels (P2X receptors) expressed in mammalian cells. *J. Physiol.* **497**: 413–422.

Falkner FG, Moss B. (1988) *Escherichia coli* gpt gene provides dominant selection for vaccinia virus open reading frame expression vectors. *J. Virol.* **62**: 1849–1854.

Falkner FG, Moss B. (1990) Transient dominant selection of recombinant vaccinia viruses. *J. Virol.* **64**: 3108–3111.

Feng Y, Broder CC, Kennedy PE, Berger EA. (1996) HIV-1 entry cofactor: functional cDNA cloning of a seven-transmembrane, G protein-coupled receptor. *Science* **272**: 872–877.

Fenner (1996) In: *Virology*. (Ed BN Fields, DM Knipe, PM Howley *et al.*) Chapter 84 pp. Lippincott-Raven Publishers, Philadelphia 1996.

Flanagan B, Pringle CR, Leppard KN. (1997) A recombinant human adenovirus expressing the simian immunodeficiency virus Gag antigen can induce long-lived immune responses in mice. *J. Gen. Virol.* **78**: 991–997.

Flexner C, Moss B. (1997) Vaccinia virus as a live vector for expression of immunogens. New Generation Vaccines 2nd edition. (Ed Levine IM) Marcel Inc, New York, pp 297–314.

Flexner C, Hugin A, Moss B. (1987) Prevention of vaccinia virus infection in immunodeficient mice by vector-directed IL-2 expression. *Nature* **330**: 259–262.

Fooks AR, Schadeck E, Liebert UG *et al.* (1995) High level expression of the measles nucleocapsid protein by using a replication-deficient adenovirus vector: induction of an MHC-1-restricted CTL response and protection in a murine model. *Virology* **210**:

Fooks AR, Jeevarajah D, Warnes A, Wilkinson GW, Clegg JC. (1996) Immunization of mice with plasmid DNA expressing the measles virus nucleoprotein gene. *Viral Immunol* **9**: 65–71.

Fooks AR, Jeevarajah D, Lee J *et al.* (1998) Oral or parenteral administration of replication-deficient adenoviruses expressing the measles virus haemagglutinin and fusion proteins: protective immune responses in rodents. *J. Gen. Virol.* **79**: 1027–1031.

Franke CA, Rice CM, Strauss JH, Hruby DE. (1985) Neomycin resistance as a dominant selectable marker for selection and isolation of vaccinia virus recombinants. *Mol. Cell. Biol.* **5:** 1918–1924.

Fuerst TR, Niles EG, Studier FW, Moss B. (1986) Eukaryotic transient-expression system based on recombinant vaccinia virus that synthesizes bacteriophage T7 RNA polymerase. *Proc. Natl Acad. Sci. USA* **83**: 8122–8126.

Fueyo J, Gomez-Manzano C, Alemany R *et al.* (2000) A mutant oncolytic adenovirus targeting the Rb pathway produces anti-glioma effect in vivo. *Oncogene* **19**: 2–12.

Gahery-Segard H, Molinier-Frenkel V, Le Boulaire C *et al.* (1997) Phase I trial of recombinant adenovirus gene transfer in lung cancer. Longitudinal study of the immune responses to transgene and viral products. *J. Clin. Invest.* **100**: 2218–2226.

Gallichan WS, Rosenthal KL. (1995) Specific secretory immune responses in the female genital tract following intranasal immunization with a recombinant adenovirus expressing glycoprotein B of herpes simplex virus. *Vaccine* **13**: 1589–1595.

Gallichan WS, Rosenthal KL. (1996) Long-lived cytotoxic T lymphocyte memory in mucosal tissues after mucosal but not systemic immunization. *J. Exp. Med.* **184**: 1879–1890.

Gallichan WS, Johnson DC, Graham FL, Rosenthal KL. (1993) Mucosal immunity and protection after intranasal immunization with recombinant adenovirus expressing herpes simplex virus glycoprotein B. *J. Infect. Dis.* **168**: 622–629.

Ganne V, Eloit M, Laval A, Adam M, Trouve G. (1994) Enhancement of the efficacy of a replication-defective adenovirus-vectored vaccine by the addition of oil adjuvants. *Vaccine* **12**: 1190–1196.

Ginsberg HS, Lundholm-Beauchamp U, Price G. (1987) Adenovirus as a model of disease. In *Molecular Basis of Virus Disease*, pp. 245–258. (Ed WC Russell, JW Almond). Cambridge: Univ. of Cambridge Press.

Gong J, Chen L, Chen D, Kashiwaba M, Manome Y, Tanaka T, Kufe D. (1997) Induction of antigen-specific antitumor immunity with adenovirus-transduced dendritic cells. *Gene Ther.* **4**: 1023–1028.

Gonin P, Fournier A, Oualikene W, Moraillon A, Eloit M. (1995) Immunization trial of cats with a replication-defective adenovirus type 5 expressing the ENV gene of feline immunodeficiency virus. *Vet. Microbiol.* **45**: 393–401.

Gonin P, Oualikene W, Fournier A, Eloit M. (1996) Comparison of the efficacy of replication-defective adenovirus and Nyvac poxvirus as vaccine vectors in mice. *Vaccine* **14**: 1083–1087.

Graham FL, Prevec, L. (1995) Methods for construction of adenovirus vectors. *Mol. Biotechnol.* **3**: 207–220.

Grunhaus A, Horwitz MS. (1992) Adenoviruses as cloning vectors. *Seminars in Virology* **3**: 237–252.

Guy B, Kieny MP, Riviere Y *et al.* (1987) HIV F/3' orf encodes a phosphorylated GTP-binding protein resembling an oncogene product. *Nature* **330**: 266–269.

Hammond JM, McCoy RJ, Jansen ES, Morrissy CJ, Hodgson AL, Johnson MA. (2000) Vaccination with a single dose of a recombinant porcine adenovirus expressing the classical swine fever virus gp55 (E2) gene protects pigs against classical swine fever. *Vaccine* **18**: 1040–1050.

Hardy PA, Mazzini MJ, Schweitzer C, Lundstrom K, Glode LM. (2000) Recombinant Semliki Forest virus infects and kills human prostate cancer cell lines and prostatic duct epithelial cells *ex vivo*. *Int. J. Molec. Med.* **5**: 241–245.

Hariharan MJ, Driver DA, Townsend K *et al.* (1998) DNA immunization against herpes simplex virus: Enhanced efficacy using Sindbis virus-based vector. *J. Virol.* **72**: 950–958.

He Z, Wlazlo AP, Kowalczyk DW, Cheng J, Xiang ZQ, Giles-Davis W, Ertl HC. (2000) Viral recombinant vaccines to the E6 and E7 antigens of HPV-16. *Virology* **270**: 146–161.

Heise C, Kirn DH. (2000) Replication-selective adenoviruses as oncolytic agents. *J. Clin. Invest.* **105**: 847–851.

Heise CC, Williams AM, Xue S, Propst M, Kirn DH. (1999) Intravenous administration of ONYX-015: a selectively replicating adenovirus, induces antitumoral efficacy. *Cancer Res.* **59**: 2623–2628.

Hendrix RM, Lindner JL, Benton FR, Monteith SC, Tuchscherer MA, Gray GC, Gaydos JC. (1999) Large, persistent epidemic of adenovirus type 4-associated acute respiratory disease in U.S. army trainees. *Emerg. Infect. Dis.* **5**: 798–801.

Hermiston T. (2000) Gene delivery from replication-selective viruses: arming guided missiles in the war against cancer. *J. Clin. Invest.* **105**: 1169–1172.

Herweijer H, Latendresse JS, Williams P, Zhang G, Danko I, Schlesinger S, Wolff JA. (1995) A plasmid-based self-amplifying Sindbis virus vector. *Hum. Gene Ther.* **6**: 1161–1167.

Hilleman MR, Werner JH. (1954) Recovery of new agent from patients with acute repiratory illness. *Proc. Soc. Exp. Biol. Med.* **85**: 183–188.

Hirsch VM, Fuerst TR, Sutter G *et al.* (1996) Patterns of viral replication correlate with outcome in simian immunodeficiency virus (SIV)-infected macaques: effect of prior immunization with a trivalent SIV vaccine in modified vaccinia virus Ankara. *J. Virol.* **70**: 3741–3752.

Hirschowitz EA, Crystal RG. (1999) Adenovirus-mediated expression of interleukin-12 induces natural killer cell activity and complements adenovirus-directed gp75 treatment of melanoma lung metastases. *Am. J. Respir. Cell. Mol .Biol.* **20**: 935–941.

Hirschowitz EA, Leonard S, Song W *et al.* (1998) Adenovirus-mediated expression of melanoma antigen gp75 as immunotherapy for metastatic melanoma. *Gene Ther.* **5**: 975–983.

Hodge JW, McLaughlin JP, Kantor JA, Schlom J. (1997) Diversified prime and boost protocols using recombinant vaccinia virus and recombinant non-replicating avian pox virus to enhance T-cell immunity and antitumor responses. *Vaccine* **15:**759–768.

Hofmann C, Loser P, Cichon G, Arnold W, Both GW, Strauss M. (1999) Ovine adenovirus vectors overcome preexisting humoral immunity against human adenoviruses in vivo. *J. Virol.* **73**: 6930–6936.

Holzer GW, Falkner FG. (1997) Construction of a vaccinia virus deficient in the essential DNA repair enzyme uracil DNA glycosylase by a complementing cell line. *J. Virol.* **71**: 4997–5002.

Holzer GW, Remp G, Antoine G *et al.* (1999) Highly efficient induction of protective immunity by a vaccinia virus vector defective in late gene expression. *J. Virol.* **73**: 4536–4542.

Horwitz MS. (1996) Adenoviruses. In *Fields Virology.*, (Eds BN. Fields, Knipe DM, Howley PM) pp. 2149–2171. Philidelphia: Lippincott-Raven.

Hovius R, Tairi A-P, Blasey H, Bernard A, Lundstrom K, Vogel H. (1998) Characterization of a mouse serotonin 5-HT3 receptor purified from mammalian cells. *J. Neurochem.* **70**: 824–834.

Hsu KH, Lubeck MD, Davis AR *et al.* (1992) Immunogenicity of recombinant adenovirus-respiratory syncytial virus vaccines with adenovirus types 4, 5, and 7 vectors in dogs and a chimpanzee. *J. Infect. Dis.* **166**: 769–775.

Hsu KH, Lubeck MD, Bhat BM *et al.* (1994) Efficacy of adenovirus-vectored respiratory syncytial virus vaccines in a new ferret model. *Vaccine* **12**: 607–612.

Hu L, Esposito JJ, Scott FW. (1996) Raccoon poxvirus feline panleukopenia virus VP2 recombinant protects cats against FPV challenge. *Virology* **218**: 248–252.

Hu SL, Klaniecki J, Dykers T, Sridhar P, Travis BM. (1991) Neutralizing antibodies against HIV-1 BRU and SF2 isolates generated in mice immunized with recombinant vaccinia virus expressing HIV-1 (BRU) envelope glycoproteins and boosted with homologous gp160. *AIDS Res. Hum. Retroviruses* **7**: 615–620.

Huang MR, Olsson M, Kallin A, Pettersson U, Totterman TH. (1997) Efficient adenovirus-mediated gene transduction of normal and leukemic hematopoietic cells. *Gene Ther.* **4**: 1093–1099.

Huckriede A, Heikema A, Wilschut J, Agsteribbe A. (1996) Transient expression of a mitochondrial precursor protein: A new approach to study mitochondrial protein import in cells of higher eukaryotes. *Eur. J. Biochem.* **237**: 288–294.

Hung PP, Morin JE, Lubeck MD *et al.* (1988) Expression of HBV surface antigen or HIV envelope protein using recombinant adenovirus vectors. *Nat. Immun. Cell Growth Regul.* **7**: 135–143.

Hung PP, Chanda PK, Natuk RJ *et al.* (1990) Adenovirus vaccine strains genetically engineered to express HIV-1 or HBV antigens for use as live recombinant vaccines. *Nat. Immun. Cell Growth Regul.* **9**: 160–164.

Irvine KR *et al.* (1997). Enhancing efficacy of recombinant anticancer vaccines with prime/boost regimens that use two different vectors. *J. Natl Cancer Inst.* **89:** 1595–1601.

Isaacs SN, Kotwal GJ, Moss B. (1990) Reverse guanine phosphoribosyltransferase selection of recombinant vaccinia viruses. *Virology* **178**: 626–630.

Jackson RJ, Hall DF, Kerr PJ. (1996) Construction of recombinant myxoma viruses expressing foreign genes from different intergenic sites without associated attenuation. *J. Gen. Virol.* **77**: 1569–1575.

Jacobs SC, Stephenson JR, Wilkinson, GWG. (1992) High-level expression of the tick-borne encephalitis virus NS1 protein by using an adenovirus-based vector: protection elicited in a murine model. *Journal of Virology* **66**: 2086–2095.

Jacobs SC, Stephenson JR, Wilkinson GWG. (1994) Protection elicited by a replication-defective adenovirus vector expressing the tick-borne encephalitis virus non-structural glycoprotein NS1. *J. Gen. Virol.* **75**: 2399–2402.

Johanning FW, Conry RM, LoBuglio AF, Wright M, Sumerel LA, Pike MJ, Curiel DT. (1995) A Sindbis virus mRNA polynucleotide vector achieves prolonged and high level heterologous gene expression *in vivo. Nucleic Acid Res.* **23**: 1495–1501.

Juillard V, Villefroy P, Godfrin D, Pavirani A, Venet A, Guillet JG. (1995) Long-term humoral and cellular immunity induced by a single immunization with replication-defective adenovirus recombinant vector. *Eur. J. Immunol.* **25**: 3467–3473.

Kalicharran KK, Springthorpe VS, Sattar SA. (1992) Studies on the stability of a human adenovirus-rabies recombinant vaccine. *Can. J. Vet. Res.* **56**: 28–33.

Kaplan JM, Yu Q, Piraino ST, Pennington SE, Shankara S, Woodworth LA, Roberts BL. (1999) Induction of antitumor immunity with dendritic cells transduced with adenovirus vector-encoding endogenous tumor-associated antigens. *J. Immunol.* **163**: 699–707.

Kauffman H, Schlom J. (2000) Vaccines for Colon Cancer. In: *Cancer Vaccines and Immunotherapy*, (Eds PL Stern, PLC Beverley, MW Carroll), Cambridge University Press UK. In Press.

Kikuchi T, Crystal RG. (1999) Anti-tumor immunity induced by in vivo adenovirus vector-mediated expression of CD40 ligand in tumor cells. *Hum. Gene Ther.* **10**: 1375–1387.

Kirn D. (2000) A phase II trial of intratumoral injection with a selectively replicating adenovirus (ONYX-015) in patients with recurrent, refractory squamous cell carcinoma of the head and neck. In: *Gene Therapy of Cancer: Methods and Protocols*, pp. 559–574. (Eds W Walther, U Stein) Totowa, New Jersey, Humana Press.

Kochneva GV, Urmanov IH, Ryabchikova EI, Streltsov VV, Serpinsky OI. (1994) Fine mechanisms of ectromelia virus thymidine kinase-negative mutants avirulence. *Virus Res.* **34**: 49–61.

Kovacs GR, Moss B. (1996) The vaccinia virus H5R gene encodes late gene transcription factor 4: purification, cloning, and overexpression. *J. Virol.* 6796–6802.

Kremer EJ, Boutin S, Chillon M, Danos O. (2000) Canine adenovirus vectors: an alternative for adenovirus-mediated gene transfer. *J. Virol.* **74**: 505–512.

Kurilla MG. (1997) Transient selection during vaccinia virus recombination with insertion vectors without selectable markers. *Biotechniques* **22**: 906–910.

Lane JM, Ruben FL, Neff JM, Millar JD. (1969) Complications of smallpox vaccination, 1968. *N. Engl. J. Med.* **281**: 1201–1208.

Langridge WH. (1983) Detection of Amsacta moorei entomopoxvirus and vaccinia virus proteins in cell cultures restrictive for poxvirus multiplication. *J. Invertebr. Pathol.* **42**: 77–82.

Lasarte JJ, Corrales FJ, Casares N *et al.* (1999) Different doses of adenoviral vector expressing IL-12 enhance or depress the immune response to a coadministered antigen: the role of nitric oxide. *J. Immunol.* **162**: 5270–5277.

Leong KH, Ramsay AJ, Ramshaw IA, Morin MJ, Robinson HL, Boyle DB. (1995) Generation of enhanced immune responses by consecutive immunisation with DNA and recombinant fowl pox vectors, Vaccines 9**5**: pp 327–33**1**: Cold Spring Harbor Press.

Le Potier MF, Monteil M, Houdayer C, Eloit M. (1997) Study of the delivery of the gD gene of pseudorabies virus to one-day-old piglets by adenovirus or plasmid DNA as ways to by-pass the inhibition of immune response by colostral antibodies. *Vet. Microbiol.* **55**: 75–80.

Leppard K. (1997) E4 function in adenovirus, adenovirus vector and adeno-associated virus infections. *J. Gen. Virol.* **78**: 2131–2138.

Levrero M, Barban V, Manteca S *et al.* (1991) Defective and nondefective adenovirus vectors for expressing foreign genes in vitro and in vivo. *Gene* **101**: 195–202.

Li K-J, Garoff H. (1996) Production of infectious recombinant Moloney murine leukemia virus particles in BHK cells using Semliki Forest virus-derived RNA expression vectors. *Proc. Natl. Acad. Sci. USA* **93**: 11658–11663.

Li K-J, Garoff H. (1998) Packaging of intron-containing genes into retrovirus vectors by alphavirus vectors. *Proc. Natl. Acad. Sci. USA* **95**: 3650–3654.

Li Y, Hall RL, Moyer RW. (1997) Transient, nonlethal expression of genes in vertebrate cells by recombinant entomopoxviruses. *J. Virol.* **71**: 9557–9562.

Liao HJ, Kobayashi R, Mathews MB. (1998) Activities of adenovirus virus-associated RNAs: purification and characterization of RNA binding proteins. *Proc. Natl Acad. Sci. USA* **95**: 8514–8519.

Liljeström P, Garoff H. (1991) A new generation of animal cell expression vectors based on the Semliki Forest virus replicon. *Bio/Technology* **9**: 1356–1361.

Linette GP, Shankara S, Longerich S *et al.* (2000) In vitro priming with adenovirus/gp100 antigen-transduced dendritic cells reveals the epitope specificity of HLA-A*0201-restricted CD8+ T cells in patients with melanoma. *J. Immunol.* **164**: 3402–3412.

Liu R, Baillie J, Sissons JG, Sinclair JH. (1994) The transcription factor YY1 binds to negative regulatory elements in the human cytomegalovirus major immediate early enhancer/promoter and mediates repression in non-permissive cells. *Nucleic Acids Res.* **22**: 2453–2459.

Lubeck MD, Davis AR, Chengalvala M *et al.* (1989) Immunogenicity and efficacy testing in chimpanzees of an oral hepatitis B vaccine based on live recombinant adenovirus. *Proc. Natl Acad. Sci. USA* **86**: 6763–6767.

Lubeck MD, Natuk RJ, Chengalvala M et al. (1994) Immunogenicity of recombinant adenovirus-human immunodeficiency virus vaccines in chimpanzees following intranasal administration. *AIDS Res. Hum. Retroviruses* **10**: 1443–1449.

Lubeck MD, Natuk R, Myagkikh M *et al.* (1997) Long-term protection of chimpanzees against high-dose HIV-1 challenge induced by immunization. *Nat. Med.* **3**: 651–658.

Lundstrom K. (1997) Alphaviruses as expression vectors. *Curr. Op. Biotech.* **8**: 578–582.

Lundstrom K. (1999) Alphaviruses as tools in neurobiology and gene therapy. *J. Receptor & Signal Transduction Res.* **19**: 673–686.

Lundstrom K, Mills A, Buell G, Allet E, Adami N, Liljeström P. (1994) High-level expression of the human neurokinin-1 receptor in mammalian cell lines using the Semliki Forest virus expression system. *Eur. J. Biochem.* **224**: 917–921.

Lundstrom K, Michel A, Blasey H, Bernard AR, Hovius R, Vogel H, Surprenant A. (1997a) Expression of ligand-gated ion channels with the Semliki Forest virus expression system. *J. Receptor & Signal Transduction Res.* **17**: 115–126.

Lundstrom K, Pralong W, Martinou JC. (1997b) Anti-apoptotic effect of bcl-2 overexpression in RIN cells infected with Semliki Forest virus. *Apoptosis* **2**: 189–191.

Lundstrom K, Richards JG, Pink JR, Jenck F. (1999a) Efficient in vivo expression of a reporter gene in rat brain after injection of recombinant replication-deficient Semliki Forest virus. *Gene Ther. Mol. Biol.* **3**: 15–23.

Lundstrom K, Schweitzer C, Richards JG, Ehrengruber MU, Jenck F, Mülhardt C. (1999b) Semliki Forest virus vectors for *in vitro* and *in vivo* applications. *Gene Ther. Mol. Biol.* **4**: 23–31.

Mackett M. (1994) Vaccinia virus recombinants: Expression vectors and potential vaccines. *Animal Cell Biotechnology* **6**: 315–371.

Mackett M, Smith GL, Moss B. (1982) Vaccinia virus: a selectable eukaryotic cloning and expression vector. *Proc. Natl Acad. Sci. USA* **79**: 7415–7419.

Mackett M, Yilma T, Rose JK, Moss B. (1985) Vaccinia virus recombinants: expression of VSV genes and protective immunization of mice and cattle. *Science* **227**: 433–435.

Mahnel H, Mayr A. (1994) Experiences with immunization against orthopox viruses of humans and animals using vaccine strain MVA. *Berl. Munch. Tierarztl. Wochenschr.* ;**107**: 253–256.

Maletic-Savatic M, Malinow R, Svoboda K. (1999) Rapid dendritic morphogenesis in CA1 hippocampal dendrites induced by synaptic activity. *Science* **283**: 1923–1927.

Marshall GS, Ricciardi RP, Rando RF, Puck J, Ge RW, Plotkin SA, Gonczol E. (1990) An adenovirus recombinant that expresses the human cytomegalovirus major envelope glycoprotein and induces neutralizing antibodies. *J. Infect. Dis.* **162**: 1177–1181.

Mason BB, Davis AR, Bhat BM *et al.* (1990) Adenovirus vaccine vectors expressing hepatitis B surface antigen: importance of regulatory elements in the adenovirus major late intron. *Virology* **177**: 452–461.

Mathias P, Wickham T, Moore M, Nemerow G. (1994) Multiple adenovirus serotypes use alpha v integrins for infection. *J. Virol.* **68**: 6811–6814.

Mayr A, Stickl H, Muller HK, Danner K, Singer H. (1978) The smallpox vaccination strain MVA:

marker, genetic structure, experience gained with the parenteral vaccination and behavior in organisms with a debilitated defence mechanism. *Zentralbl. Bakteriol. [B]* **167**: 375–390.

Mayr GA, Chinsangaram J, Grubman MJ. (1999) Development of replication-defective adenovirus serotype 5 containing the capsid and 3C protease coding regions of foot-and-mouth disease virus as a vaccine candidate. *Virology* **263**: 496–506.

McCraith S, Holtzman T, Moss B, Fields S. (2000) Genome-wide analysis of vaccinia virus protein–protein interactions. *Proc. Natl. Acad. Sci. USA* **97**: 4879–4884.

McDermott MR, Graham FL, Hanke T, Johnson DC. (1989) Protection of mice against lethal challenge with herpes simplex virus by vaccination with an adenovirus vector expressing HSV glycoprotein B. *Virology* **169**: 244–247.

Melancon P, Garoff H. (1987) Processing of the Semliki Forest virus structural polyprotein: role of the capsid protease. *J. Virol.* **61**: 1301–1309.

Merchlinsky M, Moss B. (1992) Introduction of foreign DNA into the vaccinia virus genome by in vitro ligation: recombination-independent selectable cloning vectors. *Virology* **190**: 522–526.

Merchlinsky M, Eckert D, Smith E, Zauderer M. (1997) Construction and characterization of vaccinia direct ligation vectors. *Virology* **238**: 444–451.

Meyer H, Sutter G, Mayr A. (1991) Mapping of deletions in the genome of the highly attenuated vaccinia virus MVA and their influence on virulence. *J. Gen. Virol.* **72**: 1031–1038.

Michou AI, Lehrmann H, Saltik M, Cotten M. (1999) Mutational analysis of the avian adenovirus CELO, which provides a basis for gene delivery vectors. *J. Virol.* **73**: 1399–1410.

Mittal SK, Papp Z, Tikoo SK, Baca-Estrada ME, Yoo D, Benko M, Babiuk LA. (1996) Induction of systemic and mucosal immune responses in cotton rats immunized with human adenovirus type 5 recombinants expressing the full and truncated forms of bovine herpesvirus type 1 glycoprotein gD. *Virology* **222**: 299–309.

Monastyrkaia K, Goepfert F, Hochstrasser R, Acuna G, Leighton J, Pink JR, Lundstrom K. (1999) Expression and intracellular localization of odorant receptors in mammalian cell lines using Semliki Forest virus vectors. *J. Receptor & Signal Transduction Res.* **19**: 687–702.

Monteil M, Le Potier MF, Cariolet R, Houdayer C, Eloit M. (1997) Effective priming of neonates born to immune dams against the immunogenic pseudorabies virus glycoprotein gD by replication-incompetent adenovirus-mediated gene transfer at birth. *J. Gen. Virol.* **78**: 3303–3310.

Monteil M, Le Pottier MF, Ristov AA, Cariolet R, L'Hospitalier R, Klonjkowski B, Eloit M. (2000) Single inoculation of replication-defective adenovirus-vectored vaccines at birth in piglets with maternal antibodies induces high level of antibodies and protection against pseudorabies. *Vaccine* **18**: 1738–1742.

Morgan SM, Wilkinson GW, Floettmann E, Blake N, Rickinson AB. (1996) A recombinant adenovirus expressing an Epstein-Barr virus (EBV) target antigen can selectively reactivate rare components of EBV cytotoxic T- lymphocyte memory in vitro. *J. Virol.* **70**: 2394–2402.

Morin JE, Lubeck MD, Barton JE, Conley AJ, Davis AR, Hung PP. (1987) Recombinant adenovirus induces antibody response to hepatitis B virus surface antigen in hamsters. *Proc. Natl. Acad. Sci. USA* **84**: 4626–4630.

Morral N, O'Neal W, Zhou H, Langston C, Beaudet A. (1997) Immune response to reporter proteins and high viral dose limit duration of expression with adenoviral vectors: comparison of E2a wild type and E2a deleted vectors. *Human Gene Therapy* **8**: 1275–1286.

Morrison MD, Onions DE, Nicolson L. (1997) Complete DNA sequence of canine adenovirus type 1. *J. Gen. Virol.* **78**: 873–878.

Moss B. (1968) Inhibition of HeLa cell protein synthesis by the vaccinia virion. *J. Virol.* **2**: 1028–1037.

Moss B. (1996a) Poxviridae: The viruses and their replication. In: *Virology* (eds BN Fields, DM Knipe, PM Howley, *et al.*) Chapter 83, pp 2637–2671. Lippincott-Raven Publishers, Philadelphia.

Moss B. (1996b) Genetically engineered poxviruses for recombinant gene expression, vaccination, and safety. *Proc. Natl. Acad. Sci. USA* **93**: 11341–11348.

Mossmann SP, Bex F, Berglund P *et al.* (1996) Protection against lethal simian immunodeficiency virus SIVsmmPBj14 disease by a recombinant Semliki Forest virus gp160 vaccine and by gp120 subunit vaccine. *J. Virol.* **70**: 1953–1960.

Motoi F, Sunamura M, Ding L *et al.* (2000) Effective gene therapy for pancreatic cancer by cytokines mediated by restricted replication-competent adenovirus. *Hum. Gene Ther.* **11**: 223–235.

Mueller R, Muldoon R, Jackson G. (1969) Communicability of live adenovirus type 4 in families. *J. Infect. Dis.* **119**: 60–66.

Natuk RJ, Chanda PK, Lubeck MD *et al.* (1992) Adenovirus-human immunodeficiency virus (HIV) envelope recombinant vaccines elicit high-morphogenesis in CA1 hippocampal dendrites induced by synaptic activity. *Science* **283**: 1923–1927.

Natuk RJ, Lubeck MD, Chanda PK *et al.* (1993) Immunogenicity of recombinant human adenovirus-human immunodeficiency virus vaccines in chimpanzees. *AIDS Res. Hum. Retroviruses* **9**: 395–404.

Nemerow GR, Stewart PL. (1999) Role of alpha(v) integrins in adenovirus cell entry and gene delivery. *Microbiol. Mol. Biol. Rev.* **63**: 725–734.

Ohno K, Sawai K, Iijima Y, Levin B, Meruelo D. (1997) Cell-specific targeting of Sindbis virus vectors displaying IgG-binding domains of protein A. *Nature Biotechnol.* **15**: 763–767.

Olkkonen VM, Liljestrom P, Garoff H, Simons K, Dotti CG. (1993) Expression of heterologous proteins in cultured rat hippocampal neurons using the Semliki Forest virus vector. *J. Neurosci. Res.* **35**: 445–451.

Olson KE, Higgs S, Hahn CS, Rice CM, Carlson JO, Beaty BJ. (1994) The expression of chloramphenicol acetyltransferase in *Aedes albopictus* (C6/36) cells and *Aedes triseratus* mosquitoes using a double subgenomic recombinant Sindbis virus. *Insect Biochem. Mol. Biol.* **24**: 39–48.

Ourmanov I, Brown CR, Moss B *et al.* (2000) Comparative efficacy of recombinant modified vaccinia virus Ankara expressing simian immunodeficiency virus (SIV) Gag-Pol and/or Env in macaques challenged with pathogenic SIV. *J. Virol.* **74**: 2740–2751.

Panicali D, Paoletti E. (1982) Construction of poxviruses as cloning vectors: insertion of the thymidine kinase gene from herpes simplex virus into the DNA of infectious vaccinia virus. *Proc. Natl. Acad. Sci. USA* **79**: 4927–4931.

Paolazzi CC, Perez O, De Filippo J. (1999) Rabies vaccine. Developments employing molecular biology methods. *Mol. Biotechnol.* **11**: 137–147.

Paoletti E. (1996) Applications of pox virus vectors to vaccination: an update. *Proc. Natl. Acad. Sci. USA* **93**(21): 11349–11353.

Papp Z, Middleton DM, Mittal SK, Babiuk LA, Baca-Estrada ME. (1997) Mucosal immunization with recombinant adenoviruses: induction of immunity and protection of cotton rats against respiratory bovine herpesvirus type 1 infection. *J. Gen. Virol.* **78**: 2933–2943.

Papp Z, Babiuk LA, Baca-Estrada ME. (1999) The effect of pre-existing adenovirus-specific immunity on immune responses induced by recombinant adenovirus expressing glycoprotein D of bovine herpesvirus type 1. *Vaccine* **17**: 933–943.

Pastoret PP, Brochier B. (1999) Epidemiology and control of fox rabies in Europe. *Vaccine* **26**: 1750–1754.

Paul NL, Marsh M, McKeating J, Schulz T, Liljeström P, Garoff H, Weiss RA. (1993) Expression of HIV-1 envelope glycoproteins by Semliki Forest virus vectors. *AIDS Res. Hum. Retroviruses* **9**: 963–970.

Perkus ME, Limbach K, Paoletti E. (1989) Cloning and expression of foreign genes in vaccinia virus, using a host range selection system. *J. Virol.* **63**: 3829–3836.

Pfleiderer M, Falkner FG, Dorner F. (1995) A novel vaccinia virus expression system allowing construction of recombinants without the need for selection markers, plasmids and bacterial hosts. *J. Gen. Virol.* **76**: 2957–2962.

Polo JM, Belli BA, Driver DA *et al.* (1991) Analysis of Sindbis virus promoter recognition *in vivo*, using novel vectors with two subgenomic mRNA promoters. *J. Virol.* **65**: 2501–2510.

Powell M, Wilkinson GWG. (2000) Adenovirus Cancer Gene Therapy. In: *Viruses, cell transformation and cancer.*, (ed R. Grand) pp. 479–521. Amsterdam, Elsevier.

Precious BaR WC. (1985) Growth purification and titration of adenoviruses. In: *Virology: apractical approach,* pp. 193–205. (ed BWJ Mahy). Oxford: IRL Press.

Prevec L, Schneider M, Rosenthal KL, Belbeck LW, Derbyshire JB, Graham FL. (1989) Use of human adenovirus-based vectors for antigen expression in animals [published erratum appears in J Gen Virol 1989 Sep;70(Pt 9):2539]. *J. Gen. Virol.* **70**: 429–434.

Prevec L, Campbell JB, Christie BS, Belbeck L, Graham FL. (1990) A recombinant human adenovirus vaccine against rabies. *J. Infect. Dis.* **161**: 27–30.

Prevec L, Christie BS, Laurie KE, Bailey MM, Graham FL, Rosenthal KL. (1991) Immune response to HIV-1 gag antigens induced by recombinant adenovirus vectors in mice and rhesus macaque monkeys. *J. Acquir. Immune Defic. Syndr.* **4**: 568–576.

Putzer BM, Hitt M, Muller WJ, Emtage P, Gauldie J, Graham FL. (1997) Interleukin 12 and B7-1 costimulatory molecule expressed by an adenovirus vector act synergistically to facilitate tumor regression. *Proc. Natl. Acad. Sci. USA* **94**: 10889–10894.

Ragot T, Finerty S, Watkins PE, Perricaudet M, Morgan AJ. (1993) Replication-defective recombinant adenovirus expressing the Epstein-Barr virus (EBV) envelope glycoprotein gp340/220 induces protective immunity against EBV-induced lymphomas in the cottontop tamarin. *J. Gen. Virol.* **74**: 501–507.

Ramshaw IA, Andrew ME, Phillips SM, Boyle DB, Coupar BE. (1987) Recovery of immunodeficient mice from a vaccinia virus/IL-2 recombinant infection. *Nature* **329:** 545–546.

Rapp F, Melnick JL, Butel JS, Kitahara T. (1964) The incorporation of SV40 genetic material into adenovirus 7 as measured by intranuclear synthesis of SV40 tumor antigen. *Proc. Natl Acad. Sci. USA* **52**: 1348–1352.

Rea D, Schagen FH, Hoeben RC *et al.* (1999) Adenoviruses activate human dendritic cells without polarization toward a T-helper type 1-inducing subset. *J. Virol.* **73**: 10245–10253.

Reddy JR, Kwang J, Varthakavi V, Lechtenberg KF, Minocha HC. (1999) Semliki Forest virus vector carrying the bovine viral diarrhea virus NS3 (p80) cDNA induced immune responses in mice and expressed BVDV protein in mammalian cells. *Comp. Immunol. Microbiol. Infect. Dis.* **22**: 231–246.

Reddy PS, Idamakanti N, Babiuk LA, Mehtali M, Tikoo SK. (1999a) Porcine adenovirus-3 as a helper-dependent expression vector. *J. Gen. Virol.* **80**: 2909–2916.

Reddy PS, Idamakanti N, Chen Y, Whale T, Babiuk LA, Mehtali M, Tikoo SK. (1999b) Replication-defective bovine adenovirus type 3 as an expression vector. *J. Virol.* **73**: 9137–9144.

Reddy PS, Idamakanti N, Song JY *et al.* (1998a) Nucleotide sequence and transcription map of porcine adenovirus type 3. *Virology* **251**: 414–426.

Reddy PS, Idamakanti N, Zakhartchouk AN *et al.* (1998b) Nucleotide sequence, genome organization, and transcription map of bovine adenovirus type 3. *J. Virol.* **72**: 1394–1402.

Ribas A, Butterfield LH, McBride WH *et al.* (1999) Characterization of antitumor immunization to a defined melanoma antigen using genetically engineered murine dendritic cells. *Cancer Gene Ther.* **6**: 523–536.

Robert-Guroff M, Kaur H, Patterson LJ *et al.* (1998) Vaccine protection against a heterologous, non-syncytium-inducing, primary human immunodeficiency virus. *J. Virol.* **72**: 10275–10280.

Rodriguez D, Zhou YW, Rodriguez JR, Durbin RK, Jimenez V, McAllister WT, Esteban M. (1990) Regulated expression of nuclear genes by T3 RNA polymerase and lac repressor, using recombinant vaccinia virus vectors. *J. Virol.* **64**: 4851–4857.

Rodriguez R, Schuur ER, Lim HY, Henderson GA, Simons JW, Henderson DR. (1997) Prostate attenuated replication competent adenovirus (ARCA) CN706: a selective cytotoxic for prostate-specific antigen-positive prostate cancer cells. *Cancer Res.* **57**: 2559–2563.

Rogulski KR, Freytag SO, Zhang K *et al.* (2000) In vivo antitumor activity of ONYX-015 is influenced by p53 status and is augmented by radiotherapy [In Process Citation]. *Cancer Res.* **60**: 1193–1196.

Romero CH, Barrett T, Chamberlain RW, Kitching RP, Fleming M, Black DN. (1994) Recombinant capripoxvirus expressing the hemagglutinin protein gene of rinderpest virus: protection of cattle against rinderpest and lumpy skin disease viruses. *Virology* **204**: 425–429.

Rosenberg SA. (1999) A new era for cancer immunotherapy based on the genes that encode cancer antigens. *Immunity* **10**: 281–287.

Rosenberg SA, Zhai Y, Yang JC *et al.* (1998) Immunizing patients with metastatic melanoma using recombinant adenoviruses encoding MART-1 or gp100 melanoma antigens. *J. Natl Cancer Inst.* **90**: 1894–1900.

Rowe WP, Hueber JW, Gillmore LK, Parrott RH, Ward TG. (1953) Isolation of a cytopathogenic agent from human adenoids undergoing spontaneous degeneration in tissue culture. *Proc. Soc. Exp. Biol. Med.* **84**: 570–573.

Rubin BA, Rocke LB. (1994) Adenovirus vaccines. In: *Vaccines*, 2nd edn. (ed SA Plotkin, EA Mortimer). Philadelphia: WB Saunders.

Russell WC. (1998) Adenoviruses. In: *Topley & Wilson's Microbiology and Microbial Infections*. (Ed W Topley, G Wilson, L Collier) Arnold.

Sanz-Parra A, Vazquez B, Sobrino F, Cox SJ, Ley V, Salt JS. (1999) Evidence of partial protection against foot-and-mouth disease in cattle immunized with a recombinant adenovirus vector expressing the precursor polypeptide (P1) of foot-and-mouth disease virus capsid proteins. *J. Gen. Virol.* **80**: 671–679.

Sawai K, Meruelo D. (1998) Cell-specific transfection of choriocarcinoma cells by using Sindbis virus hCG expressing chimeric vector. *Biochem. Biophys. Res. Comm.* **248**: 315–323.

Sawai K, Ohno K, Iijima Y, Levin B, Meruelo D. (1998) A novel method of cell-specific mRNA transfection. *Mol. Gen. & Metabol.* **64**: 44–51.

Scarpini C, Arthur J, Efstathiou S, McGrath Y, Wilkinson G. (1999) Herpes simplex and adenovirus vectors. In: *DNA viruses – a practical approach*, (ed AJ Cann). Oxford: Oxford University Press.

Scheer A, Björklöf K, Cotecchia S, Lundstrom K. (1999) Expression of the α1b-adrenergic receptor and G protein subunits in mammalian cell lines using the Semliki Forest virus expression system. *J. Receptor & Signal Transduction Res.* **19**: 369–378.

Scheiflinger F, Dorner F, Falkner FG. (1992) Construction of chimeric vaccinia viruses by molecular cloning and packaging. *Proc. Natl Acad. Sci. USA* **89**: 9977–9981.

Scheiflinger F, Dorner F, Falkner FG. (1998) Transient marker stabilisation: a general procedure to construct marker-free recombinant vaccinia virus. *Arch. Virol.* **143**: 467–474.

Schlaeger E-J, Lundstrom K. (1998) Effect of temperature on recombinant protein expression in Semliki Forest virus infected mammalian cell lines growing in serum-free suspension cultures. *Cytotechnology* **28**: 205–211.

Schneider J, Gilbert SC, Blanchard TJ *et al.* (1998) Enhanced immunogenicity for CD8+ T cell induction and complete protective efficacy of malaria DNA vaccination by boosting with modified vaccinia virus Ankara. *Nat. Med.* **4**: 397–402.

Schnell MJ, Mebatsion T, Conzelmann KK. (1994) Infectious rabies viruses from cloned cDNA. *EMBO J.* **13**: 4195–4203.

Scholl SM, Balloul J-M, Goc GL *et al.* (2000) A recombinant vaccinia virus encoding MUC1 and IL2 as immunotherapy in breast cancer patients. *J. Immunol. Immunother.* in press.

Schreiber S, Kampgen E, Wagner E *et al.* (1999) Immunotherapy of metastatic malignant melanoma by a vaccine consisting of autologous interleukin 2-transfected cancer cells: outcome of a phase I study. *Hum. Gene Ther.* **10**: 983–993.

Schwartz AR, Togo Y, Hornick RB. (1974) Clinical evaluation of live, oral types 1, 2 and 5 vaccines. *Am. Rev. Resp. Dis.* **109**: 233–239.

Segerman A, Mei YF, Wadell G. (2000) Adenovirus types 11p and 35p show high binding efficiencies for committed hematopoietic cell lines and are infective to these cell lines. *J. Virol.* **74**: 1457–1467.

Seong YR, Lee CH, Im DS. (1998) Characterization of the structural proteins of hepatitis C virus expressed by an adenovirus recombinant. *Virus Res.* **55**: 177–185.

Seth A, Ourmanov I, Schmitz JE *et al.* (2000) Immunization with a modified vaccinia virus expressing simian immunodeficiency virus (SIV) Gag-Pol primes for an anamnestic Gag-specific cytotoxic T-lymphocyte response and is associated with reduction of viremia after SIV challenge. *J. Virol.* **74**: 2502–2509.

Sharp IR, Wadell G. (1995) Adenoviruses. In: *Principles and Practice of Clinical Virology*, Third edn, pp. 287–308, (ed AJ Zuckerman, JE Banatvala, JR Pattison), John Wiley & Sons Ltd, London.

Shenk T. (1996) Adenoviridae: The viruses and their replication. In: *Fields Virology*, pp. 2111–2148, (eds BN Fields, Knipe DM, Howley PM), Philadelphia: Lippincott-Raven.

Shi S-H, Hayashi Y, Petralia RS, Zaman SH, Wenthold RJ, Svoboda K, Malinow R. (1999) Rapid spine delivery and redistribution of AMPA receptors after synaptic NMDA receptor activation. *Science* **284**: 1811–1816.

Shi Z, Curiel DT, Tang DC. (1999) DNA-based non-invasive vaccination onto the skin [In Process Citation]. *Vaccine* **17**: 2136–2141.

Sjöberg EM, Suomalainen M, Garoff H. (1994) A significantly improved Semliki Forest virus expression system based on translation enhancer segments from the viral capsid gene. *Bio/Technology* **12**: 1127–1131.

Smith GL, Moss B. (1983) Infectious poxvirus vectors have capacity for at least 25 000 base pairs of foreign DNA. *Gene* **25**: 21–28.

Smith GL, Symons JA, Khanna A, Vanderplasschen A, Alcami A. (1997) Vaccinia virus immune evasion. *Immunol. Rev.* **159**: 137–154.

Smith R, Huebner R, Rowe RJ, Schatten WP, Thomas WE. (1956) Studies on the use of viruses in the treatment of carcinoma of the cervix. *Cancer* **9**: 1211–1218.

Smith TAG, Mehaffey MG, Kayda DB *et al.* (1993) Adenovirus mediated expression of therapeutic plasma levels of human factor IX in mice. *Nature Genetics* **5**: 397–402.

Somogyi P, Frazier J, Skinner MA. (1993) Fowlpox virus host range restriction: gene expression, DNA replication, and morphogenesis in nonpermissive mammalian cells. *Virology* **197:** 439–444.

Song W, Kong HL, Carpenter H *et al.* (1997) Dendritic cells genetically modified with an adenovirus vector encoding the cDNA for a model antigen induce protective and therapeutic antitumor immunity. *J. Exp. Med.* **186**: 1247–1256.

Spehner D, Drillien R, Lecocq JP. (1990) Construction of fowlpox virus vectors with intergenic insertions: expression of the beta-galactosidase gene and the measles virus fusion gene. *J. Virol.* **64**: 527–533.

Stanley E, Jackson G. (1969) Spread of enteric live adenovirus type 4. *J. Infect. Dis.* **119**: 51–59.

Stevenson SC, Rollence M, White B, Weaver L, McClelland A. (1995) Human adenovirus serotypes 3 and 5 bind to two different cellular receptors via the fiber head domain. *J. Virol.* **69**: 2850–2857.

Stewart AK, Lassam NJ, Quirt IC *et al.* (1999) Adenovector-mediated gene delivery of interleukin-2 in metastatic breast cancer and melanoma: results of a phase 1 clinical trial [In Process Citation]. *Gene Ther.* **6**: 350–363.

Stewart PL, Burnett RM. (1995) Adenovirus structure by X-ray crystallography and electron microscopy. *Curr. Top. Microbiol. Immunol.* **199**: 25–38.

Stewart PL, Fuller SD, Burnett RM. (1993) Difference imaging of adenovirus: bridging the resolution gap between X-ray crystallography and electron microscopy. *Embo. J.* **12**: 2589–2599.

Strauss JH, Strauss EG. (1994) The alphaviruses: gene expression, replication and evolution. *Microbiological Reviews* **58**: 491–562.

Sutter G, Moss B. (1992) Nonreplicating vaccinia vector efficiently expresses recombinant genes. *Proc. Natl Acad. Sci. USA* **89**: 10847–10851.

Sutter G, Ohlmann M, Erfle V. (1995) Non-replicating vaccinia vector efficiently expresses bacteriophage T7 RNA polymerase. *FEBS Lett.* **371**: 9–12.

Sutter G, Wyatt LS, Foley PL, Bennink JR, Moss B. (1994) A recombinant vector derived from the host range-restricted and highly attenuated MVA strain of vaccinia virus stimulates protective immunity in mice to influenza virus. *Vaccine* **12**: 1032–1040.

Tacket CO, Losonsky G, Lubeck MD *et al.* (1992) Initial safety and immunogenicity studies of an oral recombinant adenohepatitis B vaccine. *Vaccine* **10**: 673–676.

Tartaglia J, Perkus ME, Taylor J *et al.* (1992) NYVAC: a highly attenuated strain of vaccinia virus. *Virology* **188**: 217–232.

Tartaglia J, Benson J, Cornet B *et al.* (1997) Potential improvements for poxvirus-based immunisation vehicles. In: *Onzieme Colloque des Cent Gardes,* (ed. M Girard, B Dodet), pp. 187–197. Elsevier, Paris.

Tillman BW, de Gruijl TD, Luykx-de Bakker SA *et al.* (1999) Maturation of dendritic cells accompanies high-efficiency gene transfer by a CD40-targeted adenoviral vector. *J. Immunol.* **162**: 6378–6383.

Timofeev AV, Ozherelkov SV, Pronin AV, Deeva AV, Elbert LB, Stephenson JR. (1997) [A recombinant adenovirus expressing the NS1 nonstructural protein of tick-borne encephalitis virus: some characteristics of the immunologic basis of antiviral action]. *Vopr. Virusol.* **42**: 219–222.

Timofeev AV, Ozherelkov SV, Pronin AV, Deeva AV, Karganova GG, Elbert LB, Stephenson JR. (1998) Immunological basis for protection in a murine model of tick-borne encephalitis by a recombinant adenovirus carrying the gene encoding the NS1 non-structural protein. *J. Gen. Virol.* **79**: 689–695.

Tims T, Briggs DJ, Davis RD, Moore SM, Xiang Z, Ertl HC, Fu ZF. (2000) Adult dogs receiving a rabies booster dose with a recombinant adenovirus expressing rabies virus glycoprotein develop high titers of neutralizing antibodies. *Vaccine* **18**: 2804–2807.

Tine JA, Lanar DE, Smith DM *et al.* (1996) NYVAC-Pf7: a poxvirus-vectored, multiantigen, multistage vaccine candidate for Plasmodium falciparum malaria. *Infect. Immun.* **64:** 3833–3844.

Tomko RP, Xu R, Philipson L. (1997) HCAR and MCAR: the human and mouse cellular receptors for subgroup C adenoviruses and group B coxsackieviruses. *Proc. Natl Acad. Sci. USA* **94**: 3352–3356.

Torres JM, Sanchez C, Sune C, Smerdou C, Prevec L, Graham F, Enjuanes L. (1995) Induction of antibodies protecting against transmissible gastroenteritis coronavirus (TGEV) by recombinant adenovirus expressing TGEV spike protein. *Virology* **213**: 503–516.

Torres JM, Alonso C, Ortega A, Mittal S, Graham F, Enjuanes L. (1996) Tropism of human adenovirus type 5-based vectors in swine and their ability to protect against transmissible gastroenteritis coronavirus. *J. Virol.* **70**: 3770–3780.

Townsend K, Perri S, Mento S.J *et al.* (1999) Stable alphavirus packaging cell lines for Sindbis virus- and Semliki Forest virus-derived vectors. *Proc. Natl Acad. Sci. USA* **96**: 4598–4603.

Trentin J, Yabe Y, Taylor G. (1962) The quest for human cancer viruses. *Science* **137**: 835–841.

Tripp RA, Jones L, Anderson LJ, Brown MP. (2000) CD40 ligand (CD154) enhances the Th1 and antibody responses to respiratory syncytial virus in the BALB/c mouse [In Process Citation]. *J. Immunol.* **164**: 5913–5921.

Ulmanen I, Peränen J, Tenhunen J *et al.* (1997) Expression and intracellular localization of catechol O-methyltransferase in transfected mammalian cells. *Eur. J. Biochem.* **243**: 452–459.

Usdin TB, Brownstein MJ, Moss B, Isaacs SN. (1993) SP6 RNA polymerase containing vaccinia virus for rapid expression of cloned genes in tissue culture. *Biotechniques* **14**: 222–224.

Walters RW, Grunst T, Bergelson JM, Finberg RW, Welsh MJ, Zabner J. (1999) Basolateral localization of fiber receptors limits adenovirus infection from the apical surface of airway epithelia. *J. Biol. Chem.* **274**: 10219–10226.

Wan Y, Emtage P, Zhu Q, Foley R, Pilon A, Roberts B, Gauldie J. (1999) Enhanced immune response to the melanoma antigen gp100 using recombinant adenovirus-transduced dendritic cells. *Cell Immunol.* **198**: 131–138.

Wang K, Guan T, Cheresh DA, Nemerow GR. (2000) Regulation of adenovirus membrane penetration by the cytoplasmic tail of integrin beta5 [In Process Citation]. *J. Virol.* **74**: 2731–2739.

Wang M, Bronte V, Chen PW, Gritz L, Panicali D, Rosenberg SA, Restifo NP. (1995) Active immunotherapy of cancer with a nonreplicating recombinant fowlpox virus encoding a model tumor-associated antigen. *J. Immunol.* **154**(9): 4685–4692.

Wang Y, Xiang Z, Pasquini S, Ertl HC. (1997) The use of an E1-deleted, replication-defective adenovirus recombinant expressing the rabies virus glycoprotein for early vaccination of mice against rabies virus. *J. Virol.* **71**: 3677–3683.

Ward GA, Stover CK, Moss B, Fuerst TR. (1995) Stringent chemical and thermal regulation of recombinant gene expression by vaccinia virus vectors in mammalian cells. *Proc. Natl Acad. Sci. USA*; **92**: 6773–6777.

Wesseling JG, Godeke GJ, Schijns VE, Prevec L, Graham FL, Horzinek MC, Rottier PJ. (1993) Mouse hepatitis virus spike and nucleocapsid proteins expressed by adenovirus vectors protect mice against a lethal infection. *J. Gen. Virol.* **74**: 2061–2069.

Wickham TJ, Mathias P, Cheresh DA, Nemerow GR. (1993) Integrins alpha v beta 3 and alpha v beta 5 promote adenovirus internalization but not virus attachment. *Cell* **73**: 309–319.

Wilkinson GWG, Akrigg A. (1992) Constitutive and enhanced expression from the CMV major IE promoter in a defective adenovirus vector. *Nucleic Acids Research* **20**: 2233–2239.

Wold WS, Doronin K, Toth K, Kuppuswamy M, Lichtenstein DL, Tollefson AE. (1999) Immune responses to adenoviruses: viral evasion mechanisms and their implications for the clinic. *Curr. Opin. Immunol.* **11**: 380–386.

Wu GY, Zou DJ, Koothan T, Cline HT. (1995) Infection of frog neurons with vaccinia virus permits in vivo expression of foreign proteins. *Neuron* **14**: 681–684.

Wyatt LS, Moss B, Rozenblatt S. (1995) Replication-deficient vaccinia virus encoding bacteriophage T7 RNA polymerase for transient gene expression in mammalian cells. *Virology* **210**: 202–205.

Wyatt LS, Shors ST, Murphy BR, Moss B. (1996) Development of a replication-deficient recombinant vaccinia virus vaccine effective against parainfluenza virus 3 infection in an animal model. *Vaccine* **14**: 1451–1458.

Wyatt LS, Carroll MW, Czerny CP, Merchlinsky M, Sisler JR, Moss B. (1998) Marker rescue of the host range restriction defects of modified vaccinia virus Ankara. *Virology* **251**: 334–342.

Xiang Z, Ertl HC. (1999) Induction of mucosal immunity with a replication-defective adenoviral recombinant. *Vaccine* **17**: 2003–2008.

Xiang ZQ, Yang Y, Wilson JM, Ertl HC. (1996) A replication-defective human adenovirus recombinant serves as a highly efficacious vaccine carrier. *Virology* **219**: 220–227.

Xiang ZQ, Pasquini S, Ertl HC. (1999) Induction of genital immunity by DNA priming and intranasal booster immunization with a replication-defective adenoviral recombinant. *J. Immunol.* **162**: 6716–6723.

Xiong C, Levis R, Shen P, Schlesinger S, Rice CM, Huang HV. (1989) Sindbis virus: an efficient broad host range vector for gene expression in animal cells. *Science* **243**: 1188–1191.

Xu ZZ, Both GW. (1998) Altered tropism of an ovine adenovirus carrying the fiber protein cell binding domain of human adenovirus type 5. *Virology* **248**: 156–163.

Xu ZZ, Hyatt A, Boyle DB, Both GW. (1997) Construction of ovine adenovirus recombinants by gene insertion or deletion of related terminal region sequences. *Virology* **230**: 62–71.

Yang Y, Ertl HCJ, Wilson JM. (1994) MHC class I-restricted cytotoxic T lymphocytes to viral antigens destroy hepatocytes in mice infected with E1-deleted recombinant adenoviruses. *Immunity* **1**: 433–442.

Yarosh OK, Wandeler AI, Graham FL, Campbell JB, Prevec L. (1996) Human adenovirus type 5 vectors expressing rabies glycoprotein. *Vaccine* **14**: 1257–1264.

Ye WW, Mason BB, Chengalvala M *et al.* (1991) Co-expression of hepatitis B virus antigens by a non-defective adenovirus vaccine vector. *Arch. Virol.* **118**: 11–27.

Ying H, Zaks TZ, Wang R-F *et al.* (1999) Cancer therapy using self-replicating RNA vaccine. *Nature Medicine* **5**: 823–827.

Zakhartchouk AN, Reddy PS, Baxi M, Baca-Estrada ME, Mehtali M, Babiuk LA, Tikoo SK. (1998) Construction and characterization of E3-deleted bovine adenovirus type 3 expressing full-length and truncated form of bovine herpesvirus type 1 glycoprotein gD [In Process Citation]. *Virology* **250**: 220–229.

Zhai Y, Yang JC, Kawakami Y *et al.* (1996) Antigen-specific tumor vaccines. Development and characterization of recombinant adenoviruses encoding MART1 or gp100 for cancer therapy. *J. Immunol.* **156**: 700–710.

Zhang J, Asselin-Paturel C, Bex F *et al.* (1997) Cloning of human IL-12 p40 and p35 DNA into the Semliki Forest virus vector: expression of IL-12 in human tumor cells. *Gene Ther.* **4**: 367–374.

Zheng B, Graham FL, Johnson DC, Hanke T, McDermott MR, Prevec L. (1993) Immunogenicity in mice of tandem repeats of an epitope from herpes simplex gD protein when expressed by recombinant adenovirus vectors. *Vaccine* **11**: 1191–1198.

Zhong L, Granelli-Piperno A, Choi Y, Steinman RM. (1999) Recombinant adenovirus is an efficient and non-perturbing genetic vector for human dendritic cells. *Eur. J. Immunol.* **29**: 964–972.

Zhou X, Berglund P, Rhodes G, Parker SE, Jondal M, Liljeström P. (1994) Self-replicating Semliki Forest virus RNA as a recombinant vaccine. *Vaccine* **12**: 1510–1513.

Zhou X, Berglund P, Zhao H, Liljeström P, Jondal M. (1995) Generation of cytotoxic and humoral immune responses by nonreplicative recombinant Semliki Forest virus. *Proc. Natl Acad. Sci. USA* **92**: 3009–3013.

Zolla-Pazner S, Lubeck M, Xu S *et al.* (1998) Induction of neutralizing antibodies to T-cell line-adapted and primary human immunodeficiency virus type 1 isolates with a prime-boost vaccine regimen in chimpanzees. *J. Virol.* **72**: 1052–1059.

Chapter 6

Genetic engineering of animal DNA viruses in order to study and modify pathogenesis

David A. Leib

6.1 Introduction

The study of virus genetics and the development of methodologies to alter viral genomes has led to a great deal of information regarding the interactions of viruses and their hosts. The DNA viruses have proved to be the most facile in terms of genetic manipulation, although new technologies are also now allowing the manipulation of most RNA viruses (Lai, 2000). Amongst the DNA viruses, the herpesvirus family has been perhaps the most extensively studied in terms of genetics and pathogenesis, and will therefore be the primary focus for this chapter. Reference to other DNA virus families such as the poxviruses, will also be made where appropriate.

The herpesvirus family includes almost 100 viruses which infect animals ranging from humans and non-human primates to turtles, snakes and fish (Roizman, 1996b). The members of this family exhibit tremendous biological diversity, but share two common classifying characteristics, namely structure, and the ability to establish latent infections. The virion consists of a linear double stranded DNA genome which can vary from 120 to 230 kbp with a G+C content variable from 31 to 75 percent. The DNA is enclosed within an icosahedral capsid made up of 162 capsomeres. The capsid is surrounded by an amorphous structure known as the tegument and the viral envelope which is essential for infectivity. The size of the virions varies from 120 to around 300 nm. The latent infections established by this family of viruses are a unique feature and represent a lifelong source of virus from which reactivation can occur to cause disease and spread to other susceptible hosts. The pathogenesis and latently infected cell type, however, varies greatly from virus to virus. The classification of herpesviruses has been attempted in a number of ways based upon genome type, host species, and latently infected cell type. It is the latter of these categories which has proved the most meaningful and convenient over time.

6.1.1 Classification

The *alphaherpesviruses* are characterized by generally rapid cytolytic infection in cell culture, and establishment of latency in neurons (Roizman, 1996b). Generally speaking,

---

Genetically Engineered Viruses: Development and Applications, edited by C.J.A. Ring and E.D. Blair

this subfamily has a wide host range, facilitating the development of animal models for disease. The alphaherpesviruses include the herpes simplex viruses (HSVs) 1 and 2 and varicella-zoster virus (VZV) of humans, simian B virus and pseudorabies virus (PrV) of swine. The *betaherpesviruses* generally have a slow replicative cycle with the production of enlarged cytomegalic cells in culture. These viruses can establish latency in a variety of cell types including secretory glands and lymphoreticular cells. The betaherpesviruses include both human and mouse cytomegalovirus (HCMV and MCMV) and have a highly restricted host range. The *gammaherpesviruses* replicate in lymphoblastoid cell lines in culture, although *in vivo* they may be capable of replication in epithelial and fibroblast cells. The predominant mode of these viruses in culture, however, is latency, although *in vivo* many are associated with tumors in immunosuppressed hosts. Examples include Epstein-Barr virus (EBV), Kaposi's sarcoma herpesvirus (KSHV), herpesvirus saimiri (HVS) and murine γ-herpesvirus 68 (γHV68). This group of viruses also exhibits a highly species-restricted host range.

6.2 Reverse genetics and animal models

Reverse genetics and animal model systems are clearly both necessary for the detailed study of molecular pathogenesis (*Table 6.1*). Another pivotal factor is the availability of a complete sequence for the viral genome (Davison and Scott, 1986; McGeoch and Davision, 1986; Virgin *et al.*, 1997). Viruses such as VZV, HCMV and EBV, which are significant human pathogens, lack either simple reverse genetics methods, animal model systems, or both. Recent advances have allowed some genetic manipulation in these systems allowing a better understanding of the functions of their genes at least in cell

Table 6.1. Classification, genetic systems and animal models for selected herpesviruses.

Common name	Subfamily	Natural host	Reverse genetics	Animal models
Herpes simplex 1 & 2	α	Humans	HR[1], Cosmid[2] BAC[3]	Mouse[4], rabbit[5], rat[6] guinea pig[7], tree shrew[8]
Varicella zoster	α	Humans	Cosmid[9]	Rat[10], rabbit[11], SCID/Hu mouse[12]
Pseudorabies	α	Pigs	HR[13], BAC[14]	Pig[15], rat[16], mouse[17]
Human cytomegalovirus	β	Humans	HR[18], BAC[19]	SCID/Hu mouse[20]
Murine cytomegalovirus	β	Mouse	HR[21], BAC[22]	Mouse[23]
Epstein-Barr virus	γ	Humans	HR[24], BAC[25]	None
Kaposi's sarcoma	γ	Humans	None	None
Murine γ-herpesvirus 68	γ	Bank voles	HR[26], BAC[27]	Mouse[28]

HR =, Homologous recombination; BAC =, Bacterial artificial chromosome.
[1] (Post & Roizman, 1981); [2] (Cunningham & Davision, 1993); [3] (Horsburgh *et al.*, 1999); [4] (Stevens & Cook, 1971); [5] (Nesburn *et al.*, 1972); [6] (Marks, 1975); [7] (Lausch *et al.*, 1966); [8] (Darai *et al.*, 1983); [9] (Cohen & Seidel, 1993); [10] (Sadzot-Delvaux *et al.*, 1995); [11] (Dunkel *et al.*, 1995); [12] (Moffat *et al.*, 1995); [13] (Mettenleiter *et al.*, 1994); [14] (Smith & Enquist, 2000); [15] (Becker & Bergmann, 1968); [16] (Tokumaru, 1975); [17] (Platt *et al.*, 1980); [18] (Spaete & Mocarski, 1987); [19] (Borst *et al.*, 1999); [20] (Mocarski *et al.*, 1993); [21] (Takekoshi *et al.*, 1993); [22] (Messerle *et al.*, 1997); [23] (Osborn & Medearis, 1996); [24] (Hammerschmidt & Sugden, 1989); [25] (Delecluse *et al.*, 1998); [26] (Simas *et al.*, 1998); [27] (Alder *et al.*, 1999); [28] (Sunil-Chandra *et al.*, 1992).

culture systems, but the lack of animal models remains a significant block (Borst *et al.*, 1999; Cohen and Seidel, 1993; Hammerschmidt and Sugden, 1989; Wang *et al.*, 1991). In the case of EBV, however, the availability of EBV-transformed and therefore continuously replicating latently infected B cell lines for study in cell culture has facilitated studies of the mechanisms of latent infection and cell transformation (Kaye *et al.*, 1999; Thorley-Lawson and Babcock, 1999; Woisetschlaeger *et al.*, 1991). In contrast, such lines are not available for viruses such as HSV (Wilcox and Johnson, 1987), where, despite good animal models, the molecular basis of latency and reactivation remains obscure. In the case of HCMV, which has no animal model, a great deal of effort has been devoted to its mouse homologue, murine CMV (MCMV). The ability to study MCMV in mice, its natural host, has made MCMV an attractive model system. Until relatively recently with the development of γHV68 biology (Speck and Virgin, 1999; Sunil-Chandra *et al.*, 1992), there was no satisfactory model for any form of *gammaherpesvirus* pathogenesis. Any attempts to study this area were limited either by extreme species specificity or the expense and difficulties associated with the development of non-human primate models of viral disease. It is clear therefore, that despite the large number of clinically important herpesviruses, there are only a few viruses with both a simple reverse genetics system and a tractable animal model for studying pathogenesis. The major focus of this chapter will be on HSV-1, although examples of other herpesviruses will be given where appropriate. A particular emphasis is given to how the use of recombinant viruses has elucidated the immune evasion strategies of herpesviruses, and where their use in combination with appropriate knockout mice has been particularly informative to elucidate both viral and host determinants of pathogenesis.

6.3 Herpes simplex virus

6.3.1 Overview of pathogenesis

Herpes simplex virus is a prevalent human pathogen and infection is manifested in several ways which range in severity from cold sores to life-threatening encephalitis. Studies using animal models and observations of human infections have resulted in the so-called 'classical theory' of HSV pathogenesis (Wildy *et al.*, 1982). According to this theory there are four stages which characterize an HSV infection. *Entry* into the host occurs at the time of a primary infection, and HSV replicates at peripheral sites such as the eyes, skin, or mucosae. *Spread* to the axonal terminae of sensory neurons is followed by retrograde intra-axonal transport to neuronal cell bodies in sensory ganglia where further viral replication may occur. *Establishment of latency* then occurs at which time lytic gene expression is repressed. At this stage no infectious virus can be detected but the viral genome persists in the neuron in a transcriptionally active state. *Reactivation* occurs when certain poorly defined stimuli such as stress, menstruation, or sunlight cause the controls responsible for maintaining latency to break down. This leads to the production of infectious virus in the ganglion followed by anterograde transport to the periphery which, following further replication, may be manifested by lesions at or near the site of primary infection.

6.3.2 HSV genetics and pathogenesis

HSV has long been considered as the 'prototypic' herpesvirus, and certainly remains intensively studied. The HSV genomes have been completely sequenced and consist of 152 kilobases of double-stranded DNA encoding approximately 80 genes (McGeoch and Davison, 1986). In addition, the genome contains two *cis*-acting DNA replication and packaging elements (Boehmer and Lehman, 1997). Firstly, there are origins of replication, i.e., the regions from which DNA replication is initiated, and secondly there is a signal ('a' sequence) which is required for the packaging of replicated DNA into virus particles. The naked viral genome is infectious i.e., transfection of viral DNA into cells results in production of infectious virus. Also, during viral DNA replication the genome undergoes frequent recombination. These two properties have facilitated the generation of mutant viruses for studies of viral gene function (Mocarski *et al.*, 1980; Post and Roizman, 1981). In addition, viruses with mutations in essential genes can be propagated on complementing cell lines stably transformed with these essential genes (DeLuca *et al.*, 1985). Obviously, for the study of pathogenesis, the emphasis has been upon those genes which are non-essential for viral replication in cell culture but have profound effects upon the ability of the virus to cause disease *in vivo*. The availability of a simple reverse genetics system for HSV has allowed the identification of a number of genes which play pivotal roles in pathogenesis and uncovered key aspects of virus–host interactions. The most interesting studies involve viruses with mutations in genes which do not affect replication *per se* and have thereby uncovered some unique pathway by which the virus is a successful pathogen. These genes have been shown to be important *in vivo* by many mechanisms including regulation of viral gene expression, promotion of establishment of latency or by regulation of the host's immune responses to the virus. Examples of each of these mechanisms will be discussed below.

6.3.3 Gene expression and viral replication during lytic infection

In order to fully understand the impact of reverse genetics upon our understanding of HSV pathogenesis, a brief primer on viral gene expression patterns is necessary (Roizman, 1996a). Viral gene expression is tightly regulated during the viral lifecycle. According to the manner in which they are expressed, viral genes can be divided into three temporal classes: immediate-early, early and late (Honess and Roizman, 1974; Roizman *et al.*, 1975). Broadly speaking, the immediate-early class consists of four regulatory genes and one immunomodulatory gene, the early class of genes encodes a variety of viral DNA replication functions (Boehmer and Lehman, 1997), and the products of late genes are the structural proteins of the virus. The cascade of viral gene expression is initiated and enhanced by the tegument protein VP16, which is brought into the nucleus with the viral genome (Campbell *et al.*, 1984; Pellett *et al.*, 1985). VP16 transactivates immediate-early genes as part of a complex with the cellular transcription factor oct-1 and host cell factor (LaBoissiere and O'Hare, 2000; Stern and Herr, 1991; Walker *et al.*, 1994; Wilson *et al.*, 1997). Immediate-early gene products then activate early gene expression with resultant initiation of HSV DNA replication at the viral origins. Viral DNA replication is mediated by a large number of virally encoded proteins (Boehmer and Lehman, 1997). A subset of these proteins are essential for viral DNA replication (e.g., viral DNA polymerase, origin binding protein), while others (e.g., thymidine kinase, ribonucleotide reductase) are involved in DNA metabolism and are required only in non-dividing cells. Following DNA synthesis, late genes are expressed which lead to assembly of capsids in the nucleus. A series of cleavage and packaging events occur and the capsid is filled with a unit length of linear DNA. The viral capsid then apparently acquires its envelope as it buds through the nuclear membrane. Finally, the entire lytic pathway culminates with lysis of the host cell and release of newly synthesized and assembled virus particles.

6.3.4 HSV infection of neurons

HSV has a broad host range and can lytically infect a wide variety of cell types. The peripheral nervous system (PNS) neuron, however, is unique in that it can support either a lytic or a latent mode of HSV gene expression, and it is the only cell type which has been unequivocally shown to support latency as predicted by Goodpasture in 1929 (Goodpasture, 1929). While the lytic pathway has been extensively studied and characterized in cultured non-neuronal cells, it is not known whether this conventional

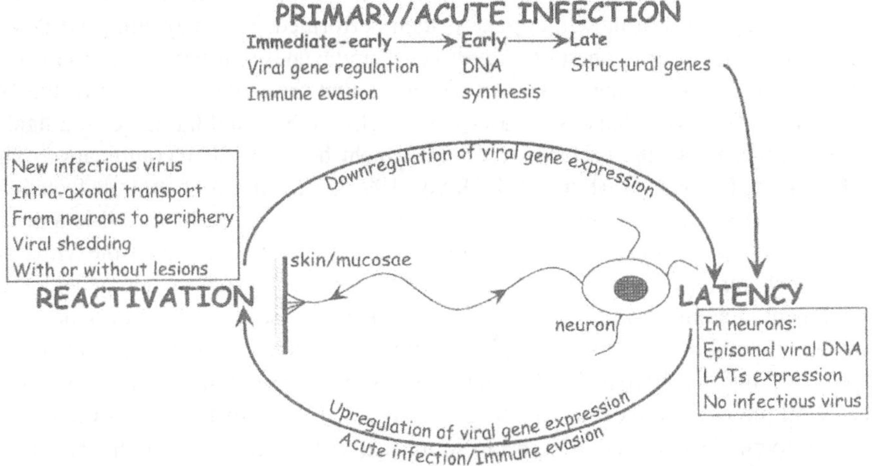

Figure 6.1. Lytic and latent pathways of HSV infection. The lytic pathway begins with activation of immediate-early gene expression. This is enhanced by, but does not require, viral structural components such as virion protein 16 (VP16). Certain immediate-early genes are required for the transactivation of early genes which function primarily during viral DNA synthesis and late genes are expressed from replicated genomes. Late gene products make up the virus structure. Latency in neurons almost certainly involves a down-regulation of immediate-early gene expression, but the mechanism is unclear. Lytic gene expression is shut-off and latency is evidenced by the persistence of episomal viral DNA, the presence of latency associated transcripts (LATs), and the absence of infectious virus. Certain ill-defined stimuli may then trigger the virus to re-enter the lytic pathway with production of new virus which moves back to the periphery by intra-axonal transport. Shedding of virus may occur with or without overt disease. Subsequent downregulation of viral gene expression leads to the re-establishment of latency.

cascade of gene expression occurs in neurons (Kosz-Vnenchak *et al.*, 1993; Nichol *et al.*, 1996). As shown in *Figure 6.1*, HSV exhibits two very different modes of gene expression during the lytic and latent phases of its life cycle. During lytic infection, nearly all viral genes are expressed. In contrast, during latency in neurons, viral gene expression is almost completely repressed. Limited transcription however does occur from one region of the viral genome and results in a family of RNA molecules referred to as latency-associated transcripts or LATs (Spivack and Fraser, 1987; Stevens *et al.*, 1987). The precise molecular mechanisms involved in the repression of lytic gene expression during the establishment and maintenance of latency and the de-repression of lytic gene expression that occurs during reactivation are not fully understood. Some of the research described below, however, has provided some clues as to the mechanisms by which HSV establishes latency and reactivates.

6.3.5 Generation of recombinant viruses

The general scheme of how the HSV genome can be manipulated to obtain a recombinant virus is shown in *Figure 6.2*. This general method has been extensively used for the mutagenesis of other herpesvirus genomes. The process of making a recombinant virus begins with the cloning of the gene of interest into a bacterial plasmid. Mutagenesis of the target gene is then performed. This may comprise deletion using endogenous restriction sites, site-directed mutagenesis, insertion of stop codons, or insertion of marker genes. A key to success is that the mutated portion of the HSV genome is flanked by at least 500 base-pairs of HSV DNA, and the larger the flanking sequences, the higher the probability of subsequent homologous recombination. Then, as shown in *Figure 6.2* (Leib and Olivo, 1993), this plasmid is transfected into permissive cells in culture along with, and in molar excess to, purified, full-length HSV DNA. Since naked viral DNA is infectious, this transfection initiates the virus lytic cycle. DNA replication and homologous recombination result in the generation of recombinant genomes which are replicated, cleaved, and packaged into new virus particles. Recombinant virus is usually detected at a frequency of 1 to 5% of the total virus yield, and can be purified by plating the mixed virus population on susceptible cells. Viral plaques are picked, and progeny viruses are then evaluated for the presence of the foreign gene by DNA hybridization, often in conjunction with use of PCR techniques. At least three rounds of plaque purification and evaluation are required to obtain a pure stock of the recombinant. Once a recombinant has been isolated and tested for a phenotype either *in vivo* or in cell culture it is *de rigeur* to perform so-called 'marker-rescue' of the recombinant. Marker-rescue involves making infectious DNA from the recombinant and cotransfecting this DNA with a plasmid containing only the wild-type allele under study. Identification and purification of this virus followed by testing it in parallel with the mutant allows the assertion that any phenotype of the mutant virus is due only to the intended mutation in the gene of interest, and not due to any unintended secondary mutation in another locus. An alternative approach is to make multiple versions of the original recombinant virus using independent transfections. If all of these independently isolated viruses have the same phenotype it is highly unlikely that secondary mutations could account for the observed differences.

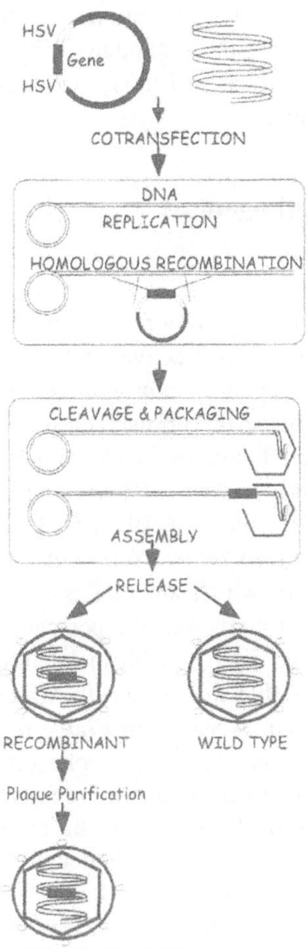

Figure 6.2. Recombinant HSV vectors (adapted from Leib and Olivo, 1993). Genomic HSV DNA is isolated and transfected into permissive cells along with a plasmid containing the mutated target gene (black rectangle) flanked by at least 0.5 kb of the HSV genome. Transfection of viral DNA initiates the virus lytic cycle and viral DNA replication. The HSV DNA sequences on the plasmid DNA undergo homologous recombination with viral DNA. Replicated and recombined viral DNA is packaged into new virus particles. Progeny virus is predominantly the parental type with a few percent (1–5) of recombinant viruses containing the mutated gene. Recombinant virus is then identified and plaque-purified away from parental virus.

A number of selection methods have been successfully employed to enable a more rapid initial isolation of recombinant viruses. Insertional inactivation of the viral thymidine kinase (*tk*) gene allows rapid selection by rendering the recombinant virus resistant to the antiviral drug acyclovir (Post and Roizman, 1981). Other methods have been employed such as 'blue/white' virus plaque selection by inserting a *lacZ* cassette into the virus (Goldstein and Weller, 1988). One other technology worthy of brief mention here that has been used to generate recombinant viruses is the use of bacterial artificial chromosomes (Horsburgh *et al*., 1999; Smith and Enquist, 2000; Wagner *et al*., 1999). Although this method has not been tested extensively for making recombinants for use in pathogenesis, it is, in principle, a method to make recombinant viruses within 7 days, without the need for multiple rounds of plaque purification described above. The only caveats are difficulty in marker-rescue of thus generated recombinants, and presence of a sufficiently large intergenic site for insertion of *loxP* sites, such that the insertion does not interfere with viral gene expression (Smith and Enquist, 2000). The marker-rescue requirement could, however, be satisfied by generation of multiple recombinant viruses in the first instance. Use of this elegant methodology by other laboratories will be needed to evaluate this technology which will likely revolutionize the way that herpesvirus genetic manipulation is performed.

6.4 Regulation of HSV gene expression and pathogenesis

6.4.1 ICP0

The observation that the levels of HSV gene expression are greatly reduced during the establishment and maintenance of latency, and increased during reactivation has led to a great deal of work on viral genes which control the gene expression programme. In cell culture, the immediate early genes have been shown to be the primary class of viral regulatory genes and therefore have been a major focus of work done using recombinant viruses *in vivo*. Of the five immediate-early genes, ICP4 and ICP27 are essential for viral replication (DeLuca and Schaffer, 1985; Preston, 1979; Sacks *et al.*, 1985), and one is immunomodulatory (ICP47) and discussed below (York *et al.*, 1994). Of the remaining two genes, ICP0 and ICP22, both of which are non-essential for growth in culture, ICP0 has attracted the most attention both from the perspective of molecular biology and pathogenesis. The interest in the role of ICP0 in latency and reactivation was significantly heightened by the observation that the LATs, which are the only abundant transcripts detected during latency, and, as shown in *Figure 6.3*, are transcribed in an 'antisense' fashion to ICP0 (Stevens *et al.*, 1987). This phenomenon will be discussed further below, but the idea that ICP0 was in some way regulated by the LATs during the establishment and reactivation of latency was an attractive hypothesis. Furthermore, the potent transactivating capacity of ICP0 fuelled a great deal of work to discover whether ICP0 could be the trigger factor that allows a latent genome to reactivate, or at the very least determines the outcome of infection of a

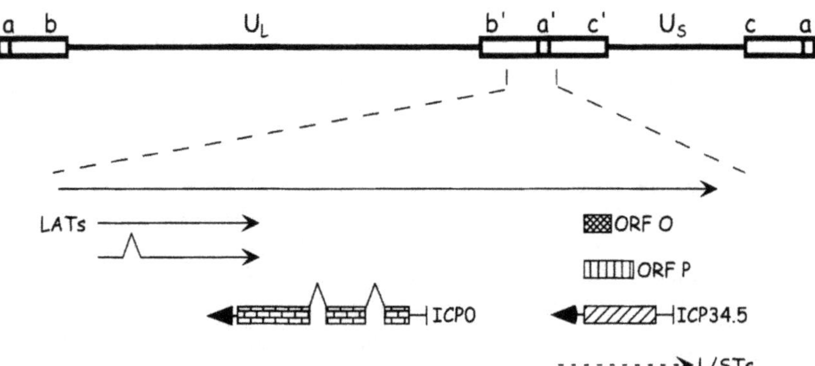

Figure 6.3. Genomic map of HSV-1 below which are shown the positions of the regulatory gene ICP0, and the virulence gene ICP34.5, relative to the latency-related gene products of the LATs, ORF O, ORF P and the L/STs. Refer to the text for details.

neuron by the virus. HSV gene regulation and latency was the subject of a recent excellent and detailed review (Preston, 2000). Some salient points are outlined below.

Early work using transient transfection assays demonstrated that ICP0 was a strong and promiscuous activator of all classes of viral genes (Everett, 1984; O'Hare and Hayward, 1985). Two laboratories then generated recombinant viruses and showed that whilst ICP0 was not essential for viral replication in cell culture, viruses lacking ICP0 showed a substantial loss of infectivity, particularly at low multiplicities of infection (Sacks and Schaffer, 1987; Stow and Stow, 1986). The reasons for this loss of infectivity were not known at that time, but it was consistent with a possible role for ICP0 in latency and reactivation since neuronal infection both entering and exiting from latency is presumed to be a low multiplicity event. Studies in mice of viruses lacking ICP0 showed that whilst the viruses were capable of replication at both peripheral and neuronal sites, the replication of the viruses was significantly reduced (Clements and Stow, 1989; Leib *et al.*, 1989b; Russell *et al.*, 1987; Stow and Stow, 1986). Interestingly, however, these ICP0-deleted viruses were able to establish latent infections but were almost incapable of reactivation. An excellent corollary to these studies was one in which HSV latency in an *in vitro* model could be reactivated by superinfection of latently infected cells with a recombinant adenovirus expressing ICP0, but not by a control adenovirus (Harris *et al.*, 1989; Zhu *et al.*, 1990). An extensive mutagenesis study was subsequently undertaken to show that the ability of mutant ICP0 constructs to transactivate reporter genes in cell culture correlated closely with the ability of these same ICP0 molecules, in the context of the viral genome, to facilitate reactivation (Cai *et al.*, 1993). This key study linked the known transactivation function of ICP0 *in vitro* with its role in pathogenesis *in vivo*. In these initial studies, the generation of well-defined recombinant viruses along with the availability of an animal model in which a stage-specific approach could be taken to pathogenesis, allowed a function for ICP0 to be discerned *in vivo*.

These initial studies, however, were phenomenological in that nothing was known of the mechanism by which ICP0 promotes growth and reactivation. Moreover, little was known of the mechanism by which ICP0 transactivates gene expression. An additional role for ICP0 was proposed in cell regulation events since it had been shown that a cellular protein could partially substitute for ICP0 at certain points in the cell cycle (Ralph *et al.*, 1994; Yao and Schaffer, 1995). More recent studies have shown ICP0 to be a RING finger protein (Everett *et al.*, 1993) which interacts with a diverse array of cellular proteins including the elongation factor 1δ, cyclin D3 which regulates cell cycle, and a ubiquitin-specific protease termed HAUSP (Everett *et al.*, 1997; Kawaguchi *et al.*, 1997a, 1997b). Disruption of this cyclin D3-binding domain of ICP0 has only a minimal effect upon the replication and neurovirulence of HSV-1, but has a significant effect upon neuroinvasion from the peritoneum, an intriguing result, and worthy of further investigation (Van Sant *et al.*, 1999). Furthermore, ICP0 has been shown to target PML (Everett *et al.*, 1998), a component of nuclear domain 10 (ND10), for degradation as well as causing the proteasome-dependent degradation of CENP-C, which plays a critical role in cell division (Everett *et al.*, 1999). Finally ICP0 can prevent progression of cells through mitosis and from G1 to S phase in the cell cycle through both p53-dependent and -independent pathways (Hobbs and DeLuca,

1999). It has been shown that neurons die by apoptosis if G0 neurons are stimulated to enter the cell cycle. It has therefore been suggested that ICP0 may function by prolonging neuronal survival, as well as promoting a permissive intracellular environment for the virus. Other hypotheses include the idea that host repression proteins which serve to silence incoming viral genomes are themselves targeted for degradation by ICP0. Some of these inhibitory proteins may be interferon inducible, so it is intriguing that ICP0 deletion mutants are hypersensitive to interferons and grow to almost wild-type levels in mice lacking interferon receptors (Leib *et al.*, 1999; Mossman *et al.*, 2000).

The proven role for ICP0 in latency and pathogenesis coupled with its complex but thematically consistent molecular properties make this particular area ripe for further study. In particular, the examination of the interactions between ICP0 and some of the above-mentioned cellular factors in neurons will allow a better understanding of the function of this protein at the mechanistic level. Whether ICP0 acts as a pivotal direct transactivation signal for viral reactivation, or plays a more indirect role through its properties described above are important questions to be resolved.

6.4.2 Latency-associated transcription and HSV pathogenesis

The discovery of latency-associated transcription for HSV in the mid to late 1980s has led to an unprecedented amount of research in HSV utilizing recombinant viruses in animal models. A literature search reveals that more than 140 papers have been written involving the study of the latency-associated transcripts (LATs). More remarkable is the fact that despite all of this research over the past 15 years, the mechanism of action of the LATs remains unknown. The LATs were originally described as alternatively spliced, abundant nuclear RNAs, approximately 2, 1.5 and 1.45 kb in length (Mitchell *et al.*, 1990; Wagner *et al.*, 1988). These transcripts are transcribed in an antisense fashion to the transcript encoding ICP0, overlapping 30% of its 3′ terminus (*Figures 6.3* and *6.4*). The LATs were therefore originally proposed to repress the expression of ICP0 during the establishment and maintenance of latency (Stevens *et al.*, 1987). For this hypothesis to be correct, a null deletion of the LATs should have resulted in a virus which was unable to establish latency, resulting in a persistent or highly neurovirulent infection. Several laboratories generated such null mutants within a relatively short timeframe and from this work it was clear that the LATs were not required for the establishment of latency in mice and that viral replication was unaffected *in vitro* and *in vivo* (Javier *et al.*, 1988; Leib *et al.*, 1989a; Steiner *et al.*, 1989). The only discernible phenotype from these initial studies was a two-fold drop in reactivation frequency from explanted latently infected ganglia. The other observation noted by one group was that ganglia infected by a LAT null mutant appeared to have more pathologic damage than ganglia infected with wild-type or marker-rescued viruses (Leib *et al.*, 1989a). A great deal of effort was put into defining the LAT promoter, particularly by groups interested in gene therapy and gene delivery to the nervous system (Ackland-Berglund *et al.*, 1995; Goins *et al.*, 1994; Radar *et al.*, 1993; Soares *et al.*, 1996; Zwaagstra *et al.*, 1991). The LATs promoter remains the only one in HSV which appears to be constitutively active during latency, providing at least a theoretical basis for its use for the regulation of therapeutic genes.

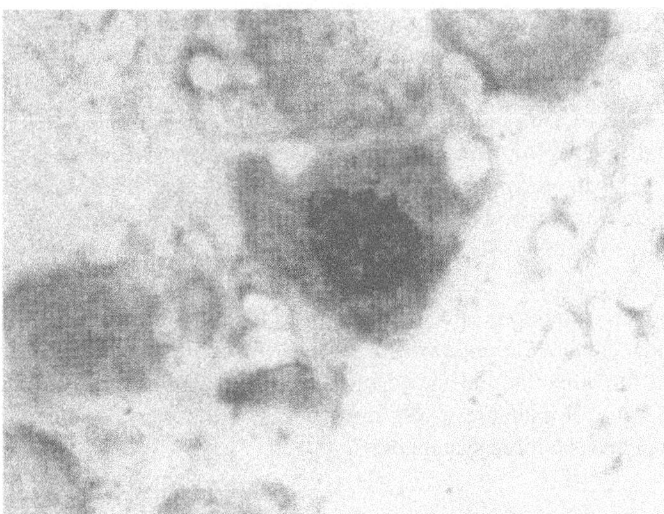

Figure 6.4. In-situ hybridization of trigeminal ganglia latently infected with HSV-1, probed for the major species of LAT (taken from Leib *et al.*, 1989a). A strong signal is notable in a neuronal nucleus indicating the presence of abundant LATs.

Advances in the mapping of the LATs led to the observation that LAT is transcribed as a primary transcript of approximately 8.5 kb and that the predominant nuclear 2 kb transcript is actually a uniquely stable intron (Farrell *et al.*, 1991; Zwaagstra *et al.*, 1990). This discovery presented profound difficulties for the design of mutations in this region because several critical pathogenesis genes, including ICP0 and ICP34.5 map to the strand opposite LAT (*Figure 6.3*). This obviously confounds the problem of assigning reactivation phenotypes specifically to the LATs. Other confounding issues have been the different animal models used and the different strains of HSV-1. Some consensus, however has emerged from this work. Firstly, in agreement with the early work using explanted trigeminal ganglia from mice, LAT null mutants have been shown to reactivate poorly by induced reactivation in mice and rabbits, and from spontaneous reactivation in rabbits (Hill *et al.*, 1990; Perng *et al.*, 1994; Sawtell and Thompson, 1992). A core issue which emerged from this work, however, was whether the LATs were required for the establishment of latency, reactivation, or both (Perng *et al.*, 2000b; Sedarati *et al.*, 1989; Thompson and Sawtell, 1997). A significant body of work has now shown that LAT mutants establish latency with reduced efficiency and that this reduction in efficiency is not due to a failure to replicate acutely. Moreover, it was shown by the construction of several recombinant viruses that the ability of LAT to promote spontaneous reactivation in rabbits was contained within the first 1.5 kb (Perng *et al.*, 1996). This region of LAT does not overlap with ICP0 or any other known HSV-1 gene thereby disproving the idea that LAT acts to promote spontaneous reactivation in rabbits through some antisense repression of ICP0. A consistent theme, however, began to emerge, that LAT was enhancing the establishment of latency, increasing the pool of latently infected neurons, thereby increasing the probability of spontaneous reactivation (Perng *et al.*, 2000b; Sawtell, 1997; Thompson and Sawtell, 1997). This idea was supported by the observation that in a partial deletion of LAT, neurovirulence was enhanced (Perng *et al.*, 1999). This was a curious result given that complete deletions of LAT showed no such increase in neurovirulence. A very recent study has shown that the HSV-1 LATs may serve to inhibit neuronal apoptosis, consistent with the idea that LAT serves to enhance establishment and maintenance of latency (Leib *et al.*, 1989a; Perng *et al.*, 2000a). This interesting result should now provide a platform from which the true mechanism of LATs' function can be uncovered.

This work exemplifies an important pitfall of overly simplistic approaches in recombinant strategies for study of pathogenesis. It is likely that the LATs and other genes have several functions and may serve to modulate the activities of other genes, both viral and cellular. Superimpose this scenario with the general complexity of an animal model for infection and the potential for misinterpretation becomes very high. The creation of null mutations can only uncover the sum of the parts and with multifunctional genes a more systematic approach is required.

6.5 Immune evasion and pathogenesis

The pathogenesis of herpesviruses with their latency and persistence strongly suggests that this family of viruses must have evolved ways to evade the host's immune response. This has become one of the most exciting areas in all of virology and many virus families have different preferred strategies. Clearly these findings have important implications not only for the basic biology of the virus itself, but also for the future design of vaccines and antiviral strategies. The herpesviruses, with reverse genetics available for HSV and MCMV have become excellent for this area of study. In addition, the ability to create recombinant mice with specific genes knocked out has allowed the elucidation of both sides of this complex equation. There are now several good examples in the literature where an attenuated virus has been restored to full virulence in mice which lack the resistance target of the viral determinant. Some of these examples will be discussed below.

6.5.1 Evasion of CD8$^+$ T and natural killer (NK) cell responses

The blockage of the MHC class I pathway is a common theme for the herpesvirus family suggesting that evasion of this particular aspect of the immune response is critical for the lifecycle of viruses which establish latent infections (Doherty *et al.*, 1992). Reactivation and infection of new susceptible hosts occurs despite the primary host's primed immune system. Indeed it is likely that this mechanism is necessary to provide the virus with a replicative window of opportunity within an immune host during reactivation events. Such mechanisms have been shown to be important for pathogenesis of HSV and MCMV as described below, and have been shown to be mediated by the EBNA1 gene of EBV although its role is pathogenesis remains unknown (Levitskaya *et al.*, 1995). Similarly, adenovirus possesses MHC class I-blocking activity through the action of the E3/19K protein, although the role of this function in pathogenesis has not been addressed directly (Andersson *et al.*, 1985; Ginsberg *et al.*, 1989). *Figure 6.5* outlines some of the events that comprise MHC

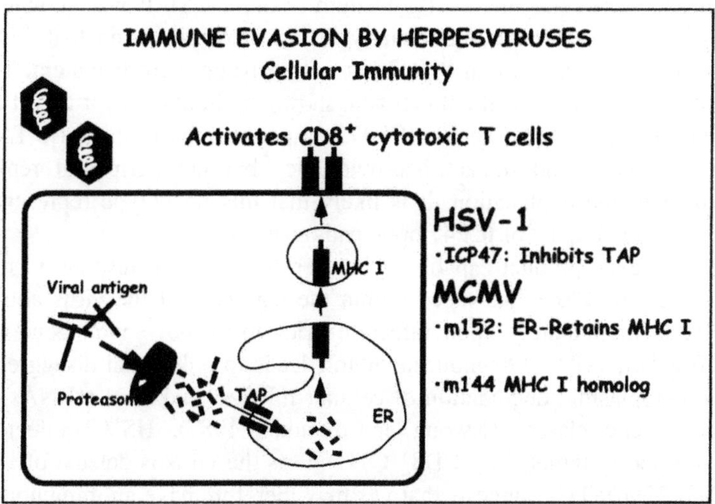

Figure 6.5. Evasion of immunity by herpesviruses by interference with MHC class I-restricted antigen presentation. Intracellular viral antigens are processed by the proteasome and the degraded peptide is delivered to the ER by the TAPs. The MHC class I heavy chains then combine with viral peptide and other cellular proteins to form a complex prior to presentation on the cell surface where recognition by CD8$^+$ T cells occurs. HSV-1 ICP47 interferes with transport of peptide into the ER by TAPs. MCMV m152 causes MHC class complexes to be retained in the ER and therefore not presented to T cells. MCMV m144 serves as an MHC class I homologue which serves as a decoy to avoid NK cell recognition due to low MHC class I levels on the cell surface.

class I-restricted antigen presentation (Heemels and Ploegh, 1995). Briefly, intracellular viral antigens are processed by the proteasome, which is a multi-subunit proteolytic complex. The degraded peptide is delivered to the endoplasmic reticulum (ER) by the transporters associated with antigen presentation (TAPs). Subsequently, the MHC class I heavy chains combine with β2 microglobulin and viral peptide to form a complex with the chaperone calnexin prior to presentation on the cell surface where recognition by CD8$^+$ (cytotoxic) T lymphocytes occurs. CD8$^+$ T cells limit viral infection through direct lysis of infected cells as well as by secretion of cytokines with antiviral activity.

6.5.1.1 HSV. Work from 2 groups using cell culture systems showed that ICP47 inhibited peptide translocation across the ER membrane with concomitant retention of MHC class I in the ER (Fruh *et al.*, 1995; Hill *et al.*, 1995; York *et al.*, 1994). It was shown that HSV-1 recombinants lacking ICP47 failed to cause this ER retention and that ICP47 colocalized and physically associated with the TAPs. What was lacking, at this point, however, was a demonstration that ICP47 was a virulence determinant *in vivo*. This issue was compounded by the fact that ICP47 bound to human TAP with a 100-fold higher affinity than mouse TAP, suggesting that it may be difficult to discern a phenotype for an ICP47 mutant in mice (Ahn *et al.*, 1996). Indeed, it was shown that ICP47-deleted viruses behaved identically to wild-type in terms of corneal replication, and stromal and periocular disease (Goldsmith *et al.*, 1998). It was shown, however, that ICP47-deleted viruses are significantly less neurovirulent and that $CD8^+$ T cell depletion restores virulence to an ICP47 but not a glycoprotein E mutant. This result, although consistent with the cell culture data, shows the limitations of use of the mouse as an animal model. As noted above, ICP47 is a poor inhibitor of mouse TAP and the reason why ICP47 would impact neurovirulence but not peripheral replication is unclear. Whatever the explanation, it is likely that this phenotype represents a gross underestimate of the impact of ICP47 upon pathogenesis.

Another HSV gene product capable of interfering with CTL responses is the virion host shutoff, or *vhs*. *Vhs* is packaged within the tegument of the virus allowing it to exert its effects immediately upon infection prior to *de novo* viral gene expression (Read and Frenkel, 1983). *Vhs*-induced shutoff leads to polysomal disaggregation and non-specific cytoplasmic degradation of cellular mRNAs and viral mRNAs belonging to all three kinetic classes (Kwong and Frenkel, 1987). HSV-2 infected human fibroblasts are poorly targeted by $CD8^+$ CTL unless the virus is deleted of *vhs* activity (Tigges *et al.*, 1996). This suggests that *vhs* may therefore have an immunomodulatory function, consistent with the observations that viruses lacking *vhs* are highly attenuated in mouse models of infection (Smith *et al.*, 2000; Strelow and Leib, 1995).

6.5.1.2 MCMV. The other system in which viral recombinants have been generated to elucidate immune evasion of T cell responses is MCMV. Indeed, this system has rightly attracted a great deal of attention due to the ability to use the natural murine host experimentally, and the availability of the vast array of reagents for exploring mouse immunology. The homology between many MCMV and HCMV genes has also allowed parallels to be drawn between the murine and human versions of CMV. Although multiple genes have been identified with immuno-evasive properties, only a subset have been deleted from the virus and their impact on pathogenesis determined. Two interesting examples of such genes, m144 and m152 are described here.

MHC class I expression not only controls cytotoxic T cell (CTL) function, but also natural killer (NK) cell function. NK and CTLs, however, respond in different ways. Recognition of the MHC class I-peptide complex results in activation of CTLs, but results in inhibition of the lytic action of NK cells (Brutkiewicz and Welsh, 1995; Karre, 1995). This mechanism allows the host with complementing defences. Presentation of viral peptide with MHC leads to CTL-mediated killing, and reduced MHC I expression leads to NK cell killing. Teleologically speaking, this presents a dilemma for viruses which downregulate MHC I expression in that they now become vulnerable to NK attack. The CMVs therefore express an MHC class I decoy molecule to potentially inhibit NK cell lysis of infected cells. The HCMV version is UL18, although the role of this gene has not been determined in the context of viral infection, although it is clearly capable of conferring resistance to NK cell attack when expressed on the surface of cells (Reyburn *et al.*, 1997). A number of studies have examined m144 of MCMV and shown that it has significant sequence homology with the MHC class I heavy chains suggesting it too may function as a decoy molecule (Farrell *et al.*, 1997). A recombinant MCMV from which the m144 gene had been deleted showed severely restricted acute stage replication in mice, and the defect in replication was restored by depletion of NK cells. This strongly suggested, and has been subsequently confirmed, that the m144 gene has an immuno-evasive function through interference with NK cell-mediated clearance.

The m152 gene of MCMV is a further excellent example of the use of recombinant viruses to elucidate pathogenesis. The product of this gene, the gp40 protein has been shown to block the export of MHC class I complexes from the ER, preventing antigen presentation to CTLs (Ziegler *et al.*, 1997). As with previous examples, one sentinel study has defined the role of this gene *in vivo* (Krmpotic *et al.*, 1999). Deletion of m152 resulted in a virus which grew normally in tissue culture, but grew with significantly reduced titres in mice, both in spleen and lungs, and was significantly reduced for virulence. Significant, however, was that the replication of the m152 mutants were restored to wild-type levels in mice which lacked either β2 microglobulin, CD8 or T cells. Moreover, it was shown that in adoptive transfer experiments the m152-deleted virus was especially sensitive to both MCMV-primed and naive lymphocyte control. Therefore, despite the presence of multiple functions to regulate MHC class I surface expression (Kleijnen *et al.*, 1997; Reusch *et al.*, 1999; Ziegler *et al.*, 1997), deletion of m152 alone is sufficient to render the virus significantly more susceptible to lymphocyte attack.

6.5.2 Herpesvirus-encoded Fc receptors: pitfalls for the unwary

The herpesviruses and coronaviruses express immunoglobulin G Fc-binding proteins on the surfaces of infected cells (Oleszak and Leibowitz, 1990; Watkins, 1964). The encoding of such receptors could clearly allow the virus to escape the effects of antibody attack in all of its manifestations. The presence of such receptors has been known for many years, but what has been lacking until recently is a clear demonstration of the importance of these receptors to pathogenesis *in vivo*. In this section some of this research will be explored highlighting some unexpected results that suggest that this function is actually dispensable for virulence of certain viruses.

HSV-encoded Fc receptor activity was first shown back in 1964 when a simple rosetting assay was performed on HSV-infected cells with IgG-coated sheep erythrocytes in culture (Watkins, 1964). A great deal of research has ultimately shown that glycoprotein E (gE) and gI form an oligomeric complex with capability of binding the Fc domain of IgG, both as a monomer and in aggregates (Bell *et al.*, 1990; Hanke *et al.*, 1990; Johnson *et al.*, 1988). gE can function alone to bind Fc with a low affinity, binding to aggregates but not monomers (Dubin *et al.*, 1990). Recombinant viruses lacking gE and gI were generated and were found to have inherent defects in cell-to-cell and neuron-to-neuron spread, with associated small plaque formation in certain cell types in culture (Balan *et al.*, 1994; Dingwell *et al.*, 1995). This clearly presented a problem with respect to the determination of the role of Fc receptor activity in pathogenesis since any attenuation seen *in vivo* would likely be due to an inherent problem with spread, not a lack of immune evasion. An extensive mutagenesis strategy was employed to show that the functional domain of gE involved in cell-to-cell spread was separable from that which was critical for Fc receptor activity (Dubin *et al.*, 1994; Weeks *et al.*, 1997). The next obvious experiment was to infect mice with a virus lacking the Fc-binding domain of gE that was phenotypically normal for cell-to-cell spread (Nagashunmugam *et al.*, 1998). The caveat with this simple approach, however, was that the HSV-1 Fc receptor does not bind murine IgG. Not surprisingly, therefore, no gross difference in virulence in mice was found. An elegant addendum to these experiments, however, was to determine whether passive immunization of mice with human anti-HSV IgG would affect the replication of wild-type or the gE-deleted virus. In these experiments, the human serum caused a significant decrease in both titre and clinical disease in mice infected with the deleted virus, but not wild-type or marker-rescued viruses. These data showed, therefore, that the HSV-1 Fc-receptor activity was important for allowing HSV to evade antibody attack and may explain, at least in part, why HSV reactivation and disease occurs in humans despite the presence of high levels of neutralizing antibodies.

Other herpesviruses such as HSV-2, varicella-zoster virus (VZV), and human and murine CMV also express Fc-binding proteins (Litwin *et al.*, 1990; MacCormac and Grundy, 1996; Thale *et al.*, 1994). VZV and HCMV have no simple animal models, so the impact of their Fc receptors on virulence has not been addressed. Murine CMV, however, has been studied and it has been shown that its Fc receptor is an 88-kDa early glycoprotein designated *fcr-1*. By extrapolation from HSV-1, it was predicted that *fcr-1* would play a role in pathogenesis by virtue of its ability to evade antibody responses (Crnkovic-Mertens *et al.*, 1998). Recombinant viruses were constructed from which *fcr-1* was deleted and it was shown that cells infected with these viruses lacked IgG-binding activity and that growth *in vitro* was unaffected. Growth *in vivo* in contrast was severely affected, consistent with the potential role for *fcr-1* in immuno-evasion. Surprisingly, however, the *fcr-1* mutants also grew poorly in μMT/μMT B-cell-deficient mice which lack antibody, demonstrating that the attenuation of *fcr-1* mutants was not due to antibody control. The one caveat with this study is that *fcr-1* may mediate viral spread in some organs *in vivo*, analogous to the situation with HSV-1 gE and gI. Nevertheless, this particular study highlights the necessity for careful *in vivo* genetic analysis of host–pathogen interactions, and that extrapolation from related viral systems can lead to false assumptions regarding viral determinants of pathogenesis.

6.5.3 Herpesviruses and the regulation of complement activity

Complement is a serum enzyme system which upon contact with microbes acts as one of the initial innate defences. There are a large number of components in the complement cascade and a small stimulus can trigger robust activity downstream. Antiviral activities include blocking of attachment, aggregation of viral particles to facilitate phagocytosis, or direct lysis of the viral envelope. Proteins with complement regulatory activities have been shown in many herpesviruses including HSV-1, pseudorabies virus, HCMV, herpesvirus saimiri, EBV and γHV68, as well as in HIV and vaccinia (Albrecht and Fleckenstein, 1992; Friedman *et al.*, 1984; Huemer *et al.*, 1992; Kapadia *et al.*, 1999; Kotwal *et al.*, 1990; Mold *et al.*, 1988; Schmitz *et al.*, 1995; Spiller *et al.*, 1997). The importance of these functions for pathogenesis of these viruses has not yet been demonstrated except for HSV-1 and vaccinia (Isaacs *et al.*, 1992; Lubinski *et al.*, 1999).

The glycoprotein gC of HSV-1 blocks activation of the complement cascade through binding of the critical complement component C3b as well as through blocking of properdin binding to C3b and C5 (Kostavasili *et al.*, 1997). In this way, gC prevents activation through both the classical and alternative pathways as well as inhibiting formation of the membrane attack complex. This interference with complement function was discovered *in vitro* prior to the more recent work done *in vivo*. Two studies have now shown that a gC null virus was significantly less virulent and grew to lower titres than wild-type or maker-rescued virus (Friedman *et al.*, 1996; Lubinski *et al.*, 1999). Moreover, when C3-deficient guinea pigs (Lubinski *et al.*, 1998) or C3 knockout mice were infected, the gC null virus grew to significantly higher titres than in wild-type animals, although its titres and disease scores were not restored exactly to that seen for control viruses. This suggested that although gC plays an immuno-evasive role *in vivo* through the complement system, gC must have other functions *in vivo* in addition to its interaction with C3. A further study has generated further recombinant viruses to examine the roles of two of the separable domains of gC, one which interacts with C3, and another which prevents C5 and properdin binding to C3. This study showed that whilst both domains are important for pathogenesis, the C3-binding domain was far more important. It was shown that a virus lacking the C3-binding domain was significantly less virulent than a virus lacking the C5/properdin domain, and equivalent in virulence to a doubly-deleted virus. These studies together have shown that gC is a major player in immune evasion through its C3-binding domain and explain why HSV is resistant to attack by complement.

Judging by the number of genes with complement-regulatory activities, resistance to attack by complement is a common theme among herpesviruses. Three of the gammaherpesviruses, EBV, herpesvirus saimiri, and γHV68 have such activities, although γHV68 is the only one among these viruses with a good model for pathogenesis and facile reverse genetics. The field is ripe for the assessment of the role of complement regulation in pathogenesis of γHV68 and other members of the herpesvirus family.

6.5.4 Herpesviruses and the regulation of cytokine activity

A large number of viruses, especially the poxviruses, retroviruses and herpesviruses are capable of interfering with the regulation of the immune response. Two common ways that viruses accomplish this is by encoding homologues or receptors for cytokines and chemokines. Some of these so-called 'virokines' and 'viroceptors' have now been identified in many large DNA viruses (McFadden *et al.*, 1998). An interesting recent discovery of a broad spectrum secreted chemokine binding protein in γHV68 (Parry *et al.*, 2000). Some of these factors have now been shown to have important roles in pathogenesis in the poxviruses, rat CMV and MCMV (Beisser *et al.*, 1998; Fleming *et al.*, 1999; McFadden *et al.*, 1998). The chemokines mediate early inflammatory responses to infection. With respect to MCMV, the chemokine homologue m131/129 was deleted from the virus resulting in significantly reduced dissemination of virus and infection of the salivary glands (Fleming *et al.*, 1999). This study strongly suggested that chemokine homologues may be important for immune system evasion early in infection, particularly from the effects of NK and T cells.

Better characterized for the herpesviruses are viral factors that inhibit responses to tumor necrosis factor and interferons, although as has arisen throughout this discussion, relatively few examples are available where the factor has a proven role in pathogenesis. This general area for other DNA viruses (herpes-, pox-, adeno- and hepatitis viruses) has been the subject of an excellent recent review (Krajcsi and Wold, 1998). One pathway that has been especially well-studied is the interference with interferon (IFN)-mediated shutoff of viral replication. Interferons are pivotal mediators of innate immunity to viruses. As shown in *Figure 6.6*, IFNs act through a number of pathways to induce the so-called 'antiviral state' within a cell. Not surprisingly, viruses have evolved a myriad of strategies to overcome these blocks. Two examples from the herpesvirus family are the Epstein-Barr virus-encoded RNAs (EBERs) of EBV and ICP34.5 of HSV-1 (Chou *et al.*, 1995; Sharp *et al.*, 1993). Both of these gene products are capable of preventing blockage of viral replication through the IFN-induced double stranded RNA-dependent protein kinase or PKR (Gale and Katze, 1998; Roberts *et al.*, 1976). Activation of PKR by IFNs and ds RNA binding leads to phosphorylation and inactivation of eukaryotic translation initiation factor 2α (eIF-2α) and a subsequent shutdown of host and viral protein synthesis and replication.

The EBERs operate by binding PKR and preventing its activation (Clemens *et al.*, 1994). The likely role of the EBERs in pathogenesis, however, remains controversial. There is good evidence from different labs that these RNAs can bind PKR and alter its function *in vitro*. One study, however, generated an EBV recombinant lacking the

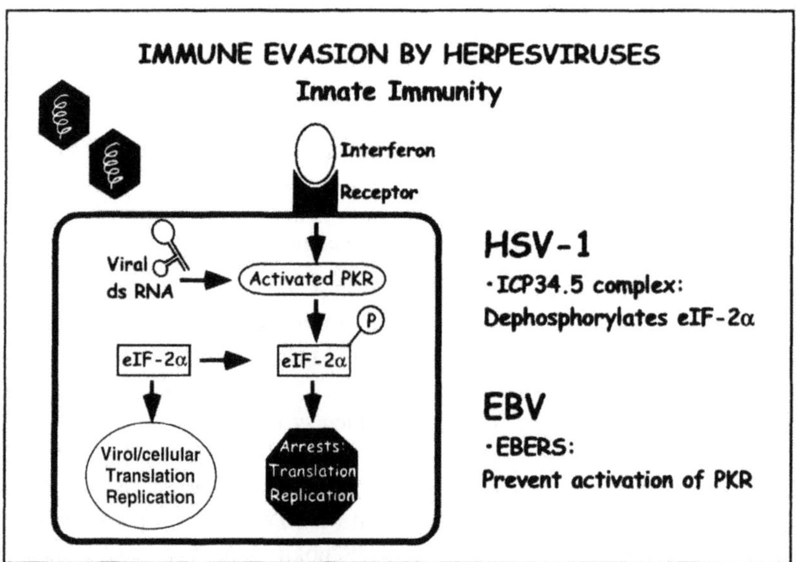

Figure 6.6. Evasion of interferon-induced antiviral by herpesviruses by interference with PKR. Interferon induces expression of PKR which is also activated through viral double-stranded RNA binding. One of the activities of PKR is to phosphorylate eIF2α leading to cessation of protein synthesis and blockage of viral replication. To evade this antiviral mechanism, HSV-1 ICP34.5 binds to protein phosphatase 1α and dephosphorylates eIF2α, lifting the block on viral replication.

EBERs and showed that EBV sensitivity to interferon was not altered (Swaminathan *et al.*, 1992). This suggests that the EBERs are actually unlikely to play a major role in downregulating IFN responses *in vivo* although the lack of an animal model for EBV precludes definitive proof. In contrast, however, it has been known for many years that ICP34.5 plays a critical role in pathogenesis of HSV-1 (Bolovan *et al.*, 1994; Chou *et al.*, 1990). Viruses mutated at the ICP34.5 locus show up to a 100 000-fold increase in LD_{50} following intracerebral injection in mice. Work done in cell culture using recombinant viruses showed that ICP34.5 binds to a protein phosphatase and causes it to dephosphorylate eIF-2α, thereby potentially negating the antiviral effects of PKR (He *et al.*, 1997, 1998). What was lacking was data linking these two observations. Recent data has shown that an HSV-1 mutant in gene ICP34.5 grew to normal levels and showed wild-type virulence in mice lacking both type 1 (IFN$\alpha\beta$) and type 2 (IFNγ) IFN receptors (Leib *et al.*, 1999). This was despite the fact that in normal mice the mutant exhibited a 10 000-fold reduction in replication and neurovirulence. Deletion of interferon receptors, however, also resulted in increased replication in mice of herpes simplex viruses with null mutations in several other genes. Increased virulence in the IFN receptor-deleted mice was therefore not a specific property of viruses lacking ICP34.5, but was consistent with the idea that ICP34.5 targeted an IFN-dependent pathway. A subsequent study showed that an ICP34.5-deleted virus exhibits wild-type replication and virulence in a host from which the PKR gene has been deleted (Leib *et al.*, 2000). The restoration of virulence was shown to be specific to ICP34.5 and PKR using additional host and viral mutants. This approach of using recombinant viruses to infect animals with null mutations in host defence genes provides a useful and rigorous genetic test for identifying *in vivo* mechanisms and targets of microbial virulence genes.

6.6 Conclusions

The exponential increase in the generation and analysis of primary DNA sequences from viruses has allowed the virtual armchair prediction of gene function, and by homology many predictions can be made regarding the possible roles of genes in pathogenesis. The increasing technology, particularly using bacterial artificial chromosomes, by which viral genomes can now be manipulated in a rational way will pave the way to a parallel exponential increase in our knowledge of the functions of these genes *in vivo*. The realism of animal models must always be considered and there are examples of null mutations made in highly conserved viral genes which, when deleted, produce no viruses with discernible phenotype in animal models (Jun *et al.*, 1998). It is likely that many determinants of pathogenesis are species-specific and models such as MCMV, rat CMV, and γHV68 are therefore highly valuable since the natural host can be infected.

The study of viral pathogenesis is currently being greatly advanced by generation and use of transgenic and knockout mice. Generation of mice expressing transgenes of viral origin has been useful in elucidating the function of the viral gene in question (Kawamura *et al.*, 1997; van Dyk *et al.*, 1999). Perhaps more significant to the study of molecular pathogenesis, a dual genetic approach, involving mutagenesis of both pathogen and host, has been and will be critical for the definition of pathways and precise molecular targets in disease. In particular, this approach will be useful for the definition of viral determinants of virulence and host resistance factors. Three general criteria have therefore been proposed for such future definitions (Krmpotic *et al.*, 1999; Leib *et al.*, 2000). First, the attenuation of the specifically mutated virus must be restored in terms of growth and virulence in a host lacking the target of the mutated viral gene. Second, the growth and virulence of the specifically mutated virus must not be restored in a host lacking other viral resistance pathways that are independent of the pathway in question. Third, attenuated viral strains with mutations in genes irrelevant to the one in question must remain attenuated in a host deleted of the resistance factor being analysed. The second and third criteria are especially important for proof of specificity. They ensure that the particular deficient host does not restore every attenuated virus to full virulence, and that the restoration of virulence in that host is specific to a particular virulence determinant. These criteria outline testable scenarios for determining mechanisms of microbial virulence *in vivo*. The data from such studies will play a critical role in the future design of vaccines and antimicrobial chemotherapies for all viruses.

Acknowledgements

David Leib is supported by National Institutes of Health grants EY10707 and EY09083 and is a recipient of a Robert McCormick Scholarship from Research to Prevent Blindness. Also acknowledged are helpful discussions with Peggy MacDonald, Lynda Morrison, Sam Speck, Skip Virgin and members of their laboratories during lab meetings.

References

Ackland-Berglund CE, Davido DJ, Leib DA. (1995) The roles of the cAMP-response element and TATA box in expression of the herpes simplex virus type 1 latency-associated transcripts. *Virology* **210**: 141–151.

Adler H, Koszinowski U, Messerle M. (1999). Unpublished data.

Ahn K, Meyer TH, Uebel S, Sempe P, Djaballah H, Yang Y, Peterson PA, Fruh K, Tampe R. (1996) Molecular mechanism and species specificity of TAP inhibition by herpes simplex virus ICP47. *EMBO J.* **15**: 3247–3255.

Albrecht JC, Fleckenstein B. (1992) New member of the multigene family of complement control proteins in herpesvirus saimiri. *J Virol.* **66**: 3937–3940.

Andersson M, Paabo S, Nilsson T, Peterson PA. (1985) Impaired intracellular transport of class I MHC antigens as a possible means for adenoviruses to evade immune surveillance. *Cell* **43**: 215–222.

Balan P, Davis-Poynter N, Bell S, Atkinson H, Browne H, Minson T. (1994) An analysis of the in vitro and in vivo phenotypes of mutants of herpes simplex virus type 1 lacking glycoproteins gG, gE, gI or the putative gJ. *J. Gen. Virol.* **75**: 1245–1258.

Beisser PS, Vink C, Van Dam JG, Grauls G, Vanherle SJ, Bruggeman CA. (1998) The R33 G protein-coupled receptor gene of rat cytomegalovirus plays an essential role in the pathogenesis of viral infection. *J .Virol.* **72**: 2352–2363.

Bell S, Cranage M, Borysiewicz L, Minson T. (1990) Induction of immunoglobulin G Fc receptors by recombinant vaccinia viruses expressing glycoproteins E and I of herpes simplex virus type 1. *J.*

Virol. **64**: 2181–2186.

Boehmer PE, Lehman IR. (1997) Herpes simplex virus DNA replication. *Annu Rev. Biochem.* **66**: 347–384.

Bolovan CA, Sawtell NM, Thompson RL. (1994) ICP34.5 mutants of herpes simplex virus type 1 strain 17syn+ are attenuated for neurovirulence in mice and for replication in confluent primary mouse embryo cell cultures. *J. Virol.* **68**: 48–55.

Borst EM, Hahn G, Koszinowski UH, Messerle M. (1999) Cloning of the human cytomegalovirus (HCMV) genome as an infectious bacterial artificial chromosome in *Escherichia coli*: a new approach for construction of HCMV mutants. *J. Virol.* **73**: 8320–8329.

Brutkiewicz RR, Welsh RM. (1995) Major histocompatibility complex class I antigens and the control of viral infections by natural killer cells. *J. Virol.* **69**: 3967–3971.

Cai W, Astor TL, Liptak LM, Cho C, Coen DM, Schaffer PA. (1993) The herpes simplex virus type 1 regulatory protein ICP0 enhances virus replication during acute infection and reactivation from latency. *J. Virol.* **67**: 7501–7512.

Campbell ME, Palfreyman JW, Preston CM. (1984) Identification of herpes simplex virus DNA sequences which encode a trans-acting polypeptide responsible for stimulation of immediate early transcription. *J. Mol. Biol.* **180**: 1–19.

Chou J, Kern ER, Whitley RJ, Roizman B. (1990) Mapping of herpes simplex virus-1 neurovirulence to gamma 134.5, a gene nonessential for growth in culture. *Science* **250**: 1262–1266.

Chou J, Chen JJ, Gross M, Roizman B. (1995) Association of a M (r) 90,000 phosphoprotein with protein kinase PKR in cells exhibiting enhanced phosphorylation of translation initiation factor eIF-2 alpha and premature shutoff of protein synthesis after infection with gamma 134.5- mutants of herpes simplex virus 1. *Proc. Natl Acad. Sci. USA* **92**: 10516–10520.

Clemens MJ, Laing KG, Jeffrey IW *et al.* (1994) Regulation of the interferon-inducible eIF-2 alpha protein kinase by small RNAs. *Biochimie* **76**: 770–778.

Clements GB, Stow ND. (1989) A herpes simplex virus type 1 mutant containing a deletion within immediate early gene 1 is latency-competent in mice. *J. Gen. Virol.* **70**: 2501–2506.

Cohen J I, Seidel KE. (1993) Generation of varicella-zoster virus (VZV) and viral mutants from cosmid DNAs: VZV thymidylate synthetase is not essential for replication in vitro. *Proc. Natl Acad. Sci. USA* **90**: 7376–7380.

Crnkovic-Mertens I, Messerle M, Milotic I *et al.* (1998) Virus attenuation after deletion of the cytomegalovirus Fc receptor gene is not due to antibody control. *J. Virol.* **72**: 1377–1382.

Cunningham C, Davison AJ. (1993) A cosmid-based system for constructing mutants of herpes simplex virus type 1. *Virology* **197**: 116–124.

Davison AJ, Scott JE. (1986) The complete DNA sequence of varicella-zoster virus. *J. Gen. Virol.* **67**: 1759–1816.

Delecluse HJ, Hilsendegen T, Pich D, Zeidler R, Hammerschmidt W. (1998) Propagation and recovery of intact, infectious Epstein-Barr virus from prokaryotic to human cells. *Proc. Natl Acad. Sci. USA* **95**: 8245–8250.

DeLuca NA, Schaffer PA. (1985) Activation of immediate-early, early, and late promoters by temperature-sensitive and wild-type forms of herpes simplex virus type 1 protein ICP4. *Mol. Cell Biol.* **5**: 1997–2208.

DeLuca NA, McCarthy AM, Schaffer PA. (1985) Isolation and characterization of deletion mutants of herpes simplex virus type 1 in the gene encoding immediate-early regulatory protein ICP4. *J. Virol.* **56**: 558–570.

Dingwell KS, Doering LC, Johnson DC. (1995) Glycoproteins E and I facilitate neuron-to-neuron spread of herpes simplex virus. *J. Virol.* **69**: 7087–7098.

Doherty PC, Allan W, Eichelberger M, Carding SR. (1992) Roles of alpha beta and gamma delta T cell subsets in viral immunity. *Annu. Rev. Immunol.* **10**: 123–151.

Dubin G, Frank I, Friedman HM. (1990) Herpes simplex virus type 1 encodes two Fc receptors which have different binding characteristics for monomeric immunoglobulin G (IgG) and IgG complexes. *J. Virol.* **64**: 2725–2731.

Dubin G, Basu S, Mallory DL, Basu M, Tal-Singer R, Friedman HM. (1994) Characterization of domains of herpes simplex virus type 1 glycoprotein E involved in Fc binding activity for immunoglobulin G aggregates. *J. Virol.* **68**: 2478–2485.

Everett RD. (1984) Trans activation of transcription by herpes virus products: requirement for two

HSV-1 immediate-early polypeptides for maximum activity. *EMBO J.* **3**: 3135–3141.

Everett RD, Barlow P, Milner A, Luisi B, Orr A, Hope G, Lyon D. (1993) A novel arrangement of zinc-binding residues and secondary structure in the C3HC4 motif of an alpha herpes virus protein family. *J. Mol. Biol.* **234**: 1038–1047.

Everett RD, Meredith M, Orr A, Cross A, Kathoria M, Parkinson J. (1997) A novel ubiquitin-specific protease is dynamically associated with the PML nuclear domain and binds to a herpesvirus regulatory protein [corrected and republished article originally printed in EMBO J 1997 Feb 3;16 (3):566–77]. *EMBO J.* **16**: 1519–1530.

Everett RD, Freemont P, Saitoh H, Dasso M, Orr A, Kathoria M, Parkinson J. (1998) The disruption of ND10 during herpes simplex virus infection correlates with the Vmw110- and proteasome-dependent loss of several PML isoforms. *J. Virol.* **72**: 6581–6591.

Everett RD, Earnshaw WC, Findlay J, Lomonte P. (1999) Specific destruction of kinetochore protein CENP-C and disruption of cell division by herpes simplex virus immediate-early protein Vmw110. *EMBO J.* **18**: 1526–1538.

Farrell HE, Vally H, Lynch DM, Fleming P, Shellam GR, Scalzo AA, Davis-Poynter NJ. (1997) Inhibition of natural killer cells by a cytomegalovirus MHC class I homologue in vivo [see comments]. *Nature* **386**: 510–514.

Farrell MJ, Dobson AT, Feldman LT. (1991) Herpes simplex virus latency-associated transcript is a stable intron. *Proc. Natl Acad. Sci. USA* **88**: 790–794.

Fleming P, Davis-Poynter N, Degli-Esposti M, Densley E, Papadimitriou J, Shellam G, Farrell H. (1999) The murine cytomegalovirus chemokine homolog, m131/129, is a determinant of viral pathogenicity. *J. Virol.* **73**: 6800–6809.

Friedman HM, Cohen GH, Eisenberg RJ, Seidel CA, Cines DB. (1984) Glycoprotein C of herpes simplex virus 1 acts as a receptor for the C3b complement component on infected cells. *Nature* **309**: 633–635.

Friedman HM, Wang L, Fishman NO, Lambris JD, Eisenberg RJ, Cohen GH, Lubinski J. (1996) Immune evasion properties of herpes simplex virus type 1 glycoprotein gC. *J. Virol.* **70**: 4253–4260.

Fruh K, Ahn K, Djaballah H *et al.* (1995) A viral inhibitor of peptide transporters for antigen presentation. *Nature* **375**: 415–418.

Gale M, Jr, Katze MG. (1998) Molecular mechanisms of interferon resistance mediated by viral-directed inhibition of PKR, the interferon-induced protein kinase. *Pharmacol. Ther.* **78**: 29–46.

Ginsberg HS, Lundholm-Beauchamp U, Horswood RL, Pernis B, Wold WS, Chanock RM, Prince GA. (1989) Role of early region 3 (E3) in pathogenesis of adebovirus disease. *Proc. Natl Acad. Sci. USA* **86**: 3823–3827.

Goins WF, Sternberg LR, Croen KD. (1994) A novel latency-active promoter is contained within the herpes simplex virus type 1 UL flanking repeats. *J. Virol.* **68**: 2239–2252.

Goldsmith K, Chen W, Johnson DC, Hendricks RL. (1998) Infected cell protein (ICP)47 enhances herpes simplex virus neurovirulence by blocking the $CD8^+$ T cell response. *J. Exp. Med.* **187**: 341–348.

Goldstein DJ, Weller SK. (1988) An ICP6::lacZ insertional mutagen is used to demonstrate that the UL52 gene of herpes simplex virus type 1 is required for virus growth and DNA synthesis. *J. Virol.* **62**: 2970–2977.

Goodpasture EW. (1929) Herpetic infection with especial reference to involvement of the nervous system. *Medicine* **8**: 223–243.

Hammerschmidt W, Sugden B. (1989) Genetic analysis of immortalizing functions of Epstein-Barr virus in human B lymphocytes. *Nature* **340**: 393–397.

Hanke T, Graham FL, Lulitanond V, Johnson DC. (1990) Herpes simplex virus IgG Fc receptors induced using recombinant adenovirus vectors expressing glycoproteins E and I. *Virology* **177**: 437–444.

Harris RA, Everett RD, Zhu XX, Silverstein S, Preston CM. (1989) Herpes simplex virus type 1 immediate-early protein Vmw110 reactivates latent herpes simplex virus type 2 in an in vitro latency system. *J. Virol.* **63**: 3513–3515.

He B, Gross M, Roizman B. (1997) The gamma (1)34.5 protein of herpes simplex virus 1 complexes with protein phosphatase 1alpha to dephosphorylate the alpha subunit of the eukaryotic translation initiation factor 2 and preclude the shutoff of protein synthesis by double-stranded RNA-activated protein kinase. *Proc. Natl Acad. Sci. USA* **94**: 843–848.

He B, Gross M, Roizman B. (1998) The gamma134.5 protein of herpes simplex virus 1 has the structural and functional attributes of a protein phosphatase 1 regulatory subunit and is present in a high molecular weight complex with the enzyme in infected cells. *J. Biol. Chem.* **273**: 20737–20743.

Heemels MT, Ploegh H. (1995) Generation, translocation, and presentation of MHC class I-restricted peptides. *Annu. Rev. Biochem.* **64**: 463–491.

Hill JM, Sedarati F, Javier RT, Wagner EK, Stevens JG. (1990) Herpes simplex virus latent phase transcription facilitates in vivo reactivation. *Virology* **174**: 117–125.

Hill A, Jugovic P, York I, *et al.* (1995) Herpes simplex virus turns off the TAP to evade host immunity. *Nature* **375**: 411–415.

Hobbs WE, 2nd, DeLuca NA. (1999) Perturbation of cell cycle progression and cellular gene expression as a function of herpes simplex virus ICP0. *J. Virol.* **73**: 8245–8255.

Honess RW, Roizman B. (1974) Regulation of herpesvirus macromolecular synthesis. I. Cascade regulation of the synthesis of three groups of viral proteins. *J. Virol.* **14**: 8–19.

Horsburgh BC, Hubinette MM, Tufaro F. (1999) Genetic manipulation of herpes simplex virus using bacterial artificial chromosomes. *Methods Enzymol.* **306**: 337–352.

Huemer HP, Larcher C, Coe NE. (1992) Pseudorabies virus glycoprotein III derived from virions and infected cells binds to the third component of complement. *Virus Res.* **23**: 271–280.

Isaacs SN, Kotwal GJ, Moss B. (1992) Vaccinia virus complement-control protein prevents antibody-dependent complement-enhanced neutralization of infectivity and contributes to virulence. *Proc. Natl Acad. Sci. USA* **89**: 628–632.

Javier RT, Stevens JG, Dissette VB, Wagner EK. (1988) A herpes simplex virus transcript abundant in latently infected neurons is dispensable for establishment of the latent state. *Virology* **166**: 254–257.

Johnson DC, Frame MC, Ligas MW, Cross AM, Stow ND. (1988) Herpes simplex virus immunoglobulin G Fc receptor activity depends on a complex of two viral glycoproteins, gE and gI. *J. Virol.* **62**: 1347–1354.

Jun PY, Strelow LI, Herman RC, Marsden HS, Eide T, Haarr L, Leib DA. (1998) The UL4 gene of herpes simplex virus type 1 is dispensable for latency, reactivation and pathogenesis in mice. *J. Gen. Virol.* **79**: 1603–1611.

Kapadia SB, Molina H, van Berkel V, Speck SH, Virgin HWT. (1999) Murine gammaherpesvirus 68 encodes a functional regulator of complement activation. *J. Virol.* **73**: 7658–7670.

Karre K. (1995) Express yourself or die: peptides, MHC molecules, and NK cells [comment]. *Science* **267**: 978–979.

Kawaguchi Y, Bruni R, Roizman B. (1997a) Interaction of herpes simplex virus 1 alpha regulatory protein ICP0 with elongation factor 1delta: ICP0 affects translational machinery. *J. Virol.* **71**: 1019–1024.

Kawaguchi Y, Van Sant C, Roizman B. (1997b) Herpes simplex virus 1 alpha regulatory protein ICP0 interacts with and stabilizes the cell cycle regulator cyclin D3. *J. Virol.* **71**: 7328–7336.

Kawakura T, Furusaka A, Koziel MJ, Chung RT, Wang TC, Schmidt EV, Liang TJ. (1997) Transgenic expression of hepatitis C virus structural proteins in the mouse. *Hepatology* **25**: 1014–1021.

Kaye KM, Izumi KM, Li H, Johannsen E, Davidson D, Longnecker R, Kieff E. (1999) An Epstein-Barr virus that expresses only the first 231 LMP1 amino acids efficiently initiates primary B-lymphocyte growth transformation. *J. Virol.* **73**: 10525–10530.

Kleijnen MF, Huppa JB, Lucin P. (1997) A mouse cytomegalovirus glycoprotein, gp34, forms a complex with folded class I MHC molecules in the ER which is not retained but is transported to the cell surface. *EMBO J.* **16**: 685–694.

Kostavasili I, Sahu A, Friedman HM, Eisenberg RJ, Cohen GH, Lambris JD. (1997) Mechanism of complement inactivation by glycoprotein C of herpes simplex virus. *J. Immunol.* **158**: 1763–1771.

Kosz-Vnenchak M, Jacobson J, Coen DM, Knipe DM. (1993) Evidence for a novel regulatory pathway for herpes simplex virus gene expression in trigeminal ganglion neurons. *J. Virol.* **67**: 5383–5393.

Kotwal GJ, Isaacs SN, McKenzie R, Frank MM, Moss B. (1990) Inhibition of the complement cascade by the major secretory protein of vaccinia virus. *Science* **250**: 827–830.

Krajcsi P, Wold WD. (1998) Inhibition of tumor necrosis factor and interferon triggered responses by DNA viruses. *Semin. Cell. Dev. Biol.* **9**: 351–358.

Krmpotic A, Messerle M, Crnkovic-Mertens I, Polic B, Jonjic S, Koszinowski UH. (1999) The immunoevasive function encoded by the mouse cytomegalovirus gene m152 protects the virus against T cell control in vivo. *J. Exp. Med.* **190**: 1285–1296.

Kwong AD, Frenkel N. (1987) Herpes simplex virus-infected cells contain a function (s) that destabilizes both host and viral mRNAs. *Proc. Natl Acad. Sci. USA* **84**: 1926–1930.

LaBoissiere S, O'Hare P. (2000) Analysis of HCF, the cellular cofactor of VP16, in herpes simplex virus-infected cells. *J. Virol.* **74**: 99–109.

Lai MM. (2000) The making of infectious viral RNA: No size limit in sight. *Proc. Natl Acad. Sci. USA* **86**: 5025-5027.

Leib DA, Olivo PD. (1993) Gene delivery to neurons: is herpes simplex virus the right tool for the job? *Bioessays* **15**: 547–554.

Leib DA, Bogard CL, Kosz-Vnenchak M, Hicks KA, Coen DM, Knipe DM, Schaffer PA. (1989a) A deletion mutant of the latency-associated transcript of herpes simplex virus type 1 reactivates from the latent state with reduced frequency. *J. Virol.* **63**: 2893–2900.

Leib DA, Coen DM, Bogard CL. (1989b) Immediate-early regulatory gene mutants define different stages in the establishment and reactivation of herpes simplex virus latency. *J. Virol.* **63**: 759–768.

Leib DA, Harrison TE, Laslo KM, Machalek MA, Moorman NJ, Virgin HW. (1999) Interferons regulate the phenotype of wild-type and mutant herpes simplex viruses in vivo. *J. Exp. Med.* **189**: 663–672.

Leib DA, Machalek MA, Williams BR, Silverman RH, Virgin HW. (2000) Specific phenotypic restoration of an attenuated virus by knockout of a host resistance gene. *Proc. Natl Acad. Sci. USA* **67**: 6097–6101.

Levitskaya J, Coram M, Levitsky V *et al.* (1995) Inhibition of antigen processing by the internal repeat region of the Epstein-Barr virus nuclear antigen-1. *Nature* **375**: 685–688.

Litwin V, Sandor M, Grose C. (1990) Cell surface expression of the varicella-zoster virus glycoproteins and Fc receptor. *Virology* **178**: 263–272.

Lubinski JM, Wang L, Soulika AM. (1998) Herpes simplex virus type 1 glycoprotein gC mediates immune evasion in vivo. *J. Virol.* **72**: 8257–8263.

Lubinski J, Wang L, Mastellos D, Sahu A, Lambris JD, Friedman HM. (1999) In vivo role of complement-interacting domains of herpes simplex virus type 1 glycoprotein gC. *J. Exp. Med.* **190**: 1637–1646.

MacCormac LP, Grundy JE. (1996) Human cytomegalovirus induces an Fc gamma receptor (Fc gammaR) in endothelial cells and fibroblasts that is distinct from the human cellular Fc gammaRs. *J. Infect. Dis.* **174**: 1151–1161.

McFadden G, Lalani A, Everett H, Nash P, Xu X. (1998) Virus-encoded receptors for cytokines and chemokines. *Semin. Cell. Dev. Biol.* **9**: 359–368.

McGeoch DJ, Davison AJ. (1986) DNA sequence of the herpes simplex virus type 1 gene encoding glycoprotein gH, and identification of homologues in the genomes of varicella-zoster virus and Epstein-Barr virus. *Nucleic Acids Res.* **14**: 4281–4292.

Messerle M, Crnkovic I, Hammerschmidt W, Ziegler H, Koszinowski UH. (1997) Cloning and mutagenesis of a herpesvirus genome as an infectious bacterial artificial chromosome. *Proc. Natl Acad. Sci. USA* **94**: 14759–14763.

Mettenleiter TC, Klupp BG, Weiland F, Visser N. (1994) Characterization of a quadruple glycoprotein-deleted pseudorabies virus mutant for use as a biologically safe live virus vaccine. *J. Gen. Virol.* **75**: 1723–1733.

Mitchell WJ, Steiner I, Brown SM, MacLean AR, Subak-Sharpe JH, Fraser NW. (1990) A herpes simplex virus type 1 variant, deleted in the promoter region of the latency-associated transcripts, does not produce any detectable minor RNA species during latency in the mouse trigeminal ganglion. *J. Gen. Virol.* **71**: 953–957.

Mocarski ES, Post LE, Roizman B. (1980) Molecular engineering of the herpes simplex virus genome: insertion of a second L-S junction into the genome causes additional genome inversions. *Cell* **22**: 243–255.

Mold C, Bradt BM, Nemerow GR, Cooper NR. (1988) Epstein-Barr virus regulates activation and processing of the third component of complement. *J. Exp. Med.* **168**: 949–969.

Mossman KL, Saffran HA, Smiley JR. (2000) Herpes simplex virus ICP0 mutants are hypersensitive to interferon. *J. Virol.* **74**: 2052–2056.

Nagashunmugam T, Lubinski J, Wang L *et al.* (1998) In vivo immune evasion mediated by the herpes simplex virus type 1 immunoglobulin G Fc receptor. *J. Virol.* **72**: 5351–5359.

Nichol PF, Chang JY, Johnson EM, Jr, Olivo PD. (1996) Herpes simplex virus gene expression in neurons: viral DNA synthesis is a critical regulatory event in the branch point between the lytic and latent pathways. *J. Virol.* **70**: 5476–5486.

O'Hare P, Hayward GS. (1985) Evidence for a direct role for both the 175,000– and 110,000-molecular-weight immediate-early proteins of herpes simplex virus in the transactivation of delayed-early promoters. *J. Virol.* **53**: 751–760.

Oleszak EL, Leibowitz JL. (1990) Immunoglobulin Fc binding activity is associated with the mouse hepatitis virus E2 peplomer protein. *Virology* **176**: 70–80.

Parry BC, Simas JP, Smith VP, Stewart CA, Minson AC, Efstathiou S, Alcami A. (2000) A broad spectrum secreted chemokine binding protein encoded by a herpesvirus [In Process Citation]. *J. Exp. Med.* **191**: 573–578.

Pellett PE, McKnight JL, Jenkins FJ, Roizman B. (1985) Nucleotide sequence and predicted amino acid sequence of a protein encoded in a small herpes simplex virus DNA fragment capable of trans-inducing alpha genes. *Proc. Natl Acad. Sci. USA* **82**: 5870–5874.

Perng GC, Dunkel EC, Geary PA, *et al.* (1994) The latency-associated transcript gene of herpes simplex virus type 1 (HSV-1) is required for efficient in vivo spontaneous reactivation of HSV-1 from latency. *J. Virol.* **68**: 8045–8055.

Perng GC, Ghiasi H, Slanina SM, Nesburn AB, Wechsler SL. (1996) The spontaneous reactivation function of the herpes simplex virus type 1 LAT gene resides completely within the first 1.5 kilobases of the 8.3- kilobase primary transcript. *J. Virol.* **70**: 976–984.

Perng GC, Slanina SM, Yukht A *et al.* (1999) A herpes simplex virus type 1 latency-associated transcript mutant with increased virulence and reduced spontaneous reactivation. *J. Virol.* **73**: 920–929.

Perng GC, Jones C, Ciacci-Zanella J *et al.* (2000a) Virus-induced neuronal apoptosis blocked by the herpes simplex virus latency-associated transcript. *Science* **287**: 1500–1503.

Perng GC, Slanina SM, Yukht A, Ghiasi H, Nesburn AB, Wechsler SL. (2000b) The latency-associated transcript gene enhances establishment of herpes simplex virus type 1 latency in rabbits. *J. Virol.* **74**: 1885–1891.

Post LE, Roizman B. (1981) A generalized technique for deletion of specific genes in large genomes: alpha gene 22 of herpes simplex virus 1 is not essential for growth. *Cell* **25**: 227–232.

Preston CM. (1979) Control of herpes simplex virus type 1 mRNA synthesis in cells infected with wild-type virus or the temperature-sensitive mutant tsK. *J. Virol.* **29**: 275–284.

Preston CM. (2000) Repression of viral transcription during herpes simplex virus latency. *J. Gen. Virol.* **81**: 1–19.

Rader KA, Ackland-Berglund CE, Miller JK, Pepose JS, Leib DA. (1993) In vivo characterization of site-directed mutations in the promoter of the herpes simplex virus type 1 latency-associated transcripts. *J. Gen. Virol.* **74**: 1859–1869.

Ralph WM, Jr, Cabatingan MS, Schaffer PA. (1994) Induction of herpes simplex virus type 1 immediate-early gene expression by a cellular activity expressed in Vero and NB41A3 cells after growth arrest-release. *J. Virol.* **68**: 6871–6882.

Read GS, Frenkel N. (1983) Herpes simplex virus mutants defective in the virion-associated shutoff of host polypeptide synthesis and exhibiting abnormal synthesis of alpha (immediate early) viral polypeptides. *J. Virol.* **46**: 498–512.

Reusch UMW, Lucin P, Burgert HG, Hengel H, Koszinowski UH. (1999) A cytomegalovirus glycoprotein re-routes MHC class I complexes to lysosomes for degradation. *EMBO J.* **18**: 1081–1091.

Reyburn HT, Mandelboim O, Vales-Gomez M, Davis DM, Pazmany L, Strominger JL. (1997). The class I MHC homologue of human cytomegalovirus inhibits attack by natural killer cells [see comments]. *Nature* **386**: 514–517.

Roberts WK, Hovanessian A, Brown RE, Clemens MJ, Kerr IM. (1976) Interferon-mediated protein kinase and low-molecular-weight inhibitor of protein synthesis. *Nature* **264**: 477–480.

Roizman B. (1996a) Herpes simplex viruses and their replication. In: *Virology*, Third edn, pp. 2231–2295, (Eds DMKBN Fields, PM Howley). Philadelphia: Lippincott-Raven.

Roizman B. (1996b) Herpesviridae. In: *Fields Virology*, pp. 2221–2230. (Eds BN Fields, Knipe DM, Howley PM), Philadelphia: Lippencott-Raven.

Roizman B, Kozak M, Honess RW, Hayward G. (1975) Regulation of herpesvirus macromolecular synthesis: evidence for multilevel regulation of herpes simplex 1 RNA and protein synthesis. *Cold Spring Harb. Symp. Quant. Biol.* **39**: 687–701.

Russell J, Stow ND, Stow EC, Preston CM. (1987) Herpes simplex virus genes involved in latency in vitro. *J. Gen. Virol.* **68**: 3009–3018.

Sacks WR, Schaffer PA. (1987) Deletion mutants in the gene encoding the herpes simplex virus type 1 immediate-early protein ICP0 exhibit impaired growth in cell culture. *J. Virol.* **61**: 829–839.

Sacks WR, Greene CC, Aschman DP, Schaffer PA. (1985) Herpes simplex virus type 1 ICP27 is an essential regulatory protein. *J. Virol.* **55**: 796–805.

Sawtell NM. (1997) Comprehensive quantification of herpes simplex virus latency at the single-cell level. *J. Virol.* **71**: 5423–5431.

Sawtell NM, Thompson RL. (1992) Herpes simplex virus type 1 latency-associated transcription unit promotes anatomical site-dependent establishment and reactivation from latency. *J. Virol.* **66**: 2157–2169.

Schmitz J, Zimmer JP, Kluxen B, Aries S, Bogel M, Gigli I, Schmitz H. (1995) Antibody-dependent complement-mediated cytotoxicity in sera from patients with HIV-1 infection is controlled by CD55 and CD59. *J. Clin. Invest.* **96**: 1520–1526.

Sedarati F, Izumi KM, Wagner EK, Stevens JG. (1989) Herpes simplex virus type 1 latency-associated transcription plays no role in establishment or maintenance of a latent infection in murine sensory neurons. *J. Virol.* **63**: 4455–4458.

Sharp TV, Schwemmle M, Jeffrey I *et al.* (1993) Comparative analysis of the regulation of the interferon-inducible protein kinase PKR by Epstein-Barr virus RNAs EBER-1 and EBER-2 and adenovirus VAI RNA. *Nucleic Acids Res.* **21**: 4483–4490.

Simas JP, Bowden RJ, Paige V, Efstathiou S. (1998) Four tRNA-like sequences and a serpin homologue encoded by murine gammaherpesvirus 68 are dispensable for lytic replication in vitro and latency in vivo. *J. Gen. Virol.* **79**: 149–153.

Smith GA, Enquist LW. (2000) A self-recombining bacterial artificial chromosome and its application for analysis of herpesvirus pathogenesis. *Proc. Natl Acad. Sci. USA* **97**: 4873–4878.

Smith TJ, Ackland-Berglund CE, Leib DA. (2000) Herpes simplex virus virion host shutoff (vhs) activity alters periocular disease in mice *J. Virol.* **74**: 3598–3604.

Soares K, Hwang DY, Ramakrishnan R, Schmidt MC, Fink DJ, Glorioso JC. (1996) cis-acting elements involved in transcriptional regulation of the herpes simplex virus type 1 latency-associated promoter 1 (LAP1) in vitro and in vivo. *J. Virol.* **70**: 5384–5394.

Spaete RR, Mocarski ES. (1987) Insertion and deletion mutagenesis of the human cytomegalovirus genome. *Proc. Natl Acad. Sci. USA* **84**: 7213–7217.

Speck SH, Virgin HW. (1999) Host and viral genetics of chronic infection: a mouse model of gamma-herpesvirus pathogenesis. *Curr. Opin. Microbiol.* **2**: 403–409.

Spiller OB, Hanna SM, Devine DV, Tufaro F. (1997) Neutralization of cytomegalovirus virions: the role of complement. *J. Infect. Dis.* **176**: 339–347.

Spivack JG, Fraser NW. (1987) Detection of herpes simplex virus type 1 transcripts during latent infection in mice [published erratum appears in J Virol 1988 Feb;62 (2):663]. *J. Virol.* **61**: 3841–3847.

Steiner I, Spivack JG, Lirette RP, Brown SM, MacLean AR, Subak-Sharpe JH, Fraser NW. (1989) Herpes simplex virus type 1 latency-associated transcripts are evidently not essential for latent infection. *EMBO J.* **8**: 505–511.

Stern S, Herr W. (1991) The herpes simplex virus trans-activator VP16 recognizes the Oct-1 homeo domain: evidence for a homeo domain recognition subdomain. *Genes Dev.* **5**: 2555–2566.

Stevens JG, Wagner EK, Devi-Rao GB, Cook ML, Feldman LT. (1987) RNA complementary to a herpesvirus alpha gene mRNA is prominent in latently infected neurons. *Science* **235**: 1056–1059.

Stow ND, Stow EC. (1986) Isolation and characterization of a herpes simplex virus type 1 mutant containing a deletion within the gene encoding the immediate early polypeptide Vmw110. *J. Gen. Virol.* **67**: 2571–2585.

Strelow LI, Leib DA. (1995) Role of the virion host shutoff (vhs) of herpes simplex virus type 1 in latency and pathogenesis. *J. Virol.* **69**: 6779–6786.

Sunil-Chandra NP, Efstathiou S, Arno J, Nash AA. (1992) Virological and pathological features of mice infected with murine gamma-herpesvirus 68. *J. Gen. Virol.* **73**: 2347–2356.

Swaminathan S, Huneycutt BS, Reiss CS, Kieff E. (1992) Epstein-Barr virus-encoded small RNAs (EBERs) do not modulate interferon effects in infected lymphocytes. *J. Virol.* **66**: 5133–5136.

Takekoshi M, Maeda-Takekoshi F, Ihara S, Sakuma S, Watanabe Y. (1993) Inducible expression of a foreign gene inserted into the human cytomegalovirus genome. *J. Gen. Virol.* **74**: 1649–1652.

Thale R, Lucin P, Schneider K, Eggers M, Koszinowski UH. (1994) Identification and expression of a murine cytomegalovirus early gene coding for an Fc receptor. *J. Virol.* **68**: 7757–7765.

Thompson RL, Sawtell NM. (1997) The herpes simplex virus type 1 latency-associated transcript gene regulates the establishment of latency. *J. Virol.* **71**: 5432–5440.

Thorley-Lawson DA, Babcock GJ. (1999) A model for persistent infection with Epstein-Barr virus: the stealth virus of human B cells. *Life Sci.* **65**: 1433–1453.

Tigges MA, Leng S, Johnson DC, Burke RL. (1996) Human herpes simplex virus (HSV)-specific CD8+ CTL clones recognize HSV-2-infected fibroblasts after treatment with IFN-gamma or when virion host shutoff functions are disabled. *J. Immunol.* **156**: 3901–3910.

van Dyk LF, Hess JL, Katz JD, Jacoby M, Speck SH, Virgin HW. (1999) The murine gammaherpesvirus 68 v-cyclin gene is an oncogene that promotes cell cycle progression in primary lymphocytes. *J. Virol.* **73**: 5110–5122.

Van Sant C, Kawaguchi Y, Roizman B. (1999) A single amino acid substitution in the cyclin D binding domain of the infected cell protein no. 0 abrogates the neuroinvasiveness of herpes simplex virus without affecting its ability to replicate. *Proc. Natl Acad. Sci. USA* **96**: 8184–8189.

Virgin HW, Latreille P, Wamsley P, Hallsworth K, Weck KE, Dal Canto AJ, Speck SH. (1997) Complete sequence and genomic analysis of murine gammaherpesvirus 68. *J. Virol.* **71**: 5894–5904.

Wagner EK, Devi-Rao G, Feldman LT, Dobson AT, Zhang YF, Flanagan WM, Stevens JG. (1988) Physical characterization of the herpes simplex virus latency-associated transcript in neurons. *J. Virol.* **62**: 1194–1202.

Wagner M, Jonjic S, Koszinowski UH, Messerle M. (1999) Systematic excision of vector sequences from the BAC-cloned herpesvirus genome during virus reconstitution. *J. Virol.* **73**: 7056–7060.

Walker S, Hayes S, O'Hare P. (1994) Site-specific conformational alteration of the Oct-1 POU domain-DNA complex as the basis for differential recognition by Vmw65 (VP16). *Cell* **79**: 841–852.

Wang F, Marchini A, Kieff E. (1991) Epstein-Barr virus (EBV) recombinants: use of positive selection markers to rescue mutants in EBV-negative B-lymphoma cells. *J. Virol.* **65**: 1701–1709.

Watkins J. (1964) Adsorption of sensitized sheep erythrocytes to HeLa cells infected with herpes simplex virus. *Nature* **202**: 1364–1365.

Weeks BS, Sundaresan P, Nagashunmugam T, Kang E, Friedman HM. (1997) The herpes simplex virus-1 glycoprotein E (gE) mediates IgG binding and cell-to-cell spread through distinct gE domains. *Biochem. Biophys. Res. Commun.* **235**: 31–35.

Wilcox CL, Johnson EM, Jr. (1987) Nerve growth factor deprivation results in the reactivation of latent herpes simplex virus in vitro. *J. Virol.* **61**: 2311–2315.

Wildy P, Field HJ, Nash AA. (1982) Classical herpes latency revisited. In: *Virus Persistence*, pp. 133–167. (Eds Mahy ACMBWJ, GK Darby), Cambridge, Cambridge University Press.

Wilson AC, Freiman RN, Goto H, Nishimoto T, Herr W. (1997) VP16 targets an amino-terminal domain of HCF involved in cell cycle progression. *Mol. Cell Biol.* **17**: 6139–6146.

Woisetschlaeger M, Jin XW, Yandava CN, Furmanski LA, Strominger JL, Speck SH. (1991) Role for the Epstein-Barr virus nuclear antigen 2 in viral promoter switching during initial stages of infection. *Proc. Natl Acad. Sci. USA* **88**: 3942–3946.

Yao F, Schaffer PA. (1995) An activity specified by the osteosarcoma line U2OS can substitute functionally for ICP0, a major regulatory protein of herpes simplex virus type 1. *J. Virol.* **69**: 6249–6258.

York IA, Roop C, Andrews DW, Riddell SR, Graham FL, Johnson DC. (1994) A cytosolic herpes simplex virus protein inhibits antigen presentation to $CD8^+$ T lymphocytes. *Cell* **77**: 525–535.

Zhu XX, Chen JX, Young CS, Silverstein S. (1990). Reactivation of latent herpes simplex virus by adenovirus recombinants encoding mutant IE-0 gene products. *J. Virol.* **64**: 4489–4498.

Ziegler H, Thale R, Lucin P *et al.* (1997) A mouse cytomegalovirus glycoprotein retains MHC class I complexes in the ERGIC/cis-Golgi compartments. *Immunity* **6**: 57–66.

Zwaagstra JC, Ghiasi H, Slanina SM *et al.* (1990) Activity of herpes simplex virus type 1 latency-associated transcript (LAT) promoter in neuron-derived cells: evidence for neuron specificity and for a large LAT transcript. *J. Virol.* **64**: 5019–5028.

Zwaagstra JC, Ghiasi H, Nesburn AB, Wechsler SL. (1991) Identification of a major regulatory sequence in the latency associated transcript (LAT) promoter of herpes simplex virus type 1 (HSV-1). *Virology* **182**: 287–297.

Chapter 7

Genetic engineering of animal RNA viruses in order to study and modify pathogenesis

Wendy S. Barclay and David J. Evans

7.1. Introduction

The animal RNA viruses include a large number of important pathogens. Their genetic modification to bring about disease control is an active area of research with significant medical and veterinary impact. The genomes of viruses in this group are comprised of either single or double-stranded RNA, the former being of either positive or negative sense. Replication of the genome involves either a virus encoded RNA-dependent RNA polymerase (RDRP) or, for the retroviruses, a virally encoded reverse transcriptase in combination with cellular polymerase. Neither RDRPs nor reverse transcriptases possess proof-reading activity and thus are inherently prone to a high rate of mutation. As a result RNA virus genomes exist as a quasispecies – a mixture of many different RNA molecules with a consensus sequence from which mutants can emerge when particular growth conditions confer a selective advantage. Historically much has been learned from the study of mutants that arise spontaneously and are then enriched or selected by altering growth conditions. Indeed, currently used live-attenuated vaccine strains of the medically important RNA viruses like poliovirus, measles and mumps have been generated by multiple passage of virus under conditions which have enriched for viruses with enhanced *in vitro* replication at the cost of their fitness *in vivo*.

More recently, techniques to generate RNA viruses from cloned DNA have been developed. The availability of reverse genetics techniques for most representatives of this group has lead to a barrage of exciting experimental data helping us to learn more about the viruses and how they cause disease. In addition it allows us to engineer virus mutants which are candidates for a new generation of live-attenuated vaccines. The ease with which each type of animal RNA viruses can be genetically manipulated varies greatly depending upon the type of genome it possesses. Thus the retroviruses, which can be recovered by simply introducing genomic cDNA alone into cultured cells, are much more amenable to such studies than are the negative sense RNA viruses for which techniques are still being refined. For viruses whose genomes are very large, like coronaviruses, researchers must rely on the inherent ability of the viral genome to undergo recombination since as yet no single cDNA molecule representing the whole

Genetically Engineered Viruses: Development and Applications, edited by C.J.A. Ring and E.D. Blair

genome has been generated. Those viruses whose genomes consist of double-stranded RNA, like reoviruses, are still refractory to genetic manipulation and are not discussed further here. In the following pages, we separately describe the techniques and latest results of genetic modification of the positive-sense RNA viruses and the negative-sense RNA viruses.

7.2 The positive-strand RNA viruses

7.2.1 The manipulation and recovery of positive-strand RNA virus genomes

Purified viral RNA from the positive-sense single-stranded RNA viruses is infectious when introduced into a suitable cell culture system *in vitro* or, in certain cases, when administered to a suitable host animal. These viruses, which include a diverse range of important human and animal pathogens, are therefore particularly amenable to the reverse genetic analysis of genome structure and function both *in vitro* and *in vivo*. Nearly two decades have elapsed since the first recovery of an animal virus from a cDNA clone, and it is therefore unsurprising that the molecular determinants of pathogenesis in some positive-strand RNA viruses are well characterized. Within the constraints of this review it is not possible to cover the full range of studies that have been conducted, and we have therefore largely focused on three representative groups, each illustrating pertinent aspects of the analysis of pathogenesis by reverse genetics.

The single-stranded positive-sense monopartite RNA viruses that infect animals range in genome length from the picornaviruses (c. 7 kb) to the coronaviruses which, at over 30 kb in length, represent some of the largest and most complex functional RNA molecules known. Despite this large size range, the practical approaches used for the cloning, maintenance, manipulation and recovery for most of these viruses are broadly similar. The major exceptions are the very large genomes of the Coronaviridae which have so far proven refractory to the standard techniques that will be outlined below, and which are therefore manipulated in a different manner. Three families of viruses form the core of this part of the review; (i) the *Picornaviridae*, which includes poliovirus, the first animal virus recovered from an infectious cDNA, and the prototype upon which many subsequent studies are based, (ii) the *Flaviviridae*, which include the major human pathogens Dengue, Yellow Fever, and Hepatitis C virus, and (iii) the *Coronaviridae* which include a number of important animal pathogens and illustrate an alternative strategy for genome manipulation.

7.2.1.1 Permissive cell culture systems. The infectious nature of purified 'naked' poliovirus RNA has been known for at least 40 years when Holland and colleagues (Holland *et al.*, 1959) demonstrated – during an analysis of poliovirus tropism – that virus RNA transfected into non-permissive cells (i.e. cells incapable of supporting virus infection) underwent a single round of replication to yield infectious progeny virus. These studies also illustrated an important tenet in the analysis of viruses *in vitro*[1] – the

[1] By which we mean cell culture, using *in vivo* to indicate within the host organism.

concept of a permissive cell culture system. The tissue tropism of poliovirus is broader *in vitro* than *in vivo*, as demonstrated by Kaplan (Kaplan, 1955) who showed that tissues not normally infected *in vivo* (specifically kidney and amnion) could be infected after culture *in vitro*. The subsequent demonstration by Holland (Holland *et al.*, 1959) that poliovirus can replicate when transfected into murine cells, even though the cells are incapable of supporting infection (due to the absence of a suitable receptor), primed our understanding of the intracellular determinants required for virus replication. These studies indicate that poliovirus will infect and replicate in cells in culture that both express a functional receptor and provide the necessary intracellular milieu for replication. In contrast, direct transfection of purified virus RNA bypasses the requirement for a receptor and so can broaden the effective tissue or host tropism.

The failure to generate infectious cDNA clones, or to observe replication, of certain positive strand RNA viruses – such as the human caliciviruses – probably reflects the lack of a suitable cell culture system, rather than a fundamental difference in the mechanism of virus replication.

7.2.2 Development of techniques for engineering positive-strand RNA virus genomes

7.2.2.1 Picornaviridae. The development of techniques for the preparation and cloning of cDNA in the 1970s resulted in the demonstration of the first infectious cDNA of an RNA virus, bacteriophage Qβ, in 1978 (Taniguchi *et al.*, 1978). This was rapidly followed by the seminal demonstration by Racaniello and Baltimore that the 7.5 kb cloned poliovirus cDNA was infectious *in vitro* (Racaniello and Baltimore, 1981). The pBR322-derived cDNA clone was introduced by calcium phosphate-mediated transfection into primate CV-1 cells and yielded approximately five plaque forming units per microgram (pfu μg^{-1}) of DNA. In the absence of any specific eukaryotic promoter in pBR322, expression of the virus was presumed to be due to fortuitous transcription from a cryptic promoter on the plasmid vector. Since the incremental development of improved infectious poliovirus cDNA clones has recently been reviewed (Evans, 1999) only the current optimized clones will be described here.

Rather than rely upon a chance transcription event from the vector sequences of a cDNA clone, the routine practice is to introduce a specific promoter for a DNA dependent RNA polymerase immediately preceding the 5′ end of the sense strand of the picornavirus cDNA (Mizutani and Colonno, 1985; Van der Werf *et al.*, 1986). Both bacteriophage T7 and SP6 promoters are in widespread use. In the case of T7, the maximal RNA yield requires the presence of three guanine nucleotides at the 3′ end of the promoter, transcriptional initiation occurring from the penultimate G residue. This results in the addition of two G nucleotides to the *in vitro* transcript – since these are not present in recovered virus RNA they are presumably removed by the action of cellular RNAses or as a consequence of the replication mechanism of the virus (Mizutani and Colonno, 1985; Van der Werf *et al.*, 1986). The 3′ end of the sense strand of the picornavirus cDNA must be followed by a suitable restriction endonuclease site with which the template can be linearized prior to *in vitro* transcription. Since the presence of non-viral sequences in this location is known to decrease the infectivity of the resulting transcript (Sarnow, 1989), optimized cDNA clones are often engineered to include a

Mlu I site (cleavage at which results in a 5′ A nucleotide) immediately following a poly(A) tail of at least 25 nucleotides (Evans, 1999). The specific infectivity of transcripts produced from such optimized cDNA constructs can be readily tested by transfection – by DEAE-Dextran, electroporation, cationic liposomes or comparable commercial transfection reagents – of permissive adherent cells, followed by overlaying the cells with semi-solid agar and monitoring the cell sheet for the production of distinct plaques. Under these conditions, *in vitro* generated RNA has between 5% and 100% the specific infectivity of purified viral RNA (Sarnow, 1989).

The improved efficiency with which virus can be recovered from cloned cDNA has been mirrored by markedly more efficient strategies for generating the initial full length cDNA clone. These were originally made by laborious assembly of several subgenomic cDNA clones – usually produced by oligo d(T) priming and GC tailing – a strategy dependent upon the availability of suitable restriction sites and commercially available enzymes. Coxsackievirus B3 was synthesized in a single piece – lacking only the two uridine residues from the 5′ end of the sense strand – using this strategy and shown to be infectious. More recently, the availability of PCR-based cloning strategies and developments in the RT-PCR amplification of transcripts of over 7 kb, means that full length infectious clones can be readily generated (Lindberg *et al.*, 1997). These technological enhancements go some way to overcome the massive variability within a population of RNA viruses that arise as a consequence of the lack of a proof-reading capability in the RNA-dependent RNA polymerase – allowing a number of full length cDNAs to be screened for infectivity or, as described below, enabling the assembly of a consensus cDNA. These technical advances also apply to the larger RNA virus genomes described below. Infectious cDNA clones for all six genera of the *Picornaviridae* have been constructed over the last two decades and extensively used for a variety of studies including genome structure and function, pathogenesis, vaccine development, and the generation of expression vectors.

7.2.2.2 Flaviviridae. Although similar in general aspects of genome organization and replication, the *Flaviviridae* family – containing the genera *Flavivirus*, *Pestivirus* and *Hepacivirus* – contains a wide range of human and animal pathogens. The genome length of individual representatives of the three genera are approximately 9.6 kb, 10.8 kb and 12.5 kb for the *Hepacivirus*, *Flavivirus* and *Pestivirus*, respectively. Like the *Picornaviridae*, the structural proteins are located at the amino-terminus of the encoded polyprotein, which is processed by virus and host proteases. In the case of the genera *Hepacivirus* and *Pestivirus*, translation is driven by an internal ribosome entry site (IRES), similar in function – but structurally divergent – from that possessed by the *Picornaviridae*. The *Flavivirus* genus has a capped 5′ terminus, and all three genera lack a poly(A) tail and are instead terminated by distinct stem-loop structures.

Infectious clones of representatives of each of the three genera of *Flaviviridae* have been generated. The combination of increased genome length and, for the Flaviviruses in particular, considerable genetic instability when maintained in prokaryotic hosts, have resulted in a number of modifications to the protocols originally developed during the production of infectious cDNA clones of picornaviruses. However, the overall strategy – the generation of a cDNA used to derive RNA *in vitro* for transfection –

remains the same. The increased size of the genome means that – whether reverse transcribed and cloned as double stranded DNA, or RT-PCR amplified – the likelihood of generating a full-length or near full-length cDNA clone is reduced. This in turn increases the chance of incorporating one or more deleterious mutations during assembly, so rendering the resulting cDNA either non-infectious or not representative of the virus population. Prior to the availability of efficient and relatively inexpensive automated DNA sequencing each full length cDNA would need to be assayed for biological function by transfection and recovery *in vitro*, and possibly even phenotypic analysis in an animal model. In contrast, nearly full-length cDNA clones can now be generated by RT-PCR amplification (Tellier *et al.*, 1996; Yanagi *et al.*, 1998) of even limiting amounts of virus RNA, and the sequence verified to resemble that of the consensus present within the virus population. This approach has recently been used for the generation of both *Pestivirus* and *Hepacivirus* infectious clones (Hong *et al.*, 1999; Kolykhalov *et al.*, 1997; Moormann *et al.*, 1996; Ruggli *et al.*, 1996).

The stable maintenance of a cDNA in the bacterial host strain is an essential aspect of an efficient reverse genetic system for the virus. In the case of the picornaviruses, the majority of infectious cDNA clones appear stable upon propagation. In contrast, early attempts to generate infectious cDNA clones of flaviviruses encountered considerable problems due to the presence of sequences within the virus genome that were toxic for the bacterial host. A number of approaches were taken to circumvent these problems ranging from the use of low copy number plasmids and the screening of a wide range of *E. coli* host strains to the propagation of the flavivirus genome on two separate plasmids, coupled with ligation *in vitro* prior to transcription. This latter approach was used to produce the infectious flavivirus cDNAs for Yellow Fever (Rice *et al.*, 1989), Japanese encephalitis (Sumiyoshi *et al.*, 1992) and tick-borne encephalitis (Mandl *et al.*, 1997). Other members of the *Flavivirus* genus can be propagated as full-length infectious cDNA clones, including Dengue (Kapoor *et al.*, 1995), Kunjin virus (Khromykh and Westaway, 1994) and Langat tick-borne encephalitis (Campbell and Pletnev, 2000). The specific infectivity of *in vitro* synthesized flavivirus RNA ranges from <1 pfu μg^{-1} to between 10^5 and 10^6 pfu μg^{-1}, in the latter cases (Yellow Fever and tick-borne encephalitis) approaching that of purified virus RNA.

In contrast to the *Flavivirus* genus, infectious full-length cDNA clones of both the *Pestivirus* and *Hepacivirus* genera can often be propagated without deleterious effects upon the host bacterial strain, even when cloned in medium and high copy number plasmid vectors. In the case of the *Hepaciviruses* the main problem lies in analysing the cDNA clone for infectivity, rather than in the generation, maintenance or *in vitro* transcription of the clone. This is due to the lack of a suitable cell culture system in which Hepatitis C virus can be reliably and reproducibly grown (Shimizu *et al.*, 1992). Following the demonstration that direct intra-cranial inoculation of mice with *in vitro* generated RNA for tick-borne encephalitis yielded infectious virus (Gritsun and Gould, 1998), several laboratories have recovered infectious Hepatitis C virus by intra-hepatic inoculation of chimpanzees (Hong *et al.*, 1999; Kolykhalov *et al.*, 1997; Yanagi *et al.*, 1997), the only known non-human host for this virus. This is a significant achievement and has enabled the *in vivo* phenotypic analysis of this major human pathogen to be initiated.

The importance of producing a full length cDNA that corresponds in sequence to the consensus present in the virus population is well illustrated by the generation of *in vivo* infectious clones of Hepatitis C virus. Two initial studies reported the construction of a HCV genotype 1a infectious cDNA from virus RNA purified from plasma – containing the H77 strain of HCV – taken from the acute phase of infection and previously shown to have high specific infectivity ($\sim 10^8$ RNA molecules and $\sim 10^{6.5}$ chimpanzee infectious doses per millilitre). In one study six full-length cDNA clones were generated by RT-PCR, sequenced and compared with the published consensus sequence of the H77 strain (Kolykhalov *et al.*, 1997). Each was shown to contain non-consensus sequence changes which were corrected during the generation of a cDNA from which uncapped T7-synthesized RNA was derived and used to inoculate chimpanzees. In a second study, 18 RT-PCR amplified cDNAs were sequenced, each containing between 6 and 28 amino acid substitutions from the H77 consensus, as well as additional mutations not present in other HCV isolates (Yanagi *et al.*, 1997). Four of these cDNAs were used to generate a chimeric near-consensus (differing only by the presence of leucine instead of phenylalanine at position 790) full-length cDNA by restriction digestion and ligation, the infectivity of which was confirmed by chimpanzee intra-hepatic transfection of RNA generated *in vitro*. The limited *in vitro* replication of HCV remains a significant restriction to our understanding of the functions of many parts of the virus genome, or the functional interactions of the virus and host proteins. The recent development of an HCV sub-genomic replicon – derived from a known chimpanzee infectious cDNA clone – provides a tool with which the ability of a cell line to support virus replication can be tested (Lohmann *et al.*, 1999). Furthermore, it is possible that a selectable replicon-based system expressing antibiotic resistance may enable the identification of cellular proteins absent from many liver-derived cell lines which could be required for HCV replication.

7.2.2.3 Coronaviridae. Although coronavirus genomic RNA is infectious when transfected into permissive cells *in vitro*, the general strategy outlined for recovery of members of the *Picornaviridae* and *Flaviviridae* has yet to be successful when applied to these, the largest RNA virus genomes. It is likely that the problems encountered in the initial construction of flavivirus infectious cDNAs – specifically the generation of representative, error-free, full-length clones – is the rate-limiting step. The capped and polyadenylated coronavirus genome is 27–31 kb in length, twice the length of the longest infectious virus cDNA (the 15.7 kb Lelystad virus; Meulenberg *et al.*, 1998), three times the size of the flaviviruses, and four times longer than the smallest picornaviruses. The necessity of using error-prone reverse transcription-based cloning strategies, possibly combined with additional errors introduced during the polymerase chain amplification, have been an insurmountable problem to the generation of an infectious coronavirus cDNA. Furthermore, the significant work involved in the identification of a consensus virus RNA sequence, and the construction and restoration of a cDNA representative of this consensus, is an arduous undertaking requiring the commitment of a considerable amount of time, personnel and reagents. Two further restrictions to this approach may apply, but have not been rigorously tested. Like certain flavivirus cDNA clones, experience has shown that large cDNA sub-fragments of the coronaviruses may not be stably maintained in *E. coli*, and the errors introduced during *in vitro* transcription (Boyer *et al.*, 1992; Sooknanan *et al.*, 1994) using bacteriophage T7 or SP6 polymerase my reduce or destroy the infectivity of the transcripts.

An alternative approach has been developed for the genetic manipulation of mouse hepatitis virus (MHV), the causative agent of a diverse range of murine diseases. This approach, designated targeted recombination, takes advantage of the high rates of homologous RNA recombination observed during natural infection. The fact that this virus replicates well *in vitro*, the availability of a tractable small animal in which to conduct pathogenesis studies, and the ability to manipulate the genome by targeted RNA recombination have made MHV the model system for the genetic analysis of the *Coronaviridae*.

The initial implementation of targeted RNA recombination involved the co-transfection of *in vitro* synthesized RNA with genomic RNA, the rare recombinants being identified using a phenotypic selection screen. The MHV *Alb4* mutant lacks a 29 amino acid sequence from the N gene and is both temperature sensitive (*ts*), forming minute plaques at the non-permissive temperature, and thermolabile, in that the titre of the mutant is reduced 100-fold more than the wild-type upon incubation at 40°C for 24 hours. Murine cells were co-transfected with *Alb4* genomic RNA and an *in vitro* generated RNA corresponding to RNA7 (the smallest subgenomic RNA containing a wild-type version of the N gene and the 3′ non-coding region of the genome) tagged with a non-coding mutation. Following selection at 40°C, viruses forming large plaques at 39°C were screened and confirmed as single-crossover recombinants due to the restoration of the 29 amino acid deletion and acquisition of the non-coding marker mutation (Koetzner *et al.*, 1992). Two significant enhancements were subsequently made to the targeted recombination technique, both of which contribute significantly to the efficiency of the process. The first was the replacement of the co-transfecting RNA with an *in vitro* synthesized defective interfering (DI) RNA which, in the presence of a suitable co-infecting helper virus replicates (Brian and Spaan, 1997), resulting in a 1000-fold increase in the frequency with which recombinants can be selected and thereby overcomes the absolute necessity for a selection system (van der Most *et al.*, 1992). This improvement in efficiency in turn allowed the use of recipient (i.e. *Alb4*) infection, rather than co-transfection with the donor molecule, making the technique much easier to use (Masters *et al.*, 1994). However, in comparison to the ease of manipulation of an infectious molecular clone, targeted recombination has a number of restrictions. With the one exception of the DI-RNA-meditated recombination of silent mutations into the 24 kb gene 1, all the introduced mutations have been restricted to the 3′ 6 kb of the genome (Koetzner *et al.*, 1992; Leparc-Goffart *et al.*, 1998; Phillips *et al.*, 1999; van der Most *et al.*, 1992). Furthermore, although other coronaviruses are known to undergo recombination (Kottier *et al.*, 1995; Stirrups *et al.*, 1998), the technique has only recently been extended to transmissible gastroenteritis virus (Sánchez *et al.*, 1999).

7.2.3 Studies and modification of pathogenicity of positive-strand RNA viruses using reverse genetics

7.2.3.1 The genetic basis of poliovirus attenuation. The Sabin vaccine strains of poliovirus were derived by repeated passage in simian extraneural cells and tissues resulting in adaptation and attenuation (Sabin and Boulger, 1973). The vaccines grow well in the gastrointestinal tract following oral administration but do not cause disease in the central nervous system. Of the three serotypes of poliovirus, the type 2 vaccine strain was already avirulent for monkeys, but the attenuation of type 1 (Mahoney) and type 3 (Leon) required the acquisition of mutations within the genome which adapted the virus to growth in the gut and resulted in the loss of the neurovirulence phenotype. The availability of relatively efficient DNA sequencing techniques in the late 1970s and early 1980s and the development of infectious molecular clones (Racaniello and Baltimore, 1981) provided the means to dissect the genetic basis of poliovirus attenuation. Since this subject has been extensively reviewed (for example Almond, 1987) only the key points and recent developments will be covered here.

The poliovirus type 3 live attenuated vaccine strain (P3/Leon/12a$_1$b) differs by 10 nucleotide substitutions from its neurovirulent progenitor (P3/Leon/37), the changes accounting for only three amino acid substitutions. Comparative sequencing of the genome of a type 3 revertant virus (P3/119) isolated from a vaccine-associated case of paralytic poliomyelitis identified a total of 8 differences from the vaccine strain. These differences, from the wild-type strain to the vaccine strain, and from the vaccine strain to P3/119, must account for the attenuation phenotype and subsequent reversion to neurovirulence. One mutation – at position 472, located in the 5′ non-coding region (NCR) – was of immediate interest, being a C in P3/Leon/12a$_1$b, a U in the vaccine strain and a C again in P3/119 (Cann *et al.*, 1984; Stanway *et al.*, 1983, 1984). The relative importance of individual substitutions or combinations of mutations was defined by the construction of recombinant genomes, the recovery of viruses *in vitro* and the *in vivo* assessment of their neurovirulent phenotype in the World Health Organization-approved vaccine safety test (Almond, 1987; Evans *et al.*, 1985) or a murine model for paralytic poliomyelitis using recombinant genomes derived from the type 2 Lansing strain (La Monica *et al.*, 1987). These studies clearly demonstrated the importance of changes at positions 472 and 2034 (the latter contributing to one of the three amino acid substitutions, at residue 91 of the capsid protein VP3) in poliovirus attenuation. Subsequent analysis identified an additional mutation (at nt. 2493) in the capsid protein VP1, previously overlooked due to the passage history of the virus used to construct the Sabin 3 cDNA, that also contributes to the attenuation phenotype of the vaccine (Tatem *et al.*, 1992; Weeks Levy *et al.*, 1991).

Similar studies have been conducted for the type 1 and 2 strains of poliovirus, though these were less straightforward due to the increased number of substitutions between attenuated and neurovirulent isolates (55 in type 1 and 27 in type 2), the lack of a neurovirulent vaccine-progenitor (type 2) or a suitable vaccine-associated isolate (type 1). Despite these limitations, major attenuating mutations were identified in the same region of the 5′ NCR to nucleotide 472 of type 3 poliovirus (Bouchard *et al.*, 1995; Macadam *et al.*, 1991, 1993; McGoldrick *et al.*, 1995; Moss *et al.*, 1989; Omata *et al.*, 1986; Pollard *et al.*, 1989; Ren *et al.*, 1991). This region of the 5′ NCR is highly structured (Skinner *et al.*, 1989) and forms an important part of the IRES, the implication being – supported by *in vitro* analysis – that attenuation is due to disruption of the internal entry of ribosomes during the initiation of polyprotein translation (Svitkin *et al.*, 1985, 1988, 1990).

An understanding of the genetic basis of poliovirus attenuation, the structure of the 5′ NCR (Skinner *et al.*, 1989) and the observation that disruption of the localized structure within the 5′ NCR affected both neurovirulence and temperature sensitivity (Macadam *et al.*, 1992) suggested that it may be possible to devise stably attenuated vaccine strains incapable of reversion to neurovirulence (Almond *et al.*, 1993; Macadam *et al.*, 1994). These would involve the construction of cDNAs in which the base-pairing of the region around nucleotide 472 was manipulated to confer an attenuation phenotype that was incapable of reversion to neurovirulence without going via a less thermally stable (and therefore more attenuated) intermediate. Although eradication of poliomyelitis may be achieved before the development, testing and licensing of such vaccine strains, the principle represents the culmination of over two decades of reverse genetic analysis of poliovirus.

7.2.3.2 The role of the coronavirus spike glycoprotein in virus tropism and pathogenesis. Distinct strains of MHV cause different patterns of pathogenesis in mice, including hepatitis, encephalitis, enteritis and chronic demyelination (Bailey *et al.*, 1949; Houtman and Fleming, 1996), depending also upon the dose, route of inoculation and immune status of the host. Comparative sequencing of different MHV strains (Parker *et al.*, 1989), and the alteration in pathogenesis induced by antigenic variants (Fleming *et al.*, 1986; Gallagher *et al.*, 1990) strongly implied that the surface envelope glycoprotein (S) was a major determinant of virus pathogenesis. The S glycoprotein is synthesized as a 180 kDa precursor and processed into two 90 kDa subunits, S1 and S2. Of these, the N-terminal derived S1 protein is presumed to form the globular head of the spike glycoprotein within which the receptor binding specificity lies, and the membrane-bound S2 protein contains the domain responsible for fusion of the host cell and virus membranes (de Groot *et al.*, 1987; Kubo *et al.*, 1994; Luytjes *et al.*, 1987). Attenuated MHV mutants derived from glial cell cultures persistently infected with MHV-A59 display altered pathogenesis, in particular reduced hepatitis and demyelination. Prior to the development of targeted recombination it was not possible to directly determine which of the two mutations, of H716D and Q159L, within the S protein were responsible for the altered phenotype. However, revertants at H716D displayed restored fusion-competence, implying that this mutation – located within the signal sequence for S cleavage – was probably not involved in pathogenesis.

Targeted recombination enabled the involvement of Q159L in MHV pathogenesis to be formally tested. Murine L2 cells infected with the *ts* mutant *Alb4* were transfected with RNA generated *in vitro* from a plasmid (pFV1) carrying the wild-type 3′ portion of the MHV genome – including the entire S gene, the wild-type N gene and the 3′ non-coding region – or an identical construct containing the Q159L mutation (pFV1-Q159L). After 24 hours selection at 40°C, recombinants were identified by their large plaque phenotype at 39°C, screened to confirm the presence of the restored N gene and sequenced to verify the encoded amino acid at position 159 of the S gene. 17% of recombinants carried both the restored wild-type N gene and the Q159L mutation indicating they had probably undergone a single crossover involving at least the 3′ 6 kb of the genome. MHV-A59 and the recombinants containing wild-type and Q159L S genes were indistinguishable *in vitro* with regard to the plaque morphology, fusion and single-step growth kinetics. Furthermore, all three viruses displayed a similar neurotropic phenotype, inducing encephalitis in C57BL/6 mice. In contrast, the Q159L recombinant virus exhibited a significantly reduced hepatotropism, replicating 2–3 $\log_{10}$ less well in the liver than MHV-A59 or the recombinant carrying a wild-type S gene. Concomitant with the reduced replication ability, the Q159L recombinant was attenuated and caused less hepatitis and less demyelination.

In addition to being a major determinant of hepatotropism, the MHV S gene also influences the neurotropism of the virus. MHV-A59 is only weakly neurovirulent, high doses being required to produce even mild encephalitis. In contrast, MHV4 is highly neurovirulent, producing a severe and often fatal encephalitis (Dalziel *et al.*, 1986; Lavi *et al.*, 1986). Antigenic variants mapped to the S protein demonstrated an association between neurovirulence and the sequence of the attachment glycoprotein (Dalziel *et al.*, 1986; Fleming *et al.*, 1986; Wege *et al.*, 1988) which could be formally tested with the availability of targeted recombination (Phillips *et al.*, 1999). As in previous experiments, the MHV *Alb4* mutant virus was used as the recipient for the introduction of the S genes of MHV-A59 and MHV4 from *in vitro* synthesized and capped RNA encoding the 3′ 6 kb of the relevant MHV genome. The standard selection strategy was inappropriate as the MHV4 S gene confers a thermolabile phenotype to the recombinant viruses. This emphasizes a major drawback of this type of reverse genetic approach when compared to the direct recovery from an infectious molecular clone. In this instance it was possible to apply an alternative selection strategy – the MHV-A59 S protein in the *Alb4* strain being selected against using two neutralizing monoclonal antibodies specific for the MHV-A59 S protein. The resulting recombinants were characterized *in vitro* – the small-plaque phenotype and low-level replication of MHV4 being similar to the *Alb4*-derived recombinant carrying the MHV4 S gene. *In vivo*, the MHV4 S gene conferred a strongly neurotropic phenotype to the recombinant, reflected in a marked increase in virulence, and viral antigen staining and inflammation in the central nervous system (Phillips *et al.*, 1999). These results confirm the importance of the S gene in MHV pathogenesis and provide the basis for an experimental approach to identify the S gene determinants that are both necessary and sufficient to confer the neurovirulent phenotype.

Although MHV has been used as a model system for coronavirus pathogenesis studies, there is considerable interest in other representatives of this family of viruses. The role of the S gene in determining the pathogenesis of transmissible gastroenteritis virus (TGEV), a porcine pathogen causing almost 100% mortality of infected newborn piglets, has recently been investigated using targeted recombination (Sánchez *et al.*, 1999). These studies – involving the analysis of the S genes from a mixed virus population containing both enteric and respiratory TGEV strains – confirmed the fundamental role the S protein has in determining virus tropism, and illustrates some additional features of the targeted recombination approach originally developed for MHV. Recombination was initiated by infecting swine testis (ST) cells with the recipient virus and subsequently electroporating the infected monolayers with an *in vitro* generated T7 runoff RNA encoding a TGEV ‘minigenome’ (Izeta *et al.*, 1999) engineered to carry the S gene of interest. The supernatants from these cultures were passaged eight times in ST cells and recombinant viruses with enteric tropism were selected *in vivo* by infection of 2-day-old piglets. Analysis of viruses isolated from the jejunum or ileum of piglets demonstrated the presence of recombinant enterotropic TGEV which resulted from a double-crossover event at the 5′ and 3′ ends of the S gene (Sánchez *et al.*, 1999). This approach shows that alternative strategies involving *in vivo* selection have potential application in the use of targeted recombination of these large RNA virus genomes. However, additional mutations were acquired during the *in vitro* passage and *in vivo* selection of these TGEV recombinants which represents a restriction to the use of this approach in the definition of the molecular determinants of virus phenotype.

7.2.3.3 Studies of the pathogenesis determinants of the Flaviviridae. The recent availability of infectious molecular clones of HCV has enabled studies on the determinants of pathogenesis to be initiated. Due to the inherent limitations of the only animal model available (chimpanzees), and the inconsistent or poor growth of HCV *in vitro*, these studies have currently been restricted to investigating the genome requirements for replication. Following discovery of a 98 nt. conserved sequence at the 3′ terminus (3′X RNA) of the genome (Kolykhalov *et al.*, 1996; Tanaka *et al.*, 1995), direct intra-hepatic inoculation of *in vitro* synthesized RNA was used to demonstrate that this sequence was required for replication and infectivity (Kolykhalov *et al.*, 2000; Yanagi *et al.*, 1999). Although infectious molecular clones of genotype 1a and 2a have been constructed, intertypic chimeras between these genotypes were non-infectious in chimpanzees (Yanagi *et al.*, 1999) suggesting that there may be a required functional complementarity between the regions of the genome encoding the structural and non-structural proteins. These results illustrate both similarities and differences with other members of the *Flaviviridae* where the 3′ end sequences and structures are of critical importance in replication (see Kolykhalov *et al.*, 1996), but in which intertypic chimeras – for example between tick-borne encephalitis and dengue virus – retain viability (Pletnev *et al.*, 1992, 1993). The importance of the four HCV-encoded enzymatic activities – the NS2–3 protease, NS3–4A serine protease, NS3 NTPase/helicase and the NS5B polymerase – was confirmed by introducing point mutations into the active site of each in a chimpanzee infectious clone, and monitoring infection following intra-hepatic inoculation (Kolykhalov *et al.*, 2000). The failure to detect any signs of HCV infection, including viraemia, seroconversion or elevation of serum alanine aminotransferase, confirmed that these enzymatic activities are required for infection, and that they therefore form valid targets for therapeutic intervention.

Bovine viral diarrhoea virus (BVDV), a member of the *Pestivirus* genus, generally causes either subclinical infections or mild diarrhoea, fever and immunosuppression in the majority of cattle. Persistent infection of cattle can also occur following infection of pregnant animals and vertical transmission, which involves a non-cytopathic (non-cp) isolate of BVDV. However, BVDV is also the causative agent of mucosal disease (MD), a sporadic disease of persistently infected calves which is usually fatal and is typically characterized by extensive ulceration of the gastrointestinal tract. BVDV isolates from MD cases are either non-cp or cytopathic (cp) – the latter replicating to very high levels in damaged tissues (in contrast to the non-cp isolate which is found in all tissues of infected animals). Sequence and polyprotein analysis of cp and non-cp isolates demonstrated that cytopathic isolates are characterized by the presence of a separate NS3 protein – which has serine protease, NTPase and helicase activity. In contrast, non-cp isolates have an NS2–3 fusion protein, which is not processed during virus replication (reviewed in Meyers and Thiel, 1996). NS3 production in cp (MD-inducing) isolates appeared to be related to the acquisition of mutations or RNA recombination events – including the incorporation of cellular sequences – thereby allowing the processing of NS3 (Meyers *et al.*, 1991, 1998; Tautz *et al.*, 1994, 1996). The importance of NS3 production, and relationship to cp isolates, has been tested by reverse genetic analysis of two infectious molecular cp clones, CP7 and NADL, which respectively contain a 9 residue duplication in NS2 and a 90 residue insertion of a novel cellular protein. In both cases, deletion of the inserted sequence resulted in a non-cp phenotype and the restored production of an unprocessed NS2–3 fusion (Mendez *et al.*, 1998; Meyers *et al.*, 1996a, b). Analysis of the Oregon strain of BVDV showed an alternative mechanism for NS3 production, namely the presence of certain point mutations in NS2 (Kümmerer *et al.*, 1998). The importance of these was confirmed by the mutagenesis of a chimeric infectious molecular clone consisting of the Oregon-strain polyprotein encoding region and 5′ and 3′ sequences derived from CP7, and demonstration that there was again a direct correlation between NS3 production and the cytopathic phenotype (Kümmerer and Meyers, 2000).

Within the *Flavivirus* genus, reverse engineering of infectious clones has been used to generate chimeric genomes with attenuated phenotypes with a view to deriving novel vaccines. These studies have largely focused on the use of a dengue type 4 backbone into which the structural proteins for dengue types 1–3 (Bray and Lai, 1991; Chen *et al.*, 1995) or the tick-borne encephalitis viruses (Pletnev *et al.*, 1992, 1993; Pletnev and Men, 1998) were engineered. Alternatively, the well characterized 17D vaccine strain of Yellow Fever (YF 17D) has been used as the backbone for expression of Japanese encephalitis structural proteins (Chambers *et al.*, 1999; Monath *et al.*, 1999). Such chimeras induce protective immunity in small animals and non-human primates (Bray and Lai, 1991; Bray *et al.*, 1996; Monath *et al.*, 1999; Pletnev *et al.*, 1993). Alternative strategies for generating attenuated vaccine candidates have included the site-directed mutagenesis of proteolytic cleavage or glycosylation sites in the genome (Muylaert *et al.*, 1996; Pletnev *et al.*, 1993), or the deletion of sequences from the 3′ non-coding region (Mandl *et al.*, 1998). These studies indicate the value of reverse genetic approaches in the analysis of virus pathogenesis and the use of the technique for the development of novel potential vaccines.

7.3 The negative-strand RNA viruses

7.3.1 Development of techniques for engineering negative-strand RNA virus genomes

Naked negative-sense RNA is not infectious. Indeed, RNA genomes of negative-strand RNA viruses exist in virions as a ribonucleoprotein (RNP) complex in association with other virally encoded proteins. In other words, these viruses carry the viral transcriptase with them into the infected cell. Thus in order to rescue virus-like synthetic RNAs, the transcriptase must also be introduced into the cell along with the engineered RNA.

Ground-breaking studies with influenza A virus, an orthomyxovirus whose genome comprises eight segments of negative-sense RNA, were the first to achieve this. The viral polymerase complex, consisting of a three subunit polymerase and viral nucleoprotein, was purified from concentrated virions and mixed with synthetic model influenza RNAs before introduction into cells as an RNP complex (Luytjes *et al.*, 1989). The model RNA consisted of a synthetic T7 polymerase transcript with termini identical to those found on authentic influenza RNAs. These termini, which constituted the whole of the non-coding sequences from segment 8 RNA, flanked an antisense sequence of the reporter gene encoding the bacterial enzyme chloramphenicol acetyl transferase (CAT). Following introduction of the synthetic RNP complex into cells, the model RNA was expressed, replicated and packaged when the cells were coinfected with a 'helper' virus whose own transcriptase and replicase recognized the promoters of the synthetic RNA as influenza-like. Following on from this, full-length copies of viral RNA segments have been reconstituted as RNPs and rescued in cells (Enami and Palese, 1991). Virions which assemble in such cells might contain progeny RNAs derived from the synthetic one in place of their own 'helper' segment. Providing there is a means of selecting the former from the very high background of progeny helper viruses, then viruses with specifically altered genotypes can be recovered. A strong selection is afforded by the ability of a gene to confer growth in a certain cell type. For example, the neuraminidase (NA) of influenza A/WSN/33 allows multicycle growth in cultured cells in the absence of trypsin. Many of the early studies with genetically engineered influenza A viruses have taken advantage of this observation to manipulate RNA segment 6 which encodes NA (see below). However, for most other RNA segments and also for the less well studied influenza types B and C viruses, strong selection systems are not available and this has severely limited the use of the helper-dependent system.

Following the reconstitution of active influenza RNP from synthetic RNA, several groups began to develop reverse genetics systems for the other negative-strand RNA viruses whose genomes consist of fewer segments, for example, the bunyaviridae, or of a single RNA molecule, the mononegavirales or non-segmented negative strand (NNS) viruses. Model RNAs of the type described above, which contained relevant promoter sequences at their termini flanking reporter genes, were used to establish the technology of reconstituting the RNP *in vivo*. Plasmids encoding the NNS virus nucleocapsid (N) and polymerase proteins (L and P) under the control of bacteriophage T7 promoters were transfected into cells previously infected with a vaccinia virus recombinant expressing T7 RNA polymerase. An additional plasmid containing a cDNA copy of the model RNA between a T7 promoter and a self-cleaving hepatitis delta virus (HDV) ribozyme sequence was also transfected. T7 transcription of this plasmid gave rise to an RNA, which, following self-cleavage by the ribozyme, possessed authentic 3′ and 5′ termini. In studies with the rhabdovirus vesicular stomatitis virus (VSV), cotransfection of plasmids which encoded the structural proteins, M and G, resulted in the packaging and propagation of the model RNA to new cell monolayers (Pattniak *et al.*, 1992). However, for some time the application of this technology for the recovery of recombinant viruses from full length cDNA representing the entire NNS genome remained elusive. Interestingly, recovery of NNS viruses from cDNA has only been achieved by supplying a positive sense copy of the genome. Conzelmann and co-workers were the first to isolate a NNS virus entirely from cloned DNA by reconstituting the RNP of rabies virus *in vivo* (Schnell *et al.*, 1994). The anti-genomic RNA was replicated by the N, P and L proteins supplied *in trans*, and the genomic RNA transcribed to produce the full complement of viral proteins allowing the assembly of progeny virus. The same approach has led to the successful recovery of nearly all the members of the NNS group entirely from cloned cDNA (reviewed by Conzelmann (1996) and Roberts and Rose (1998)). This technology has also been successfully applied to the rescue of a negative strand RNA virus with a tripartite genome. Bridgen and Elliott recovered Bunyamwera virus by co-transfecting three plasmids encoding all the virus proteins (N, NSs, G1, G2, NSm, and L) under T7 control and three plasmids containing cDNAs of the antigenomic RNAs (L, M, and S) into cells infected with vTF7-3 (Bridgen and Elliott, 1996). Each antigenomic RNA was expressed from a T7 promoter with a HDV ribozyme at the 3′ end. Cultivation of the progeny on insect cells permitted the recovery of infectious Bunyamwera virus from contaminating vaccinia virus.

The systems used for recovery of the NNS and bunyaviridae are independent of helper virus, and even highly attenuating mutations have been recovered in live viruses. As a result a barrage of interesting information concerning the pathogenesis of these viruses is beginning to emerge and some examples will be discussed below. Meanwhile, it became obvious that the helper-dependent reverse genetics technology used for engineering orthomyxoviruses was limiting, and that a helper-independent methodology similar to that developed for the NNS group was required.

Sure enough, two independent laboratories have recently reported the helper-independent rescue of recombinant influenza A virus (Fodor *et al.*, 1999; Neumann *et al.*, 1999). Both relied upon human polymerase I promoters to drive *in vivo* transcription of influenza vRNAs. This strategy was first described by Hobom and coworkers for influenza A model RNAs (Neumann *et al.*, 1994) and then adapted by Pleschka *et al.*, (1996) who successfully rescued recombinant virus in a helper-independent manner using pol I transcription to generate RNA *in vivo*. Thus, unlike recovery of the NNS viruses, recombinant vaccinia vTF7-3 or cells which express T7 polymerase were not required. Fodor *et al.* attained authentic 3′ termini via ribozyme cleavage whereas Neumann *et al.* used polymerase I terminator. Both groups have used negative sense RNAs in contrast to the anti-genomic RNA of the NNS methodologies. Interestingly Neumann *et al.* also showed the recovery of virus was enhanced by co-expression of all of the influenza structural proteins to form virus-like particles which would assemble in the transfected cells and package the newly formed vRNAs. The assembly of virus-like influenza particles was first described by Portela and coworkers (Mena *et al.*, 1996) and was used by them to show that all nine influenza A virus structural proteins were necessary for assembly and encapsidation of a model RNA. Since at least four of the viral structural proteins (the polymerase components) must be expressed to enable recombinant influenza virus recovery, the number of plasmids to be transfected could be reduced by utilizing the same cDNAs to generate both genomic and antigenomic (messanger sense) RNAs. Indeed, the Kawaoka laboratory have recently demonstrated that this ‘ambisense’ approach is possible (Hoffmann *et al.*, 2000).

Although similar strategies to those described above are being applied to the filoviruses, Ebola and Marburg virus, no significant advances in understanding the extreme pathogenicity of these agents have yet been published. Likewise, the newly identified NNS, Borna disease virus, (BDV), has not yet been successfully recovered from cDNA, and the bunyavirus remains the only other segmented genome to be manipulated: no publications have yet described arenavirus reverse genetics. However, one suspects that these related viruses will not remain refractory to genetic manipulation for long. Meanwhile, we describe below some examples which demonstrate the power of reverse genetics applied to the negative-sense RNA viruses which are human pathogens.

7.3.2 Studies and modification of pathogenicity of negative-strand RNA viruses using reverse genetics

7.3.2.1 Influenza virus. The genes of influenza virus which have been genetically manipulated to date have been limited by the availability of selection systems to separate viruses derived from the helper virus from those which contain the engineered RNA segment. Nonetheless experiments in which the neuraminidase gene, the hemagglutinin gene and the non-structural protein NS1 have been specifically altered have confirmed that this approach could lead to the design and generation of attenuated viruses suitable as vaccine candidates.

Manipulation of influenza neuraminidase. The sequences at the extreme termini of influenza A virus RNA segments contain *cis*-acting signals which direct the transcription and replication of the RNA. These sequences are completely conserved for all eight RNA segments, but differ from those found at the equivalent position of influenza B virus RNAs. Muster and coworkers engineered a recombinant influenza A virus in which the promoter sequences on RNA segment 6 were replaced by promoter sequences derived from an influenza B virus segment 8 RNA (Muster *et al.*, 1991). The resulting virus showed attenuated growth *in vitro* which was due to less efficient transcription of segment 6 RNA (Luo *et al.*, 1992). Certainly this and other recombinant viruses which have reduced levels of neuraminidase show attenuated phenotypes in mice (Solorzano *et al.*, 2000). These results have lead to the suggestion that this strategy could be considered for the generation of live attenuated influenza virus vaccines.

Manipulation of influenza haemagglutinin. Influenza A viruses are pathogens of poultry as well as of man. A major determinant of virulence of the avian influenza viruses is the amino acid sequence at the site where the precursor of the hemagglutinin protein HA0 is cleaved by host cell protease into the subunit proteins HA1 and HA2. In some strains of the H5 and the H7 subtype, the HA cleavage site has acquired several basic amino acid residues which render the protein susceptible to cleavage by ubiquitous intracellular enzymes such as furin. For these viruses, replication is no longer limited by the availability of extracellular protease and a systemic disease ensues followed by rapid death of the birds. Direct evidence for the correlation between HA cleavability and virulence of an H5N9 virus was provided by generating a series of mutants containing different numbers of basic amino acids at the HA cleavage site using reverse genetics (Horimoto and Kawaoka, 1994). In 1997 a strain of H5N1 virus which possessed a multi-basic HA cleavage site emerged in southern China and was also epidemic in the Hong Kong poultry markets. Unusually, 18 humans became infected with this avian influenza strain and 6 people died as a result. This virus could potentially have emerged as a new and highly virulent pandemic strain. Production of a vaccine for this strain is problematic. Not only is there the hazard due to the extreme virulence of this pathogen for both man and chickens, but also the yield of virus from eggs, the conventional medium for vaccine production, is low. A virus specifically engineered to remove all but one of the basic amino acids from the HA cleavage site was found to be a safe and immunogenic vaccine candidate, protecting chickens against wild-type H5N1 virus challenge (Li *et al.*, 1999).

Manipulation of influenza NS1. Reverse genetics systems allow the study of viral gene functions in the context of infection, and can give insight into the interactions of the pathogen with the host which were not accessible using other methodologies. A good example is the generation of a recombinant influenza A virus lacking the NS1 gene (Garcia-Sastre *et al.*, 1998). Several putative functions for the non-structural protein NS1 in the virus life cycle have been proposed following work largely based on over-expression of NS1 from cDNAs. Muster and coworkers were successful in isolating mutant viruses with either very large or total deletions of the NS1 coding sequences. Fortuitously, the cell line they chose to use for the virus isolation was the Vero (African Green Monkey kidney) line which lacks the ability to mount an interferon response. It was subsequently shown that a major function of NS1 is to overcome the host cell interferon response to virus infection. Thus the Vero line is one of only a few cells which allow replication of the NS1 deletant. Furthermore, a virus which lacks NS1, is highly attenuated for growth and virulence in normal cell systems and in IFN-competent mice and yet does elicit an immune response. The engineered virus may thus represent a good future candidate for a live-attenuated vaccine.

7.3.2.2 Rhabdovirus. The *lyssavirus*, rabies virus, a member of the rhabdovirus family, is a human and animal pathogen which was the first in the mononegavirales order to be genetically manipulated. Vesicular stomatitis virus (VSV) can be propagated to very high titres in cell culture and has served as a classical model of NNS virus replication both before and after reverse genetics technology became available.

Manipulation of VSV gene order. The mononegavirales have a tandem array of genes whose order is remarkably conserved across the different groups within this virus family. The gene order controls their expression since transcription proceeds by initiation at a single transcriptional promoter, and genes proximal to this site are transcribed at higher levels than those distal from the promoter. Thus the most abundant protein in the infected cell is the viral nucleoprotein which is the first to be transcribed, and the least abundant is the polymerase protein, L, which comes last in the gene order. Wertz and Ball postulated that by rearranging the gene order, they could manipulate the amounts of each gene produced and that this might provide a means for attenuating the virus. They tested their hypothesis using the rhabdovirus, VSV, and created a series of recombinant viruses in which gene orders were altered. They showed that as the N gene was moved progressively down the genome, the viruses became attenuated both for replication in cell culture and also *in vivo* in mouse pathogenicity studies (Wertz *et al.*, 1998). Moreover, the recombinant attenuated viruses still induced a protective immunity. Thus they argue that, since homologous recombination events have never been reported for this type of virus, these manipulations might represent a mechanism for engineering a stable vaccine which is unable to revert to virulence.

Manipulation of rhabdovirus tropism. Other workers in the rhabdovirus field have used reverse genetics techniques to alter the tropism of a recombinant virus in order to target the virus to a novel cell or tissue type. Conzelmann's group first showed that rabies virus particles will bud from cells in the absence of their spike protein, G, by deleting the entire gene for G from the virus genome (Mebatsion *et al.*, 1996). Then, independently, Rose and coworkers, and Conzelmann's laboratory engineered rhabdoviruses which express heterologous envelope proteins from their genomes in place of their own G protein (Mebatsion *et al.*, 1997; Schnell *et al.*, 1997). They generated recombinant viruses that encoded the genes for CD4 and the chemokine receptor CCR5, the two membrane proteins that serve as the cellular receptors for HIV. These viruses were able to bind to, infect and kill cells which are themselves infected with HIV and consequently expressing the HIV envelope protein gp120 at their surface. The generation of the pseudotyped viruses is possible because rhabdoviruses are quite promiscuous for the membrane proteins they will incorporate into their envelope.

7.3.2.3 Measles virus. Within the paramyxoviridae family reside several important human pathogens, amongst them measles virus which displays a complex pathogenicity where infection is accompanied by immunosuppression and sometimes neuropathology. The genes responsible for these traits are thus far poorly understood, and the study of measles pathogenicity lends itself to dissection using reverse genetics technology.

Manipulation of measles F protein. Like the influenza virus HA protein, several of the NNS viruses have fusion proteins which require activation by cellular proteases. The measles virus fusion protein, F, is usually cleaved inside the infected cell by the ubiquitous protease, furin. By mutating the amino acid sequence at the cleavage site, a virus was generated which required the addition of exogenous protease for fusion and infectivity in cell culture. The engineered virus did not induce neural disease after intracerebral inoculation whereas it could still replicate in the lung, where other host proteases abound (Maisner *et al.*, 2000). Thus, mutagenesis of fusion protein cleavage sites may serve as a general strategy for altering tissue tropism and pathogenicity.

Manipulation of measles H protein. Exchange of the measles virus surface protein, H, has been carried out in experiments which attempt to map genetic determinants of measles neurovirulence. In these experiments the H gene alone from a rodent brain-adapted measles virus was inserted into a full length infectious clone of the attenuated measles vaccine strain, Edmonston (Duprex *et al.*, 1999). The chimeric recombinant virus caused neuropathology following inoculation into mice, although the extent of disease was less than for the parental neurovirulent strain suggesting that further determinants in other positions of the genome are required for full neurovirulence.

Deletion of measles M protein. A different form of measles neuropathology is that associated with subacute sclerosing panencephalitis (SSPE). The cell-associated viruses which are isolated from patients with this disease often lack coding sequences for the matrix gene, M, and also contain mutations in the cytoplasmic tail of the fusion protein, F. Virus recombinants were generated which harboured deletions of the M gene alone or in combination with F cytoplasmic tail deletions (Cathomen *et al.*, 1998). These viruses were highly fusogenic in cell cultures but grew to lower titres than wild-type virus. Within infected cells, the lack of matrix protein seemed to prevent proper assembly of virus particles. However, following inoculation into mice which carry the measles virus receptor, CD46, the viruses which lacked matrix protein penetrated more deeply into the brain suggesting that the enhanced cell fusion demonstrated by these strains contributes to the phenotype of viruses associated with human cases of SSPE.

Deletion of measles C or V proteins. The P gene of measles virus encodes the phosphoprotein which is an essential component of the viral RNP complex. This region of the genome also contains open reading frames which encode small nonstructural proteins, C and V, with no previously assigned functions. Using reverse genetics, viruses were engineered which no longer encoded either C or V proteins and these viruses showed no obvious change in their replication in cultured Vero cells (Radecke and Billeter, 1996). However, the C deletant, but not the V deletant, showed decreased growth in human peripheral blood cells which are natural hosts for measles infection (Escoffier *et al.*, 1999). Further experiments using a mouse model suggest that both C and V are accessory proteins required for efficient virus survival *in vivo*. Interestingly, the equivalent region of the paramyxovirus genome also encodes small nonstructural proteins which may also serve as accessory factors *in vivo* (see Section 7.3.2.4. below).

Expression of IL-12 by measles virus. Measles virus infection is associated with immunosuppression. Neither the C nor V proteins referred to above seem to be responsible for this effect (Escoffier *et al.*, 1999). It has been previously suggested that the decrease in cell-mediated immune responses is due to a decreased production of interleukin-12. To address this hypothesis, Singh and Billeter have recently generated a recombinant measles virus which expresses biologically active IL-12 (Singh and Billeter, 1999). The foreign sequences are maintained and expressed through multiple passages of virus in cell culture. Studies to address the pathogenicity of this virus will be of great interest.

7.3.2.4 Sendai virus. The mouse pathogen, Sendai virus, which has been assigned to the respirovirus genus, is the prototype of the paramyxoviridae family.

Deletion of Sendai virus C and V genes. It is very likely that the mononegavirales, like Sendai virus, possess accessory proteins which enable them to counteract the host's interferon-mediated antiviral response. Indeed, cells infected with wild-type Sendai virus exhibit little or no increase in levels of Stat-1 – a primary interferon-induced gene (Garcin *et al.*, 1999). This interferon-resistant state could also be induced by expression of wild-type C protein, a small protein encoded from the P region of the genome. Recombinant viruses which lacked C protein or with point mutations in C were unable to inhibit the increase in Stat-1. Likewise, deletion of the V protein, which is produced from the same region of the genome following RNA editing of the P gene transcript, resulted in a virus which replicated like wild-type virus *in vitro* but was remarkably attenuated for replication *in vivo* and showed decreased pathogenicity in mice (Kato *et al.*, 1997).

The ability to inhibit the interferon response in different cell types or species may well contribute to differences in host range displayed by members of this group. For example, work by Randall and colleagues suggest that in murine cells, Sendai virus, but not the canine paramyxovirus, SV5, was able to inhibit IFN-alpha/beta responsive promoters, whereas both viruses were active in this respect in human cells (Didcock *et al.*, 1999a). More recently, this group has shown that the V protein of Sendai virus expressed alone can mimic the effect of virus infection in this respect (Didcock *et al.*, 1999b). A logical extension of this work will be to apply reverse genetics techniques in order to resolve whether interferon sensitivity can truly account for the difference in host range of these two NNS viruses. Lamb and coworkers have succeeded in establishing the cDNAs required for manipulation of the SV5 genome and such studies are now possible (He *et al.*, 1997).

Loss of phosphorylation of Sendai virus M protein. In another reverse genetics study with Sendai virus, the role of M protein phosphorylation was assessed by engineering virus mutants in which the phosphorylation sites were destroyed. Viruses which totally lacked M phosphorylation showed no phenotype change *in vitro* or *in vivo* (Sakaguchi *et al.*, 1997). Interestingly, a similar result was found in studies which assessed the role of phosphorylation of the small influenza A virus protein, M2 (Thomas *et al.*, 1998). Viruses which lacked the target sequences for the modification showed no change *in vitro* and only slight attenuation in an *in vivo* mouse model, even though, as for the Sendai M protein, the sequences for phosphorylation are totally conserved in all virus strains.

7.3.2.5 Respiratory syncytial virus. Respiratory syncytial virus (RSV) falls in the pneumovirus genus of the paramyxoviridae. RSV is a major respiratory pathogen of young children and the elderly and much effort has been invested in the development and testing of vaccine candidates.

Deletion of RSV SH and NS-2 genes. Following the leads set by studies with other members of the NNS virus groups, recombinant 'knock out' RS viruses have been recovered in order to understand the roles of individual viral genes and also to attempt to create stably attenuated strains. Sequences encoding small proteins of RSV, such as SH and NS-2, can be deleted from the genome without loss of viability and some of these engineered viruses are currently under investigation as candidate vaccines (Whitehead *et al.*, 1999).

Insertion of immunomodulatory genes into RSV genome. As an alternative method of attenuating virus pathogenicity, the gene for interferon gamma was inserted into an intergenic region of the RSV genome such that it would be co-expressed with viral proteins from infected cells (Bukreyev *et al.*, 1999). IFN gamma has both intrinsic antiviral activity as well as the ability to stimulate and regulate immune responses. The results showed that, in mice, this recombinant virus was attenuated but still retained immunogenicity.

7.4 Concluding remarks

The availability of infectious molecular clones, enabling the application of the considerable power of recombinant DNA technology, has revolutionized our understanding of the biology of RNA viruses. However, a final point should be made regarding the usefulness of the general approach. Evolution has applied an exquisite selection pressure to the virus population, resulting in the genome structure, organization and sequence conservation we see today. Our understanding of the structure and function of these genomes using a reverse genetic approach is restricted by our ability to observe a phenotype for any mutation we introduce. At either end of the possible spectrum – alive and indistinguishable from the unmodified virus or dead and completely unrecoverable – it becomes difficult if not impossible to define a function. Examples of the former are cited in section 7.3.2.4 which describes both an orthomyxovirus and a paramyxovirus mutated at highly conserved phosphorylation sites with no detectable alteration in phenotype. Examples of the latter – the 'dead virus' – abound in any laboratory which has attempted the reverse genetics techniques. Moreover, whilst these techniques are in their infancy and recovery of recombinant viruses is less than optimal, it is often difficult to decide whether the mutation introduced truly abrogates viability or is instead so attenuating that the mutant virus cannot be recovered under the present conditions. However, as the techniques become more refined and recovery efficiency rises to approximate or surpass the rate of mutation of the viral RDRP, then a new 'string' to the reverse genetics 'bow' becomes available. Thus if the mutation introduced into an infectious molecular clone would alone be sufficiently debilitating to render the clone non-infectious, it is possible (and a not-uncommon occurrence) that a compensatory second-site mutation may restore sufficient replication competence to enable recovery of a virus (see for example Macadam *et al.*, 1994a; Meredith *et al.*, 1999). Identification and characterization of such viruses can provide important information about the molecular interactions during the virus replication cycle.

References

Almond JW. (1987) The attenuation of poliovirus neurovirulence. *Ann. Rev. Microbiol.* **41**: 153–180.

Almond JW, Stone D, Burke K *et al.* (1993) Approaches to the construction of new candidate poliovirus type 3 vaccine strains. *Dev. Biol. Stand.* **78**: 161–169.

Bailey OT, Pappenheimer AM, Cheever FS, Daniels JB. (1949) A murine virus (JHM) causing disseminated encephalomyelitis with extensive destruction of myelin. II. Pathology. *J. Exp. Med.* **90**: 195–212.

Bouchard MJ, Lam DH, Racaniello VR. (1995) Determinants of attenuation and temperature sensitivity in the type-1 poliovirus sabin vaccine. *J. Virol.* **69**: 4972–4978.

Boyer JC, Bebenek K, Kunkel TA. (1992) Unequal human immunodeficiency virus type 1 reverse transcriptase error rates with RNA and DNA templates. *Proc. Natl Acad. Sci. USA* **89**: 6919–6923.

Bray M, Lai CJ. (1991) Construction of intertypic chimeric dengue viruses by substitution of structural protein genes. *Proc. Natl Acad. Sci. USA* **88**: 10342–10346.

Bray M, Men RH, Lai CJ. (1996) Monkeys immunized with intertypic chimeric dengue viruses are protected against wild-type virus challenge. *J. Virol.* **70**: 4162–4166.

Brian DA, Spaan WJM. (1997) Recombination and defective interfering RNAs. *Sem. Virol.* **8**: 101–111.

Bridgen A, Elliott RM. (1996) Rescue of a segmented negative-strand RBA virus entirely from cloned complementary DNAs. *Proc. Natl Acad. USA* **93:** 15400–15404.

Bukreyev A, Whitehead SS, Bukreyev N, Murphy B, Collins P. (1999) Interferon gamma expressed by a recombinant respiratory syncytial virus attenuates virus replication in mice without compromising immunogenicity. *Proc. Natl Acad. Sci. USA* **96:** 2367–2372.

Campbell MS, Pletnev AG. (2000) Infectious cDNA clones of Langat Tick-Borne Flavivirus that differ from their parent in peripheral neurovirulence. *Virology* **269**: 225–237.

Cann AJ, Stanway G, Hughes PJ, Minor PD, Evans DM, Schild GC, Almond JW. (1984) Reversion to neurovirulence of the live-attenuated Sabin type 3 oral poliovirus vaccine. *Nucl. Acids Res.* **12**: 7787–7792.

Cathomen T, Mrkic B, Spehner D *et al.* (1998) A matrix-less measles virus is infectious and elicits extensive cell fusion: consequences for propagation in the brain. *EMBO J.* **17:** 3899–3908.

Chambers TJ, Nestorowicz A, Mason PW, Rice CM. (1999) Yellow fever/Japanese encephalitis chimeric viruses: construction and biological properties. *J. Virol.* **73**: 3095–3101.

Chen W, Kawano H, Men R, Clark D, Lai CJ. (1995) Construction of intertypic chimeric dengue viruses exhibiting type 3 antigenicity and neurovirulence for mice. *J. Virol.* **69**: 5186–5190.

Conzelmann K-K. (1996) Genetic manipulation of non-segmented negative-strand RNA viruses. *J. Gen. Virol.* **77:** 381–389.

Dalziel RG, Lampert PW, Talbot PJ, Buchmeier MJ. (1986) Site-specific alteration of murine hepatitis virus type 4 peplomer glycoprotein E2 results in reduced neurovirulence. *J. Virol.* **59**: 463–471.

de Groot RJ, Luytjes W, Horzinek MC, van der Zeijst BA, Spaan WJ, Lenstra JA. (1987) Evidence for a coiled-coil structure in the spike proteins of coronaviruses. *J. Mol. Biol.* **196**: 963–966.

Didcock L, Young DF, Goodbourn S, Randall R. (1999a) Sendai virus and simian virus 5 block activation of interferon-responsive genes: Importance for virus pathogenesis. *J. Virol.* **73:** 3125–3133.

Didcock L, Young DF, Goodbourn S, Randall R. (1999b) The V protein of simian virus 5 inhibits interferon signaling by targeting STAT1 for proteasome-mediated degradation. *J. Virol.* **73:** 9928–9933.

Duprex WP, Duffy I, McQuaid S *et al.* (1999) The h gene of rodent brain-adapted measles virus confers neurovirulence to the Edmonston vaccine strain. *J. Virol.* **73:** 6916–6922.

Enami M, Palese P. (1991) High efficiency formation of influenza virus transfectants. *J. Virol.* **65:** 2711–2713.

Escoffier C, Manie S, Vincent S, Muller CP, Billeter AA, Gerlier D. (1999) Nonstructural C protein is required for efficient measles virus replication in human peripheral blood cells. *J. Virol.* **73:** 1695–1698.

Evans DJ. (1999) Reverse genetics of picornaviruses. *Adv. Virus Res.* **53**: 209–228.

Evans DMA, Dunn G, Minor PD *et al.* (1985) Increased neurovirulence associated with a single nucleotide change in a noncoding region of the sabin type-3 poliovaccine genome. *Nature* **314**: 548–550.

Fleming JO, Trousdale MD, el-Zaatari FA, Stohlman SA, Weiner LP. (1986) Pathogenicity of antigenic variants of murine coronavirus JHM selected with monoclonal antibodies. *J. Virol.* **58**: 869–875.

Fodor E, Devenish L, Engelhardt OG, Palese P, Brownlee GG, Garcia-Sastre A. (1999) Rescue of influenza A virus from recombinant DNA. *J. Virol.* **73:** 9679–9682.

Gallagher TM, Parker SE, Buchmeier MJ. (1990) Neutralization-resistant variants of a neurotropic coronavirus are generated by deletions within the amino-terminal half of the spike glycoprotein. *J. Virol.* **64**: 731–741.

Garcia-Sastre A, Egerov A, Matassov D *et al.* (1998) Influenza A virus lacking the NS1 gene replicates in interferon-deficient systems. *Virology.* **252:** 324–330.

Garcin D, Itoh M, Kolakofsky D. (1997) A point mutation in the Sendai virus accessory C protein attenuates virulence for mice, but not virus growth in cell culture. *Virology* **238:** 424–431.

Garcin D, Latorre P, Kolakofsky D. (1999) Sendai virus C proteins counteract the interferon-mediated induction of an antiviral state. *J. Virol.* **73:** 6559–6565.

Gritsun TS, Gould EA. (1998) Development and analysis of a tick-borne encephalitis virus infectious clone using a novel and rapid strategy. *J. Virol. Methods* **76**: 109–120.

He B, Paterson RG, Ward CD, Lamb RA. (1997) Recovery of infectious SV5 from cloned DNA and expression of a foreign gene. *Virology* **237:** 249–260.

Hoffmann E, Neumann G, Hobom G, Webster RG, Kawaoka Y. (2000) "Ambisense" approach for the generation of influenza A virus vRNA and mRNA synthesis from one template. *Virology* **267:** 310–317.

Holland JJ, McLaren LC, Syverton JT. (1959a) The mammalian cell–virus interaction: IV Infection of naturally insusceptible cells with enterovirus ribonucleic acid. *J. Exp. Med.* **110**: 65–80.

Holland JJ, McLaren LC, Syverton JT. (1959) Mammalian cell-virus relationhip. III. Poliovirus production by non-primate cells exposed to poliovirus ribonucleic acid. *Proc. Soc. Exp. Biol. Med.* **100**: 843.

Hong Z, Beaudet-Miller M, Lanford RE *et al.* (1999) Generation of transmissible hepatitis C virions from a molecular clone in chimpanzees. *Virology* **256**: 36–44.

Horimoto T, Kawaoka Y. (1994) Reverse genetics provides direct evidence for a correlation of hemagglutinin cleavability and virulence of an avian influenza virus. *J. Virol.* **68:** 3120–3128.

Houtman JJ, Fleming JO. (1996) Pathogenesis of mouse hepatitis virus-induced demyelination. *J. Neurovirol.* **2**: 361–376.

Izeta A, Smerdou C, Alonso S, Penzes Z, Mendez A, Plana-Duran J, Enjuanes L. (1999) Replication and packaging of transmissible gastroenteritis coronavirus-derived synthetic minigenomes. *J. Virol.* **73**: 1535–1545.

Kaplan AS. (1955) The susceptibility of monkey kidney cells to poliovirus *in vivo* and *in vitro*. *Virology* **1**: 377–392.

Kapoor M, Zhang L, Mohan PM, Padmanabhan R. (1995) Synthesis and characterization of an infectious dengue virus type-2 RNA genome (New Guinea C strain). *Gene* **162**: 175–180.

Kato A, Kiyotani K, Sakai Y, Yoshida T, Nagai Y. (1997) The paramyxovirus, Sendai virus, V protein encodes a luxury function required for viral pathogenesis. *EMBO J.* **16:** 578–587.

Khromykh AA, Westaway EG. (1994) Completion of Kunjin virus RNA sequence and recovery of an infectious RNA transcribed from stably cloned full-length cDNA. *J. Virol.* **68**: 4580–4588.

Koetzner CA, Parker MM, Ricard CS, Sturman LS, Masters PS. (1992) Repair and mutagenesis of the genome of a deletion mutant of the coronavirus mouse hepatitis virus by targeted RNA recombination. *J. Virol.* **66**: 1841–1848.

Kolykhalov AA, Feinstone SM, Rice CM. (1996) Identification of a highly conserved sequence element at the 3′ terminus of hepatitis C virus genome RNA. *J. Virol.* **70**: 3363–3371.

Kolykhalov AA, Agapov EV, Blight KJ, Mihalik K, Feinstone SM, Rice CM. (1997) Transmission of hepatitis C by intrahepatic inoculation with transcribed RNA. *Science* **277**: 570–574.

Kolykhalov AA, Mihalik K, Feinstone SM, Rice CM. (2000) Hepatitis C virus-encoded enzymatic activities and conserved RNA elements in the 3′ nontranslated region are essential for virus replication in vivo. *J. Virol.* **74**: 2046–2051.

Kottier SA, Cavanagh D, Britton P. (1995) Experimental evidence of recombination in coronavirus infectious bronchitis virus. *Virology* **213**: 569–580.

Kubo H, Yamada YK, Taguchi F. (1994) Localization of neutralizing epitopes and the receptor-binding site within the amino-terminal 330 amino acids of the murine coronavirus spike protein. *J. Virol.* **68**: 5403–5410.

Kümmerer BM, Meyers G. (2000) Correlation between point mutations in NS2 and the viability and cytopathogenicity of bovine viral diarrhea virus strain Oregon analyzed with an infectious cDNA clone. *J. Virol.* **74**: 390–400.

Kümmerer BM, Stoll D, Meyers G. (1998) Bovine viral diarrhea virus strain Oregon: a novel mechanism for processing of NS2-3 based on point mutations. *J. Virol.* **72**: 4127–4138.

La Monica N, Almond JW, Racaniello VR. (1987) A mouse model for poliovirus neurovirulence identifies mutations that attenuate the virus for humans. *J. Virol.* **61**: 2917–2920.

Lavi E, Gilden DH, Highkin MK, Weiss SR. (1986) The organ tropism of mouse hepatitis virus A59 in mice is dependent on dose and route of inoculation. *Lab. Anim. Sci.* **36**: 130–135.

Leparc-Goffart I, Hingley ST, Chua MM, Phillips J, Lavi E, Weiss SR. (1998) Targeted recombination within the spike gene of murine coronavirus mouse hepatitis virus-A59: Q159 is a determinant of hepatotropism. *J. Virol.* **72**: 9628–9636.

Li S, Liu C, Klimov A, Subbarao K *et al.* (1999) Recombinant influenza V virus vaccines for the pathogenic human A/Hong Kong/97 (H5N1) viruses. *J. Inf. Dis.* **179:** 1132–1138.

Lindberg AM, Polacek C, Johansson S. (1997) Amplification and cloning of complete enterovirus genomes by long distance PCR. *J. Virol. Meth.* **65**: 191–199.

Lohmann V, Körner F, Koch J, Herian U, Theilmann L, Bartenschlager R. (1999) Replication of subgenomic hepatitis C virus RNAs in a hepatoma cell line. *Science* **285**: 110–113.

Luo G, Bergmann M, Garcia-Sastre A, Palese P. (1992) Mechanism of attenuation of a chimeric influenza A/B transfectant virus. *J. Virol.* **66:** 4679–4685.

Luytjes W, Sturman LS, Bredenbeek PJ, Charite J, van der Zeijst BA, Horzinek MC, Spaan WJ. (1987) Primary structure of the glycoprotein E2 of coronavirus MHV-A59 and identification of the trypsin cleavage site. *Virology* **161**: 479–487.

Luytjes W, Krystal M, Enami M, Parvin JD, Palese P. (1990) Amplification, expression, and packaging of a foreign gene by influenza virus. *Cell* **59:** 1107–1113.

Macadam AJ, Pollard SR, Ferguson G, Dunn G, Skuce R, Almond JW, Minor PD. (1991) The 5′ noncoding region of the type-2 poliovirus vaccine strain contains determinants of attenuation and temperature sensitivity. *Virology* **181**: 451–458.

Macadam AJ, Ferguson G, Burlison J, Stone D, Skuce R, Almond JW, Minor PD. (1992) Correlation of RNA secondary structure and attenuation of Sabin vaccine strains of poliovirus in tissue culture. *Virology* **189**: 415–422.

Macadam AJ, Pollard SR, Ferguson G, Skuce R, Wood D, Almond JW, Minor PD. (1993) Genetic basis of attenuation of the Sabin type-2 vaccine strain of poliovirus in primates. *Virology* **192**: 18–26.

Macadam AJ, Stone DM, Almond JW, Minor PD. (1994a) The 5′ noncoding region and virulence of poliovirus vaccine strains. *Trends. Microbiol.* **2**: 449–454.

Macadam AJ, Ferguson G, Fleming T, Stone DM, Almond JW, Minor PD. (1994) Role of poliovirus protease 2A in cap independent translation. *EMBO J.* **13**: 924–927.

Maisner A, Mrkic B, Herrler G, Moll M, Billeter MA, Cattanei R, Klenk H-D. (2000) Recombinant measles virus requiring an exogenous protease for activation of infectivity. *J. Gen. Virol.* **81:** 441–449.

Mandl CW, Ecker M, Holzmann H, Kunz C, Heinz FX. (1997) Infectious cDNA clones of tick-borne encephalitis virus European subtype prototypic strain Neudoerfl and high virulence strain Hypr. *J. Gen. Virol.* **78**: 1049–1057.

Mandl CW, Holzmann H, Meixner T, Rauscher S, Stadler PF, Allison SL, Heinz FX. (1998) Spontaneous and engineered deletions in the 3′ noncoding region of tick-borne encephalitis virus: construction of highly attenuated mutants of a flavivirus. *J. Virol.* **72**: 2132–2140.

Masters PS, Koetzner CA, Kerr CA, Heo Y. (1994) Optimization of targeted RNA recombination and mapping of a novel nucleocapsid gene mutation in the coronavirus mouse hepatitis virus. *J. Virol.* **68**: 328–337.

McGoldrick A, Macadam AJ, Dunn G *et al.* (1995) Role of mutations G-480 and C-6203 in the attenuation phenotype of Sabin type 1 poliovirus. *J. Virol.* **69**: 7601–7605.

Mebatsion T, Konig M, Conzelmann K-K. (1996) Budding of rabies virus particles in the absence of the spike glycoprotein. *Cell* **84:** 941–951.

Mebatsion, T, Finke S, Weiland F, Conzelmann KK. (1997) A CXCR4/CD4 pseudotype rhabdovirus that selectively infects HIV-1 envelope protein-expressing cells. *Cell* **90:** 841–847.

Mena I, Vivo A, Perez E, Portela A. (1996) Rescue of a synthetic chloramphenicol acetyltransferase RNA into influenza virus-like particles obtained from recombinant plasmids. *J. Virol.* **70:** 5016–5024.

Mendez E, Ruggli N, Collett MS, Rice CM. (1998) Infectious bovine viral diarrhea virus (strain NADL) RNA from stable cDNA clones: a cellular insert determines NS3 production and viral cytopathogenicity. *J. Virol.* **72**: 4737–4745.

Meredith JM, Rohll JB, Almond JW, Evans DJ. (1999) Similar interactions of the poliovirus and rhinovirus 3D polymerases with the 3′ untranslated region of rhinovirus 14. *J. Virol.* **73**: 9952–9958.

Meulenberg JJ, Bos-de Ruijter JN, van de Graaf R, Wensvoort G, Moormann RJ. (1998a) Infectious transcripts from cloned genome-length cDNA of porcine reproductive and respiratory syndrome virus. *J. Virol.* **72**: 380–387.

Meulenberg JJ, Bos-de Ruijter JN, Wensvoort G, Moormann RJ. (1998b) An infectious cDNA clone of porcine reproductive and respiratory syndrome virus. *Adv. Exp. Med. Biol.* **440**: 199–206.

Meyers G, Thiel H-J. (1996) Molecular characterization of Pestiviruses. *Adv. Virus Res.* **47**: 53–118.

Meyers G, Tautz N, Dubovi EJ, Thiel HJ. (1991) Viral cytopathogenicity correlated with integration of ubiquitin-coding sequences. *Virology* **180**: 602–616.

Meyers G, Tautz N, Becher P, Thiel HJ, Kümmerer BM. (1996a) Recovery of cytopathogenic and noncytopathogenic bovine viral diarrhea viruses from cDNA constructs. *J. Virol.* **70**: 8606–8613.

Meyers G, Thiel HJ, Rumenapf T. (1996b) Classical swine fever virus: recovery of infectious viruses from cDNA constructs and generation of recombinant cytopathogenic defective interfering particles. *J. Virol.* **70**: 1588–1595.

Meyers G, Stoll D, Gunn M. (1998) Insertion of a sequence encoding light chain 3 of microtubule-associated proteins 1A and 1B in a pestivirus genome: connection with virus cytopathogenicity and induction of lethal disease in cattle. *J. Virol.* **72**: 4139–4148.

Mizutani S, Colonno RJ. (1985) In vitro synthesis of an infectious RNA from cDNA clones of human rhinovirus type 14. *J. Virol.* **56**: 628–632.

Monath TP, Soike K, Levenbook I *et al.* (1999) Recombinant, chimaeric live, attenuated vaccine (ChimeriVax (TM)) incorporating the envelope genes of Japanese encephalitis (SA14-14-2) virus and the capsid and nonstructural genes of yellow fever (17D) virus is safe, immunogenic and protective in nonhuman primates. *Vaccine* **17**: 1869–1882.

Moormann RJ, van Gennip HG, Miedema GK, Hulst MM, van Rijn PA. (1996) Infectious RNA transcribed from an engineered full-length cDNA template of the genome of a pestivirus. *J. Virol.* **70**: 763–770.

Moss EG, O Neill RE, Racaniello VR. (1989) Mapping of attenuating sequences of an avirulent poliovirus type 2 strain. *J. Virol.* **63**: 1884–1890.

Muster T, Subbarao K, Enami M, Murphy B, Palese P. (1991) An influenza A virus containing influenza B virus 5′ and 3′ noncoding regions on the neuraminidase gene is attenuated in mice. *Proc. Natl Acad. Sci. USA* **88**: 5177–5181.

Muylaert IR, Chambers TJ, Galler R, Rice CM. (1996) Mutagenesis of the N-linked glycosylation sites of the yellow fever virus NS1 protein: Effects on virus replication and mouse neurovirulence. *Virology* **222**: 159–168.

Neumann G, Zobel A, Hobom G. (1994) RNA-polymerase I-mediated expression of influenza viral-RNA molecules. *Virology* **202:** 477–479.

Neumann G, Watanabe T, Ito H *et al.* (1999) Generation of influenza A viruses entirely from cloned cDNAs. *Proc. Natl Acad. Sci. USA* **96:** 9345–9350.

Omata T, Kohara M, Kuge S *et al.* (1986) Genetic analysis of the attenuation phenotype of poliovirus type 1. *J. Virol.* **58**: 348–358.

Parker SE, Gallagher TM, Buchmeier MJ. (1989) Sequence analysis reveals extensive polymorphism and evidence of deletions within the E2 glycoprotein gene of several strains of murine hepatitis virus. *Virology* **173**: 664–673.

Pattnaik AK, Ball LA, Legrone AW, Wertz GW. (1992) Infectious defective interfering particles of VSV from transcripts of a cDNA clone. *Cell* **69:** 1011–1020.

Phillips JJ, Chua MM, Lavi E, Weiss SR. (1999) Pathogenesis of chimeric MHV4/MHV-A59 recombinant viruses: the murine coronavirus spike protein is a major determinant of neurovirulence. *J. Virol.* **73**: 7752–7760.

Pleschka S, Jaskunas SR, Engelhardt OG, Zurcher T, Palese P, Garcia-Sastre A. (1996) A plasmid-based reverse genetics system for influenza A virus. *J. Virol.* **70:** 4188–4192.

Pletnev AG, Men R. (1998) Attenuation of the Langat tick-borne flavivirus by chimerization with mosquito-borne flavivirus dengue type 4. *Proc. Natl Acad. Sci. USA* **95**: 1746–1751.

Pletnev AG, Bray M, Huggins J, Lai CJ. (1992) Construction and characterization of chimeric tick-borne encephalitis/dengue type 4 viruses. *Proc. Natl Acad. Sci. USA* **89**: 10532–10536.

Pletnev AG, Bray M, Lai CJ. (1993) Chimeric tick-borne encephalitis and dengue type 4 viruses: effects of mutations on neurovirulence in mice. *J. Virol.* **67**: 4956–4963.

Pollard SR, Dunn G, Cammack N, Minor PD, Almond JW. (1989) Nucleotide sequence of a neurovirulent variant of the type 2 oral poliovirus vaccine. *J. Virol.* **63**: 4949–4951.

Racaniello VR, Baltimore D. (1981) Cloned poliovirus complementary DNA is infectious in mammalian cells. *Science* **214**: 916–919.

Radecke F, Billeter MA. (1996) The nonstructural C protein is not essentail for multiplication of Edmonston B strain measles virus in cultured cells. *Virology* **217:** 418–421.

Radecke F, Spielhofer P, Schneider H *et al.* (1995) Rescue of measles virus from cloned DNA. *EMBO J.* **14:** 5773–5784.

Ren RB, Moss EG, Racaniello VR. (1991) Identification of two determinants that attenuate vaccine-related type 2 poliovirus. *J. Virol.* **65**: 1377–1382.

Rice CM, Grakoui A, Galler R, Chambers TJ. (1989) Transcription of infectious yellow fever RNA from full-length cDNA templates produced by in vitro ligation. *New Biol.* **1**: 285–296.

Roberts A, Rose JK. (1998) Recovery of negative-strand RNA viruses from plasmid DNAs: A positive approach revitalizes a negative field. *Virology* **247:** 1–6.

Ruggli N, Tratschin JD, Mittelholzer C, Hofmann MA. (1996) Nucleotide sequence of classical swine fever virus strain Alfort/187 and transcription of infectious RNA from stably cloned full-length cDNA. *J. Virol.* **70**: 3478–3487.

Sabin AB, Boulger LR. (1973) History of Sabin attenuated poliovirus oral live vaccine strains. *J. Biol. Standard.* **1**: 115–118.

Sakaguchi T, Kiyotani K, Kato A, Asakawa M, Fujii Y, Nagai Y, Yoshida T. (1997) Phosphorylation of the Sendai virus M protein is not essential for virus replication either *in vitro* or *in vivo*. *Virology* **235:** 360–366.

Sánchez CM, Izeta A, Sánchez-Morgado JM *et al.* (1999) Targeted recombination demonstrates that the spike gene of transmissible gastroenteritis coronavirus is a determinant of its enteric tropism and virulence. *J. Virol.* **73**: 7607–7618.

Sarnow P. (1989) Role of 3′-end sequences in infectivity of poliovirus transcripts made in vitro. *J. Virol.* **63**: 467–470.

Schnell MJ, Mebatsion T, Conzelmann K-K. (1994) Infectious rabies viruses from cloned cDNA. *EMBO J.* **13:** 4195–4203.

Schnell MJ, Johnson JE, Buonocore L, Rose JK. (1997) Construction of a novel virus that targets HIV-1-infected cells and controls HIV-1 infection. *Cell* **90:** 849–857.

Shimizu YK, Iwamoto A, Hijikata M, Purcell RH, Yoshikura H. (1992) Evidence for in vitro replication of hepatitis C virus genome in a human T-cell line. *Proc. Natl Acad. Sci. USA* **89**: 5477–5481.

Singh M, Billeter MA. (1999) A recombinant measles virus expressing biologically active human interleukin-12. *J. Gen. Virol.* **80:** 101–106.

Skinner MA, Racaniello VR, Dunn G, Cooper J, Minor PD, Almond JW. (1989) New model for the secondary structure of the 5′ non-coding RNA of poliovirus is supported by biochemical and genetic data that also show that RNA secondary structure is important in neurovirulence. *J. Mol. Biol.* **207**: 379–392.

Solorzano A, Zheng H, Fodor E, Brownlee GG, Palese P, Garcia-Sastre A. (2000) Reduced levels of neuraminidase of influenza A viruses correlate with attenuated phenotypes in mice. *J. Gen. Virol.* **81:** 737–742.

Sooknanan R, Howes M, Read L, Malek LT. (1994) Fidelity of nucleic acid amplification with avian myeloblastosis virus reverse transcriptase and T7 RNA polymerase. *Biotechniques* **17**: 1077–1080, 1083–1085.

Stanway G, Cann AJ, Hauptmann R *et al.* (1983) The nucleotide sequence of poliovirus type 3 leon 12 a1b: comparison with poliovirus type 1. *Nucl. Acids Res.* **11**: 5629–5643.

Stanway G, Hughes PJ, Mountford RC, Reeve P, Minor PD, Schild GC, Almond JW. (1984) Comparison of the complete nucleotide sequence of the genomes of the neurovirulent poliovirus P3/Leon/37 and its attenuated Sabin vaccine derivative P3/Leon 12a1b. *Proc. Natl Acad. Sci. USA* **81**: 1539–1543.

Stirrups K, Shaw K, Evans S, Dalton K, Cavanagh D, Britton P. (1998) Rescue of IBV D-RNA by heterologous helper virus strains. *Adv. Exp. Med. Biol.* **440**: 259–264.

Sumiyoshi H, Hoke CH, Trent DW. (1992) Infectious Japanese encephalitis virus RNA can be synthesized from in vitro-ligated cDNA templates. *J. Virol.* **66**: 5425–5431.

Svitkin YV, Maslova SV, Agol VI. (1985) The genomes of attenuated and virulent poliovirus strains differ in their in vitro translation efficiencies. *Virology* **147**: 243–252.

Svitkin YV, Pestova TV, Maslova SV, Agol VI. (1988) Point mutations modify the response of poliovirus RNA to a translation initiation factor: a comparison of neurovirulent and attenuated strains. *Virology* **166**: 394–404.

Svitkin YV, Cammack N, Minor PD, Almond JW. (1990) Translation deficiency of the Sabin type 3 poliovirus genome: association with an attenuating mutation C472–U. *Virology* **175**: 103–109.

Tanaka T, Kato N, Cho MJ, Shimotohno K. (1995) A novel sequence found at the 3′ terminus of hepatitis C virus genome. *Biochem. Biophys. Res. Commun.* **215**: 744–749.

Taniguchi T, Palmieri M, Weissmann C. (1978) QB DNA-containing hybrid plasmids giving rise to QB phage formation in the bacterial host. *Nature* **274**: 223–228.

Tatem JM, Weekslevy C, Georgiu A *et al.* (1992) A Mutation Present in the Amino Terminus of Sabin-3 Poliovirus VP1 Protein Is Attenuating. *J. Virol.* **66**: 3194–3197.

Tautz N, Thiel HJ, Dubovi EJ, Meyers G. (1994) Pathogenesis of mucosal disease: a cytopathogenic pestivirus generated by an internal deletion. *J. Virol.* **68**: 3289–3297.

Tautz N, Meyers G, Stark R, Dubovi EJ, Thiel HJ. (1996) Cytopathogenicity of a pestivirus correlates with a 27-nucleotide insertion. *J. Virol.* **70**: 7851–7858.

Tellier R, Bukh J, Emerson SU, Miller RH, Purcell RH. (1996) Long PCR and its application to hepatitis viruses: amplification of hepatitis A, hepatitis B, and hepatitis C virus genomes. *J. Clin. Microbiol.* **34**: 3085–3091.

Thomas JM, Stevens MP, Percy N, Barclay WS. (1998) Phosphorylation of the M2 protein of influenza A virus is not essential for virus viability. *Virology* **252:** 54–64.

van der Most RG, Heijnen L, Spaan WJM, de Groot RJ. (1992) Homologous RNA recombination allows efficient introduction of site-specific mutations into the genome of coronavirus MHV-A59 via synthetic co-replicating RNAs. *Nucl. Acids Res.* **20**: 3375–3381.

Van der Werf S, Bradley J, Wimmer E, Studier FW, Dunn JJ. (1986) Synthesis of infectious poliovirus RNA by purified T7 RNA polymerase. *Proc. Natl Acad. Sci. USA* **83**: 2330–2334.

Weeks Levy C, Tatem JM, Dimichele SJ, Waterfield W, Georgiu AF, Mento SJ. (1991) Identification and characterization of a new base substitution in the vaccine strain of Sabin 3 poliovirus [published erratum appears in Virology 187:845 (1992)]. *Virology* **185**: 934–937.

Wege H, Winter J, Meyermann R. (1988) The peplomer protein E2 of coronavirus JHM as a determinant of neurovirulence: definition of critical epitopes by variant analysis. *J. Gen. Virol.* **69**: 87–98.

Wertz GW, Perepelista VP, Ball LA. (1998) Gene rearrangement attenuates expression and lethality of a nonsegmented negative strand RNA virus. *Proc. Natl Acad. Sci. USA* **95:** 3501–3506.

Whitehead SS, Bukreyev A, Teng MN *et al.* (1999) Recombinant respiratory syncytial virus bearing a deletion of either the NS2 or SH gene is attenuated in chimpanzees. *J. Virol.* **73:** 3438–3442.

Yanagi M, Purcell RH, Emerson SU, Bukh J. (1997) Transcripts from a single full-length cDNA clone of hepatitis C virus are infectious when directly transfected into the liver of a chimpanzee. *Proc. Natl Acad. Sci. USA* **94**: 8738–8743.

Yanagi M, St Claire M, Shapiro M, Emerson SU, Purcell RH, Bukh J. (1998) Transcripts of a chimeric cDNA clone of hepatitis C virus genotype 1b are infectious in vivo. *Virology* **244**: 161–172.

Yanagi M, Purcell RH, Emerson SU, Bukh J. (1999a) Hepatitis C virus: an infectious molecular clone of a second major genotype (2a) and lack of viability of intertypic 1a and 2a chimeras. *Virology* **262**: 250–263.

Yanagi M, St Claire M, Emerson SU, Purcell RH, Bukh J. (1999b) In vivo analysis of the 3′ untranslated region of the hepatitis C virus after in vitro mutagenesis of an infectious cDNA clone. *Proc. Natl Acad. Sci. USA* **96**: 2291–2295.

Chapter 8

Therapeutic applications of viral vectors

Pedro Lowenstein, Daniel Stone, Anne David, Alan Melcher, Richard Vile, Clare Thomas, Dominique Bataille and Maria Castro

8.1 Viruses as vectors: what's new?

8.1.1 Viral vectors and their attraction

Any virus can, in principle, be adapted as a vector for gene transfer and eventual gene therapy. To do so, minimal changes are necessary to inhibit or eliminate virus replication and pathogenicity, and to insert the transgene to be expressed under appropriate regulatory control elements. Thus, any virus can be converted into a vector.

Whether a vector will be limited to its use in gene transfer experiments, or could eventually be used in clinical gene therapy will depend on vector pathogenicity, efficiency for gene transfer *in vivo*, and ease of production at high concentrations and clinical grade purity. Many viruses have been adapted for their use as vectors for gene therapy, the most popular ones remaining the RNA viruses (retroviruses and lentiviruses), and select DNA viruses (i.e. adeno-associated virus, herpes simplex virus type 1, and adenovirus). Many other viruses have been adapted for gene transfer, and could one day become vectors for gene therapy (i.e. feline immune deficiency virus, Semliki Forest virus, influenza virus, baculovirus, etc.).

The perceived attractions of viruses as vectors for gene transfer are their high efficiency of gene transfer into target cells, high levels of transgene expression, and for some viruses, the existence of mechanisms to achieve long-term expression. High efficiency of gene transfer into target cells depends on the ubiquity of receptors used for entry into cells. Once inside cells, efficient mechanisms allow the eventual delivery of vector DNA into target cell nuclei. Long-term expression will then depend on the virus used as vector, and the tissue transduced. The basic mechanisms of vector genome transfer to the nucleus and its subsequent fate is still being analysed, even for the most common vectors in generalized use.

Some viruses genomes integrate into those of the target cells. Type C retroviruses can integrate their genome into dividing cells only, while lentiviruses and AAV are able to do so in non-dividing postmitotic cells also. HSV-1 can establish latency in dorsal root ganglion neurons *in vivo*, and these mechanisms could eventually be harnessed to achieve longevity of expression in other areas of the brain, as well as in other organs. Adenoviruses, on the other hand, while not being known to establish standard latency as

Genetically Engineered Viruses: Development and Applications, edited by C.J.A. Ring and E.D. Blair

per HSV-1 or lentiviruses, can nevertheless provide up to at least 1 year of transgene expression in an immune-privileged organ such as the brain. Importantly, the new high-capacity adenoviral vectors also allow up to 12 months transgene expression in the periphery. While persistent infections can result from infection with non-integrating RNA viruses, it is not thought that such persistence mechanisms could be useful to achieve long-term transgene expression in a therapeutic context. The status of viral vector genomes during long-term expression in various target cells remains to be determined.

8.1.2 What are the limitations of viral vectors?

Which factors contribute to vector transgene expression, toxicity, or its induced inflammatory and immune responses? Although this remains a poorly explored area, essentially any component of viral vectors, i.e. the virion particle itself, the vector genome backbone, the sequences and functions of the encoded transgenes, as well as the promoters and transcriptional elements used, will influence how the vectors interact with the target cells and immune and inflammatory cells *in vivo*.

Recently, it has become possible to engineer viral vector particles to retarget vectors to specific receptors. Also, vector genomes can be completely redesigned and replaced by non-viral sequences of known functions, and many new promoter elements and transgenes have been made available through the rapid progress of the Human Genome Project (and the sequencing of various other genomes). Many groups are now exploring in detail the direct cytotoxic effects, as well as the inflammatory and immune responses to viral vectors.

8.1.3 Why are therapeutic applications taking so long, and which is the way forward?

Viral vectors can be made to express transgenes efficiently. However, the main limitations still hampering the full-scale implementation of clinical trials is the direct and indirect toxicity associated with viral vectors. Viral vector toxicity depends on the particular vector being used, the mode of administration, the administered dose, its direct cytotoxic activity, its potential for acute inflammation, its immunogenicity and the development of chronic inflammatory and immune responses.

What does toxicity depend on? Viral vectors can be directly cytotoxic, even in the absence of gene expression. The mechanisms of these phenomena remain poorly understood. Whether this depends on specific interactions of virions with membrane receptors, viral entry, or other intracellular responses to viral vector infection is currently being addressed. Intrinsic toxicity varies between vectors, and is also crucially dependent on the dose of vector injected. To some extent comparative toxicity will depend on viral doses used for each type of vector, i.e. usually retroviral vectors and AAV vectors are used at much lower doses than adenovirus vectors. Importantly, many molecular and biological aspects of the direct cytotoxic effects of infection with high doses of virus vectors remain to be examined in detail in most tissues.

Targeting viral vectors to specific receptors may reduce at least some of the non-specific toxicity. Many advances have been made recently in the retargeting of RNA and DNA viral vectors. How this affects early responses to virus interactions with target cells has not yet been determined. Importantly, new virus mutants are now available in which specific interactions with cellular receptors are completely abolished. How this influences early inflammatory responses will be a fascinating topic to investigate.

The use of novel powerful, cell type specific, and inducible promoters is also likely to lead to important reductions in vector cytotoxicity. In the brain, viral vectors only cause acute inflammatory responses when injected above a certain threshold dose. The use of novel very powerful promoters allows a substantial reduction (up to 3 logs) in the dose of vector needed to achieve high-level transgene expression.

While safety aspects continue to be improved, the use of transcriptional targeting elements has facilitated cell type specific, inducible, and recently, combined cell type specific and inducible transgene expression. A major limitation, nevertheless, has been the low levels of transgene expression. Promiscuous, constitutively active, viral promoters, continue to be the strongest. Even so, very high doses of virus are needed to achieve any physiologically or clinically relevant transgene expression, independently of the viral vector used. Thus, although it has not been published formally, various laboratories have evidence that with most vectors, multiple infection of target cells not only occurs, but is necessary to achieve high enough levels of transgene expression to detect physiologically relevant consequences. Recent results from our laboratory have now shown that using a more recently described viral promoter, recombinant adenoviruses can be engineered, allowing a cell which is infected by a single viral particle to produce enough transgene to enable detection by immunohistochemistry. We believe it will be very important to improve the strength of promoter function, and allow this to become cell type specific and inducible, while retaining the high levels of transgene expression.

In spite of the perception of viral vectors being very efficient, most *in vitro*, as well as most *in vivo* experiments have necessitated very high doses of vector. This automatically increases vector-induced side effects, since, although in theory one infectious viral particle should be capable of expressing its genes in a target cell, in gene therapy experiments very high multiplicities of infection need to be employed in order to see any effects. Although most gene therapy groups have accepted this matter of fact, the reasons behind the need for these high MOIs *in vivo*, and *in vitro*, are not well understood.

8.1.4 In vivo problems in humans: antibodies, toxicity, and inhibition by vector carriers

The recent death of a supposedly control patient subjected to gene therapy has highlighted the general issues on the safety of the conduct of clinical gene therapy trials. Given that many inflammatory and immune responses to viral vectors, as well as the status of the human immune system remain uncharacterized, more basic work will certainly be done in the near future. Although a high percentage of the human population may have been previously infected with adenoviruses, the level, specificity and activity of such antibodies in humans remain unknown.

It will be important to assess the role of circulating antibodies on viral vectors administered to patients. The issue has been raised recently whether much of the reported lack of toxicity in human patients has been due to the inactivation of the viral vectors before ever reaching the target tissues. There is certain concern that circulating antibodies could have masked toxicities expected to occur at lower doses. Such a false sense of security would have prompted the use of higher doses of viral vectors.

Recent data have also highlighted the possibility that certain tools used to deliver viruses into humans, like various catheters used for intravascular gene vector delivery, could inactivate viruses. If true, much lower viral titres would have been injected into tissues to be transduced. As Phase I clinical trials are limited to a detection of possible toxicity and adverse effects, the lack of these would have prompted an increase in the dose of vector administered.

8.1.5 What is being done clinically?

Table 8.1. Number of clinical trials in gene therapy

Category	Protocols		Patients	
	Number	%	Number	%
Cancer therapy	252	63.6	2269	69.2
Gene marking	41	10.4	227	6.9
Healthy volunteers	2	0.5	6	0.2
Infectious diseases	33	8.3	412	12.6
Monogenic diseases	53	13.4	298	9.1
Other	15	3.8	66	2
Total	396	100	3278	100

Data for these tables was reproduced with permission from the *Journal of Gene Medicine* website (http://www.wiley.co.uk/genmed). © John Wiley & Sons Limited

Table 8.2. Transgenes being used in clinical gene therapy trials

Gene type	Gene
Antibody	ScFv against erbB2, ScFv against rev, ScFv against -env, specific anti-idiotype, TR Ab
Antigen	CEA, env/rev, gp100, HIV-IT(V) env/rev, HLA-B7, HLA-B7/b2m, MART-1, MUC-1, PSA, HLA-A2 or HLA-B13 or H-2K(k)
Antigen + marker	CEA/KanaR, HLA-B7/neoR
Antisense	c-fos or c-myc, IGF-1, TAR, Pol 1, tat and rev, TGF-β2
Antisense + drug resistance	Antisense BCR/ABL + DHFR
Antisense or decoy + negative transdominant	Antisense TAR/revTD, RSV-TAR/revM10
Antisense + tumour suppressor	Antisense k-ras/p53
Antitumoural	E1A
Cytokine	GM-CSF, IFN-α, IFN-β, IFN-γ, IFN-γ/IL-2, IL-12, IL-2, IL-2/GM-CSF, IL-2/HLA-B7, IL-2/lymphotactin, IL-4, IL-6/sIL-6R, IL-7/IL-12/GM-CSF, IL-7/IL-2
Cytokine + marker	IL-12/NeoR, IL-2/NeoR, LacZ/IL-2, TNF/NeoR
Decoy	RRE/polyTAR
Decoy + marker	NeoR or RRE decoy, RRE/polyTAR/NeoR
Deficiency	AAT, ADA, ASPA, CD18, CD154, CFTR, common gamma chain, FACC, factor IX, GC, gp91phox, IDS, IDUA, LDLR, OTC, p47phox, PNP
Deficiency + marker	ADA/NeoR
Drug resistance	MDR-1, MGMT
Drug resistance + marker	MDR-1/NeoR
Marker	Hy, LacZ, NeoR
Multiple	IL-7/TK/Hy, TK-NeoR + delta LNGFR

Table 8.2. (cont)

Gene type	Gene
Negative transdominant	Trev
Oncogene regulator	HER-2/neu
Other	CNTF, FGF-4, hIGF-1, IRAP, VEGF
Other + marker	LacZ/VEGF
Receptor	CC49-Zeta TcR, CD4-Zeta chimera, CD80(B7-1), Mov-gamma
Ribozyme	Anti HIV-tat and rev, anti HIV-Tat ribozyme, anti-HIV ribozyme
Suicide	CD, Staphylococcus endotoxin B, TK
Suicide + marker	Hy/TK, LacZ/TK, TK/NeoR
Tumor suppressor	BRCA-1, p53, Rb Drug resistance

Data for these tables was reproduced with permission from the *Journal of Gene Medicine* website (http://www.wiley.co.uk/genmed).

Table 8.3. Clinical trials in gene therapy

Phase	Protocols		Patients	
	Number	%	Number	%
I	261	65.9	1656	50.5
I/II	97	24.5	857	26.1
II	36	9.1	514	15.7
III	2	0.5	251	7.7
Total	396	100	3278	100

Data for these tables was reproduced with permission from the *Journal of Gene Medicine* website (http://www.wiley.co.uk/genmed).

Table 8.4. Vectors used in clinical gene therapy trials

Phase	Protocols		Patients	
	Number	%	Number	%
AAV	3	0.8	36	1.1
Adenovirus	71	17.9	437	13.3
Electroporation	2	0.5	20	0.6
Gene gun	4	1	35	1.1
Herpesvirus	1	0.3	0	0
Lipofection	73	18.4	735	22.4
Lipofection/AAV	2	0.5	0	0
Lipofection/Adenovirus	1	0.3	3	0.1
Naked DNA	16	4	69	2.1
Poxvirus	26	6.6	130	4
Retroviral vector producing cells	20	5.1	408	12.4
Retrovirus	158	39.9	1217	37.1
Retrovirus/gene gun	1	0.3	6	0.2
RNA transfer	1	0.3	30	0.9
Other transfections	7	1.8	101	3.1
N/C	9	2.3	51	1.6
AAV	1	0.3	0	0
Total	396	100	3278	100

Data for these tables was reproduced with permission from the *Journal of Gene Medicine* website (http://www.wiley.co.uk/genmed).

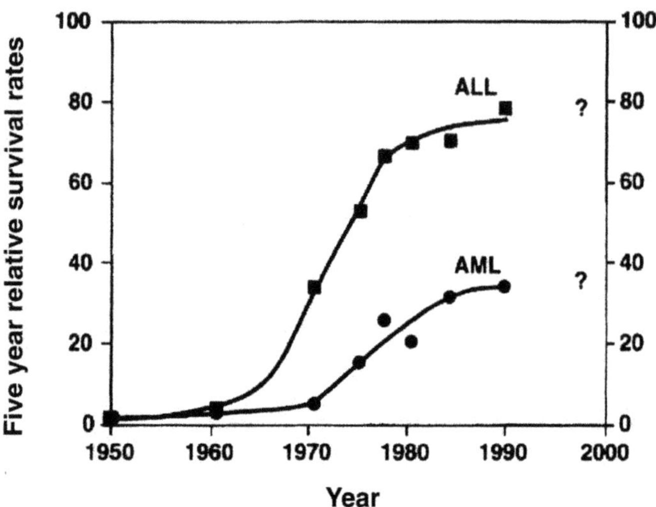

Figure 8.1. Even for the development of effective treatments for acute lymphoblastic leukaemia, it took classical cancer chemotherapy close to 50 years to achieve a high rate of success in at least some cancers. Reproduced from Kersey, 1997, Fifty years of studies of the biology and therapy of childhood leukemia, *Blood* 90(11): 4243–4251, © American Society of Hematology.

Although it is difficult to predict when gene therapy will provide clear evidence of 'cure' in large numbers of patients, it is useful to recall that even the development of classical pharmacotherapy for cancers that are now considered 'curable' took more than 50 years to come to fruition (*Figure 8.1*).

8.2 Retroviral vectors

8.2.1 Retroviral biology

Retroviral vectors have been identified as attractive candidates for gene transfer and they are currently the vehicle of choice in many ongoing clinical gene therapy trials. The ability of retroviruses to stably integrate their own genome into the host genome of many cell types has led to the production of vector systems capable of long-term gene transfer. Current vector systems based on type C retroviruses, such as the Moloney murine leukaemia virus (MoMLV), have proved very popular as they are relatively non-pathogenic. These vectors are devoid of all viral genes, are replication-defective, can carry up to 8 kilobases of foreign DNA and can be engineered to have varying cell tropism.

Retroviruses are RNA viruses, which show a common virion structure amongst all isolates. The virion structure contains a lipid envelope, derived from the host cell, which is modified by the insertion of viral envelope proteins in place of normal cellular components. The envelope protein complex is comprised of two sub-units held together by di-sulphide bonds and non-covalent interactions. The smaller transmembrane (TM) sub-unit acts as an anchor and the larger SU sub-unit, which is often heavily glycosylated, contains the receptor binding function.

Within the viral envelope lies a shell-like capsid, which is primarily made up of capsid (CA) proteins. This capsid is attached to the outer envelope by the matrix protein (MA), which is thought to be chemically cross-linked to lipid. The capsid houses two copies of the single stranded RNA genome as well as the non-structural protease, reverse transcriptase, integrase and RNA associated nucleoprotein (NC).

Retroviruses are the only diploid viruses, in that they contain two copies of their genome. This RNA genome is plus stranded, 7–10 kilobases long, and is capped at the 5′ end and polyadenylated at the 3′ end by the cellular transcription machinery. The non-coding regulatory elements are located in the terminal regions. These *cis*-acting regions contain the promoter/enhancer and transcriptional control regions which regulate viral gene expression of the integrated DNA provirus within a sequence which is also repeated at the 3′ end of the genome and is, therefore, called the Long Terminal Repeat (LTR). Just downstream of the 5′ LTR lie sequences, which include the Ψ packaging signal, flanking the four central coding regions termed *gag*, *pro*, *pol* and *env*. The *gag* gene produces a polyprotein, which is cleaved to yield the capsid proteins MA, CA and NC. The *pro* gene encodes the protease responsible for cleavage of *gag*, *pol* and *env* polyproteins. The *pol* gene encodes a polyprotein, which is cleaved to produce the reverse transcriptase and integrase proteins. The *env* gene encodes a polyprotein, which is cleaved to produce the two envelope proteins TM and SU (For a general review of retrovirus biology see Coffin, 1996).

The retrovirus life cycle (*Figure 8.2a*) begins with the process of cell attachment, which occurs when the SU sub-unit of the envelope protein interacts with a specific cell surface receptor (Sommerfelt, 1999; Weiss, 1993; Weiss and Tailor 1995). The cellular tropism depends on the native envelope glycoprotein, for example ecotropic MoMLV (MoMLV-E) has a strictly murine host range and the native receptor for MoMLV-E is the cationic amino acid transporter CAT-1 (Sommerfelt, 1999), whilst amphotropic MoMLV (MoMLV-A) has a murine and human host range and the native receptor for MoMLV-A is the transmembrane phosphate transporter PIT-1 (Sommerfelt, 1999). After attachment the viral envelope fuses with the cell membrane enabling the capsid core to be released into the cytoplasm. The process of reverse transcription can then begin within the capsid, as the RNA genome is converted to double stranded proviral DNA by the viral reverse transcriptase within the infected cell's cytoplasm (For review see Coffin, 1996). A pre-integration complex, including the proviral genome and integrase, is then transported to the nuclear membrane, where entry into the nucleus can only occur after breakdown of the nuclear membrane during mitosis (Roe *et al.*, 1993).

Upon entry into the nucleus the proviral genome, consisting of *gag*, *pro*, *pol* and *env* genes flanked by two identical long terminal repeats (LTRs) containing the *cis*-acting regions, is randomly integrated into the host genome by the viral integrase (For review see Coffin, 1996). Synthesis of viral RNA by the cellular RNA polymerase, can then be initiated from the LTR promoter. A single RNA species is produced which can then be processed into genomic RNA or multiply spliced, giving rise to individual mRNA transcripts. Translation of viral proteins can then begin and this is followed by the formation of virion particles at the cell surface. The capsid and envelope are thought to assemble at the same time and once the viral genome is packaged into the nascent

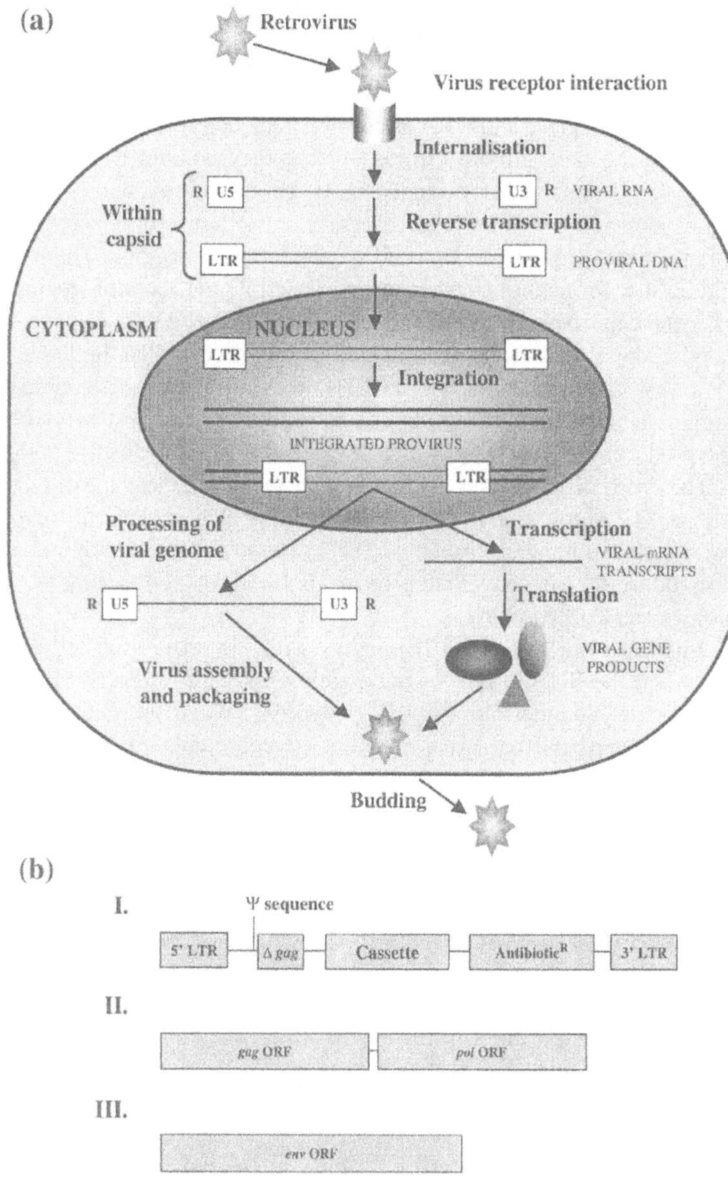

Figure 8.2. Schematic representation of the retrovirus life cycle showing internalization, reverse transcription, integration, virus assembly and budding (a). Representation of the constructs used to make Mo-MLV derived retroviral vectors (b). Construct I is transfected into a cell line, which stably expresses constructs II and III. Following transfection the cells are placed under antibiotic selection. Individual clones are then selected and assessed for their ability to produce recombinant retrovirus vectors. Within these retroviral producer cells constructs II and III produce the packaging functions in *trans* and the genome is derived from construct I, which contains the packaging signal Ψ and LTR regions. The viral integrase and reverse transcriptase, produced by the *pol* orf of construct II, are also packaged into infective virions (modified from Stone *et al.*, 2000).

virion, along with the non-structural viral proteins, budding of the newly formed virion can occur.

8.2.2. Retroviral vectors

Initial research into the use of retroviruses for gene transfer resulted in the production of first generation retroviral vectors, derived from the MoMLV, which were designed to provide stable gene transfer without producing replication competent retroviruses (RCRs). Packaging cell lines, derived from NIH3T3 cells (mouse fibroblasts), were developed containing a single plasmid incorporating the MoMLV genome deleted of the packaging signal Ψ (Cone and Mulligan, 1984; Mann *et al.*, 1983). As all other viral proteins are produced by the packaging cell line *in trans*, transfection of a vector plasmid, containing the MLV LTRs, packaging signal Ψ and therapeutic gene, results in the production of replication-defective virus. By changing the *env* gene within the first generation packaging plasmids it was possible to change the tropism of a recombinant retrovirus. Envelope proteins derived from ecotropic (able to infect rodent cells) (Mann *et al.*, 1983) or amphotropic (able to infect human, and rodent cells) (Cone and Mulligan, 1984) MLV variants were used.

The first generation packaging cell lines have however been shown to produce RCRs at regular intervals as only a single recombination event between packaging plasmid and vector plasmid is required to produce wild-type virus (Muenchau *et al.*, 1990). This led to the development of second generation retroviral vectors, where further deletions have been made within the packaging cell line (Miller and Buttimore, 1986). The Ψ deletion has been extended into part of the 5′ LTR and the 3′ LTR has been replaced by the SV40 polyadenylation signal. Thus the risk of RCR formation is reduced as a minimum of two recombinations would need to occur to enable reversion to wild-type genome. The protoype second generation packaging line is the PA317 cell line, which pseudotypes MoMLV core particles with amphotropic (4070A) envelope proteins. This line was used for the production of the recombinant retroviruses for transfer to human cells *ex vivo* and then also *in vivo* with direct injection of packaging cell lines.

Despite the increased safety of second generation retroviral vectors, RCR formation has also been detected in second generation packaging cell lines (Scarpa *et al.*, 1991). Subsequent development of packaging cell lines, where the packaging function is split between two plasmids, has reduced the likelihood of RCR formation even further (Markowitz *et al.*, 1988a, 1988b). These third generation packaging cell lines have the *gag* and *pol* gene regions on one plasmid and the *env* gene on another (*Figure 8.2b*). The number of recombinations needed to generate wild-type virus is again increased although RCR formation has still been observed within these newer packaging cell lines (Chong and Vile, 1996, 1998). Murine retroviral vectors contaminated with RCR have been shown to be tumorigenic in non-human primates raising concerns over the risk to humans (Donahue *et al.*, 1992). These data however, can be viewed in either, or both, of two different ways. Clearly it is ominous that replicating murine retroviruses can cause disease in primates (and, therefore, presumably also in patients) if administered inadvertantly. On the other hand, in order to see disease at all, the investigators in this study had to spike the recombinant stocks deliberately with levels of RCR at least two logs higher than those which can be easily detected by current quality control screening procedures. In this latter light, such experiments give high levels of confidence that administration of recombinant retroviral vector stocks should be safe even if extremely low levels of RCR did go undetected.

Retroviruses are good candidate vectors for gene therapy as they have a broad host range and they are largely non-injurious to infected cells. The primary advantage however, is their ability to stably introduce genes into cells. Genes integrated into host cell DNA are regulated as if belonging to the cell and this enables stable transfer of the gene to subsequent daughter cells following mitosis. Levels of gene expression however, will depend on the site of vector insertion. In a clinical application, where long-term correction of a genetic defect is required, retroviral vectors offer us a means to achieve this goal. The ability of retroviral vectors to easily transduce cells *in vitro* has led to the development of *ex vivo* therapeutic strategies, whereby retroviral vectors are used to infect peripheral blood lymphocytes or bone marrow cells. In phase one clinical trials using such a procedure, long-term gene expression has been demonstrated (Blaese *et al.*, 1995, Bordignon *et al.*, 1995). One potential problem has been noted, however. In several *in vitro* and *in vivo* studies, it has been shown that long-term integration of the viral vector persists in infected cells but that this can be accompanied by shut down of expression either of the viral LTR promoter or of internal viral promoters used within the vector. This shut off appears to be due in part at least to methylation of the foreign promoter/enhancers introduced into cells by the retroviral vector. Therefore, although persistent expression for long-term correction of disease is clearly a potential advantage of retroviruses as vectors, there is increasingly an effort to use internal, cellular, tissue specific promoters to avoid such shut off problems.

The need for a safe gene transfer system *in vivo* provides retroviral vectors with advantages over other currently used viral vectors. Current adenovirus or herpes virus vectors are based on human viruses and pre-exposure to wild-type adenovirus or herpes virus may result in the production of replication competent virus, through recombination with endogenous virus. As the current retroviral vectors are based on murine viruses the likelihood of producing replication competent virus, through recombination with endogenous human retroviruses, is reduced. An alternative viewpoint is that the introduction of a vector derived from a non-human pathogen – i.e. a mouse virus – may be more dangerous than from a virus already within the human arena. Thus, if a recombination were to occur *in vivo* with human endogenous viral elements, the resultant recombinant would have properties that the human immune system had never previously experienced – thereby potentially making the agent highly pathogenic with no known treatments. It is difficult to decide which of these two arguments is of greater concern and continued research into the possibilities of recombination between vectors and endogenous viruses remains a high priority – as evidenced by the xenotransplantation field at present. Despite this fact, the possibility that RCR formation may still occur is a drawback to the use of recombinant retroviruses. Recombination events between vector sequences, packaging constructs or endogenous viral sequences may lead to RCR formation (Chong *et al.*, 1998). As RCRs are able to undergo many cycles of infection and integration, there is a risk of integration occurring within a tumor suppressor gene. This risk becomes higher than when using a replication-defective virus.

Another limitation to the use of retroviral vectors for gene transfer *in vivo* is the inactivation of MLV based vectors by human serum (Welsh *et al.*, 1975). Inactivation occurs through a non-lytic process, which relies upon complement activation (Rother *et al.*, 1995) and the presence of natural antibodies (Takeuchi *et al.*, 1996). MLV virus produced in mouse cells is readily inactivated by human serum whereas MLV produced in human or mink cells is not (Takeuchi *et al.*, 1994). The process of galactosyl (α1–3) galactosylation of envelope proteins, which occurs in many non-human packaging cells, enables the virus to be recognized by anti-gal (α1–3) gal antibodies present in human serum. Retroviral vectors generated in packaging cell lines, which do not express an (α1–3) galactosyltransferase, are resistant to human serum (Cosset *et al.*, 1995) and the construction of vector systems based on these retroviruses would be advantageous for *in vivo* applications.

A further limitation to the use of retroviruses *in vivo* is the current inability to produce viral titres greater than 10^7 cfu ml^{-1}. The instability of MLV envelope proteins has prevented ultracentrifugation and therefore recombinant retroviruses cannot be concentrated to higher titres. Recently, more gentle concentration techniques have been described using a calcium phosphate precipitation technique which can achieve titres in excess of 10^8 cfu ml^{-1}. However, whether such physical techniques will ever allow recombinant retroviral titres to approach those of adenoviral vectors is difficult to assess. For an *in vivo* gene transfer application, where a high efficiency of gene transfer is demanded, limited viral titres will be a disadvantage. Strategies targeting more efficient gene transfer have included the development of higher titre recombinant viruses. By pseudotyping MLV-based vectors, with the more stable vesicular stomatitis virus envelope protein (VSV-G) (Burns *et al.*, 1993), it is possible to ultracentrifuge the virus and generate titres greater than 10^9 cfu ml^{-1}. Alternative strategies for increasing gene transfer efficiency include the modification of envelope proteins to express N-terminal binding domains or functional antibody fragments in order to target retroviral vectors to specific cell types (For review see Cosset and Russell, 1996).

MLV vectors have an inability to stably integrate proviral DNA into non-dividing cells (Roe *et al.*, 1993). For the therapeutic treatment of cancer this can be deemed advantageous, as dividing cancer cells can be transduced with a toxic or drug-activating gene, which will not be expressed in normal non-dividing cells. However, recent clinical experience has shown that very few human tumor cells are cycling at any given time *in vivo*. Thus, the discrepancy between the rates of division of rodent tumors in animal models and human tumors in patients is probably one of the principal reasons why the former can be readily cured by retroviral vectors whilst the latter have not so far been treatable in a similar manner. For the purpose of gene therapy in non-dividing cell populations however, such as post-mitotic ventral mesencephalic neurons in Parkinson's disease, MLV-based vectors would appear to be inappropriate.

8.2.3 Lentiviral vectors

Despite the fact that type C retroviruses have already been used in clinical settings, their inability to transduce non-proliferating cells has limited the number of potential applications. In a strategy that requires gene delivery into non-dividing, or very sparse, cell populations, such as neurons, monocytes or haematopoietic stem cells, retroviral vectors have been unsuccessful. The recent development of lentiviral vector systems (for reviews see Federico, 1999; Klimatcheva *et al.*, 1999; Naldini, 1998) may change this train of thought as lentiviruses are retroviruses capable of integrating their proviral genome into the genome of non-dividing host cells (Lewis and Emerman 1994; Lewis *et al.*, 1992; Weinberg *et al.*, 1991).

Initially a three plasmid expression system was developed to generate lentiviral vectors, in 293T cells, which are pseudotyped with the VSV-G envelope glycoprotein (Naldini *et al.*, 1996a). The first packaging plasmid contains all the viral genes needed in *trans* and is deleted of the LTRs, viral envelope and *Vpu* accessory protein. The second vector plasmid contains the viral LTRs, packaging signal and rev responsive element. The final plasmid contains the VSV-G envelope protein which is produced in *trans*. Further deletions, within the packaging plasmid, have facilitated the generation of infectious recombinant lentivrus produced, without *Vif*, *Vpr* and *Nef* coding regions, thought to be essential for HIV virulence (Zuffrey *et al.*, 1997). Vectors have been further developed in that they may now be generated independently of the *Tat* transactivator (Dull *et al.*, 1998; Kim *et al.*, 1998). Also, they may include a self-inactivating system by which transcriptional inactivation of the LTRs is achieved through deletion of the U_3 LTR region (Miyoshi *et al.*, 1998; Zuffrey *et al.*, 1998).

Safety issues concerning HIV-1 derived lentiviral vector systems have been raised so recently systems have been developed using the less pathogenic HIV-2 strain (Arya *et al.*, 1998; Poeschla *et al.*, 1998a) or elements of the primate simian immunodeficiency virus (SIV) lentivirus (White *et al.*, 1999). These vectors may in theory be 'safer' but they are still based on primate lentiviruses so lentiviral vectors based on non-primate viruses would reduce these concerns. Vector systems based on feline immunodeficiency virus (FIV), equine infectious anaemia virus (EIAV) and caprine arthritic encephalitis virus (CAEV) have been developed. Both FIV- (Johnston *et al.*, 1999; Poeschla *et al.*, 1998b) and EIAV- (Mitrophanous *et al.*, 1999; Olsen, 1998) based vectors have shown reasonable transduction efficiency in human cells, but CAEV vectors have shown very poor levels of transduction (Mselli-Lakhal *et al.*, 1998). It is felt that the lack of knowledge of molecular biology within these non-primate species may impede further vector development in the short term.

By pseudotyping lentiviral vectors with the envelope glycoprotein of a broad host range virus such as VSV-G, the vector tropism can be increased. To date sustained gene expression has been shown in neurons (Naldini *et al.*, 1996a, b), retina (Miyoshi *et al.*, 1997), liver (Kafri *et al.*, 1997), muscle (Kafri *et al.*, 1997), airway epithelium (Goldmann *et al.*, 1997), pancreatic beta cells (Ju *et al.*, 1998), haematopoietic stem cells (Uchida *et al.*, 1998) and cardiac myocytes (Rebolledo *et al.*, 1998).

8.3 HSV-1 vectors

8.3.1 HSV-1 biology

Herpes simplex viruses (HSV) and especially HSV-1 (HSV type 1), a human neurotropic virus, have a number of features that suggest that they can be adapted for use as vectors for gene transfer to the nervous system. These are: (i) the ability to persist in a latent state in neurons without apparent interference with nerve cells' function, or the induction of autoimmunity; (ii) the ability to express transgenes from the viral genome both transiently and long-term, and (iii) the possibility of deleting large amounts of HSV-1 sequences from the genome to provide space for the transgenes.

To obtain HSV-1 derived vectors useful for gene and cancer therapy, some problems remain to be overcome. These are to (i) eliminate or reduce the cytotoxicity and the immune responses induced by the virus vector, (ii) correctly deliver and express transgenes encoded by the vectors, and (iii) express the eventual transgene in a stable, efficient and controllable manner.

The HSV-1 genome is a double strand DNA molecule of 152 kbp in size and contains unique sequences (UL and US), flanked by repeat regions (Sheldrick and Berthelot, 1975). The genome contains also three origins of DNA replication (ori) (Stow and McMonagle, 1983; Weller *et al.*, 1985), cleavage-packaging signals (the *a* sequence), and encodes for approximately 80 different proteins. The function of most of them has been determined (for a review, see Roizman and Sears, 1993).

During the lytic cycle, the viral proteins are sequentially expressed in a cascade fashion, from the three main classes of genes, the immediate early genes (α) genes, the early (β) genes and the late (γ) genes, classified accordingly to their kinetics of expression during the replication cycle of HSV-1.

The first proteins to be expressed are the five α proteins, ICP0, 4, 22, 27 and 47, which are involved in regulatory functions or prevention of host response to infection, and are necessary for the expression of the β and γ proteins. The β proteins are mainly DNA synthesis factors, such as the viral DNA polymerase and proteins involved in nucleic acid metabolism. The γ proteins are expressed following the replication of viral DNA and encode the structural components of the virion. Of these proteins, more than half seem to be dispensable for viral replication in tissue culture (for a list, see Roizman, 1996). This could theoretically allow replacement of more than 30 kbp of viral DNA with transgenic sequences, although the insertion of such large transgenic sequences remains to be demonstrated.

HSV-1 infects humans usually by entry through mucosal membrane contacts between an infected and an uninfected person. The infected cells then undergo a lytic infection, which leads to the production of new progeny virions and the destruction of productively infected cells. The nucleocapsids can be transported retrogradely, from the axonal terminals innervating the infection site, to the nucleus of sensory neurones in the dorsal root ganglia. A lytic cycle can be initiated in these neurones, but in some of them, the virus can establish a latent infection, in which the viral genome remains episomally (Rock and Frazer, 1983). During latency, no viral protein expression is detected, the viral activity being restricted to the transcription of a set of RNAs, called LATs, and mapped in the inverted repeats flanking UL (Deatly *et al.*, 1987; Stevens *et al.*, 1987).

Various stimuli, such as stress or exposure of the skin to UV, prime the reactivation of the latent virus, which then enters a new lytic cycle, leading to the production of further infectious particles. These virions can be transported anterogradely from the cell bodies of dorsal root neurons to the axonal ends, where they produce a benign lytic infection in immunocompetent individuals.

The LATs seem to have a role in HSV-1 reactivation from latency, but not on the establishment or maintenance of the latent state (Javier *et al.*, 1988; Steiner *et al.*, 1989). Of central interest are the LAT promoters (Batchelor and O'Hare, 1990; Dobson *et al.*, 1989; Zwaagstra *et al.*, 1990), which allows the possibility of expressing a transgene during the viral latency.

8.3.2 HSV-1 derived vectors

As for other viral vectors, HSV-1 vectors can be used (i) to replace or stimulate a defective function, (ii) to deliver a drug sensitive gene, (iii) to increase the immune response against tumours, and finally (iv) to kill specific cells, such as tumour cells.

To be useful as a vector, the virus used must be non-pathogenic for the cells or the organism. Therefore, viral pathogenicity and the cellular immune response induced by the virus must be reduced or eliminated. Further, expression of the therapeutic transgene must be maintained at a sufficient level to be efficient, in the correct cells, and ideally, in a controllable manner and for the required therapeutic period. It is with these objectives in mind that the HSV-1 genome can be genetically modified.

Because of the large size of the HSV-1 genome, there are three major procedures used to construct the HSV-1 derived vectors. The first one involves homologous recombination between the viral genome and a plasmid bearing the sequence to be inserted flanked by targeted genome sequences. The DNA can be inserted into the cells by transfection or by infection, and if the inserted sequence contains also a selective marker, it is possible to specifically select the recombinant virus (Post and Roizman, 1981).

Another method to produce HSV mutants uses a set of overlapping cosmids which between them contain the entire complement of DNA sequences encompassing all of the HSV genome (Cunningham and Davison, 1993). By genetic modification on the appropriate cosmid, it becomes thus possible by co-transfection with the other cosmids to produce the required mutant. This procedure does not need the selection step for the isolation of the desired recombinant.

However, genome recombination during the generation of recombinants can make the task of producing HSV-1 vectors difficult. A new procedure has been proposed to overcome these problems. The whole HSV-1 genome, except the cleavage-packaging signal, has recently been inserted in a bacterial artificial chromosome (BAC) (Stravropoulos and Strathdee, 1998). This system promises to allow easier manipulation of the virus.

8.3.2.1 Replication defective vectors. To deliver transgenes of therapeutic interest using HSV-1-based vectors, replication defective HSV-1 have been produced to reduce or eliminate their inherent pathogenicity. This first category of vector includes HSV-1 virus in which the genome is partially modified to render them incompetent for replication and HSV-1 defective virus, deleted of almost all genomic sequences called amplicons.

The construction of HSV-1 vectors deficient for replication involves the inactivation of one or more genes essential for virus replication. Such viruses can then be propagated by growth in cell lines that provide the missing gene function in *trans.*

Historically, the first replication incompetent vectors were deleted of the essential immediate early gene encoding ICP4, the principal HSV-1 transactivator (De Luca *et al.*, 1985). However, despite reduced pathogenicity, they remained toxic to neurones in culture (Johnson *et al.*, 1992) and *in vivo*. It has been suggested that this toxicity could be the result of the expression of the few HSV-1 genes which can be expressed in the absence of ICP4, the four other α proteins, ICP 0, 27, 22 and 47, the β protein ICP6, which is a subunit of the ribonucleotide reductase, and also some HSV-1 constituents of the virion's tegument, such as the VP16 transactivator. Among them, ICP0, 27 and 22 have been shown to be indeed toxic in stable transfection assays. More recently, mutants simultaneously deleted in ICP4, ICP27 and ICP22 have been constructed and exhibit a prolonged gene expression and cell survival after infection (Wu *et al.*, 1996). In the near future, more deleted HSV-1 mutants will be produced. Care must be taken however, because the more the virus is deleted, the greater the probability that they will be inefficient as transfer gene vectors.

HSV-1 amplicons are derived from HSV-1 defective interfering particles, which are spontaneously found during serial passage of wild-type virus (Spaete and Frenkel, 1982). The minimal requirement for an amplicon to grow consists of an origin of DNA replication (ori) and the *a* sequence for cleavage-packaging (Spaete and Frenkel, 1985). Several copies of this defective genome are arranged head-to-tail in the virions, forming a concatemer up to approximately 150 kbp. Because the amplicon genome does not code for any proteins necessary for HSV-1 replication, it usually needs the presence of a helper virus to grow.

HSV-1 amplicons are widely used in gene transfer. An easy way to produce it is to construct a plasmid with the ori and *a* sequences, and also the transgene with the correct promoter, then to transfect it into cells and then super-infect cells with a helper HSV-1 virus, to provide in *trans* the factors necessary for amplicon replication and packaging into virions (Geller and Breakefield, 1988; Spaete and Frenkel, 1982). The main limitations in these kinds of vectors come from the necessity to provide a helper virus that can be pathogenic, the low vector:helper ratio usually obtained, and from the instability of the amplicon genomes, which tend to rearrange. Most of the studies with amplicons vectors have used a helper virus defective in the ICP4 function. However, recent results have shown that ICP27 defective vectors result in higher vector titres, a more favourable vector:helper ratio, and a lower rate of reversion to the wild-type helper (Lim *et al.*, 1996). Recently, a new system of amplicon has been reported, in which the helper virus does not have the capacity to be packaged, thereby avoiding the contamination of amplicon preparations with helper virus (Fraefel *et al.*, 1996). The titre of the amplicon in this system remains low (*Figure 8.3*).

A large number of studies have already reported the use of HSV-1 replication incompetent vector system, both *in vitro* and *in vivo*. Here we only outline some examples of applications. Thus, in the context of single gene therapy, the HSV-1 vectors appear promising for the treatment of neurological disorders, such as Parkinson's or Alzheimer's diseases. In order to find alternative therapy approaches to Parkinson's disease, it has been shown that a defective HSV-1 vector which, expresses the human tyrosine hydroxylase (TH) gene increases the release of catecholamines from cultured striatal neurons (Geller *et*

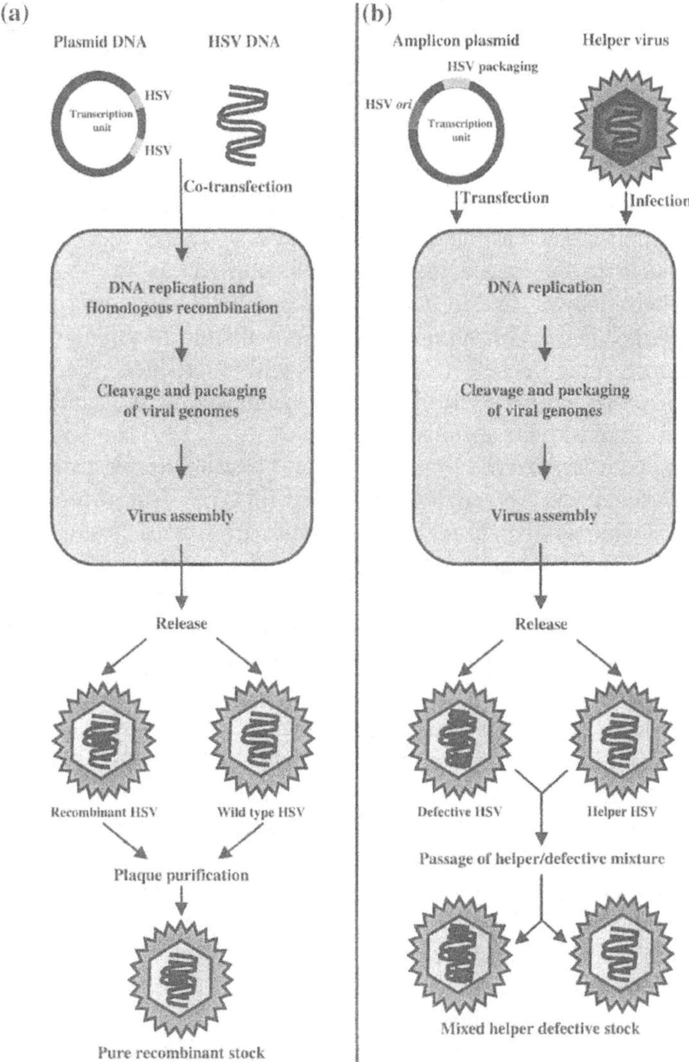

Figure 8.3. Schematic representation of recombinant and amplicon HSV vector production. Recombinant HSV vectors (a). HSV genomic DNA is co-transfected into permissive cells along with a plasmid containing the transcription unit inserted in a fragment of HSV genome. During mitosis homologous recombination occurs between the HSV sequences within the plasmid and the viral DNA. Replicated and recombined viral genomes are then processed and packaged into new virus particles. Recombinant virus can then be plaque purified from the parental virus. Amplicon derived HSV vectors (b). The amplicon plasmid containing the transcription unit (blue) is transfected into permissive cells, which are subsequently infected with a helper virus that provides all the regulatory and structural genes needed for viral growth. Following helper infection DNA replication is initiated. Helper virus DNA and linear head-to-tail concatemers of the amplicon plasmid are then packaged into separate viral particles, which are both released from the cells. The amplicon-derived virus cannot be physically separated from the helper virus and cannot replicate in the absence of helper virus. This figure has been modified from Leib and Olivo (1993), Wilkinson *et al.* (1994), and Stone *et al.* (2000).

al., 1995). Moreover, introduction of this vector into the striatum of a rat model of Parkinson's disease results in long-term behavioral recovery for up to 1 year, probably an effect of the persistent expression of the TH gene and of the increase in striatal dopamine levels which are detected in these animals (During *et al.*, 1994).

Because delivery of Nerve Growth Factor (NGF) to patients with Alzheimer's disease may be beneficial, HSV-1 vectors, expressing the NGF gene have been constructed. They have been shown to increase choline acetyltransferase activity in cultured neurons (Geschwind *et al.*, 1994) and reverse specific neurodegenerative effects of axotomy (Federoff *et al.*, 1992). Moreover, the HSV-1 vectors expressing NGF can reduce the local cytopathogenic effect of these viral vectors in the rat striatum (Pakzaban *et al.*, 1994), which add another advantage to this type of approach.

Finally, analysis of a defective HSV-1 vector expressing the gene for the glucose transporter has shown that the vector (i) can enhance the glucose uptake in hippocampal culture and in the rat hippocampus (Ho *et al.*, 1993), (ii) can protect hippocampal neurons in culture against hypoglycaemia and reduce seizure-induced damage to the rat hippocampus *in vivo* (Lawrence *et al.*, 1995) and (iii) can protect cultured neurons from necrotic insults (Ho *et al.*, 1995).

8.3.2.2 Replication conditional or oncolytic vectors. A unique property of some HSV-1 mutants is that they are able to replicate only in dividing cells. This makes the use of such herpetic vectors very interesting for the treatment of glioblastomas and possibly other tumours. In the nervous system, these mutants, can replicate only within glioma cells, and because the virus is lytic, they specifically kill these tumor cells.

At least two kinds of mutations generate HSV-1 recombinants either unable to replicate in non-dividing cells, or attenuated for neurovirulence. First, a deletion of a herpetic function involved in nucleic acid metabolism generates HSV-1 virus such as neurones in which the level of nucleic acid is low. The second kind of mutation comes from other studies, which have shown that the deletion of both copies of the ICP34.5 gene, generates mutants unable to replicate in the brain, or to cause encephalitis (Bolovan *et al.*, 1994; Chou *et al.*, 1990; McLean *et al.*, 1991).

The first generation HSV-1 vectors used introduced mutations in a single gene, thus restricting viral replication to dividing cells, and becoming avirulent after infection of the brain. Results show that deletion of the TK gene (Jia *et al.*, 1994, Martuza *et al.*, 1991, Boviastis *et al.*, 1994), the ICP6 gene (Mineta *et al.*, 1994, Marbert *et al.*, 1993, Boviastis *et al.*, 1994), or the ICP34.5 gene (Chambers *et al.*, 1995) generates recombinants with a selective ability to destroy gliomas in immunocompetent rats. Such vectors are able to decrease tumour growth and to prolong survival of nude mice implanted with human glioma cells, by a mechanism of direct oncolysis. Such viruses are also avirulent upon intracerebral inoculation in normal rodent, thus they are interesting tools for the treatment of glioblastomas. Moreover, the expression of the herpetic TK gene in a cell renders it susceptible to the gancyclovir (GCV) drug, which induces the death of the cell. Thus, if the HSV-1 oncolytic vector retains an intact TK gene, it has been shown that the addition of GCV enhances the oncolytic effect of the vector (Miyatake *et al.*, 1997a).

In order to increase the safety of these vectors, the second generation of HSV-1 vectors combine two deletions, e.g. the ICP34.5 and ICP6 gene (Mineta *et al.*, 1995, Kramm *et al.*, 1997). These vectors have shown promising results because they show powerful oncolytic ability towards *in vivo* glioblastoma models. This is why three different phase I clinical trials have already begun in the United States and United Kingdom. The preliminary reports show no evidence of encephalitis or adverse complications from the viral infection in the patients receiving increasing doses of the viral vectors.

However, the mutations inserted in such vectors do not affect any gene essential for viral replication. Thus, these vectors remain theoretically able to replicate. This could lead to important cytopathogenic effects, as was reported in immunosuppressed mice (Lasner *et al.*, 1998). Of particular interest are some recent works that demonstrated that if the oncolytic vectors also express a cytokine that can increase the immune response, such as interleukins 4 (Andreansky *et al.*, 1998) or 12 (Toda *et al.*, 1998), the antitumour activity mediated by these vectors is synergized.

8.3.2.3 Chimaeric vectors. In order to combine the advantages of different viruses, some hybrid vectors have been constructed, using an HSV-1 amplicon backbone. Thus, Johnson *et al.* reported a HSV/AAV (Adeno-Associated Virus) vector system which use the HSV-1 amplicon properties of entry and nuclear localisation and the AAV properties of transgene amplification and integration into a nonessential locus in nondividing humans cells (Johnson *et al.*, 1997). The hybrid vector, in comparison with traditional HSV-1 amplicon vectors, has been shown to extend transgene expression in dividing human glioma cells, humans and rodent hepatic cell lines and non-dividing cells in the rodent liver (Fraefel *et al.*, 1997).

Another hybrid vector combines the HSV-1 amplicon vector with the Epstein-Barr virus (EBV) element oriP and EBV Nuclear Antigen 1 (EBNA1) system, which allow episomal retention and replication in dividing cells (Wang and Vos, 1996). Finally, an HSV-1 amplicon vector harbouring the *gag*, *pol* and *env* genes of MLV has been shown to rescue defective retrovirus vectors (Savard *et al.*, 1997).

8.3.2.4 Control of transgene expression in HSV-1 vectors. The expression of the vector (for the oncolytic vectors) or of the transgene is critical to the success of the viral therapy. This means that the expression of the transgene must be as stable and as efficient as possible and that the expression should be controllable or targeted to specific tissues or cells. It is worth noting that depending on its position within the vector genome, the type of infected cells (Roemer *et al.*, 1991), or the viral state (i.e., lytic or latent) (Lachmann *et al.*, 1996), the activity of the promoter can differ.

What made HSV-1 an attractive candidate as a vector for expression of foreign genes in the nervous system is the capacity to establish long-term latent infection in neurons. The latent infections are characterized by the absence of viral gene expression (which is potentially cytotoxic), except for the synthesis of the latency-associated transcripts (LATs). It has been shown that the LATs promoter, in association with elements like the MLVLTR (Moloney murine leukemia virus long terminal repeat) (Carpenter and Stevens, 1996; Lonkensgard *et al.*, 1994) or IRES (internal ribosomal entry site) (Lachmann and Efstathiou, 1997) can express a transgene in latently infected cells for long periods ranging from 6 to 18 months.

Many efforts have been made to target the transgenic expression to specific tissues or cell types. For example, transgene expression has been targeted to a specific region of the brain, specifically to cells expressing rat preproenkephalin, by using the preproenkephalin promoter (Kaplitt *et al.*, 1994a), to catecholaminergic neurons by using the rat tyrosine hydroxylase promoter (Song *et al.*, 1997), or to glial cells by using the glial fibrillary acidic protein promoter (McKie *et al.*, 1998). Outside the brain expression has been targeted to albumin expressing cells (in the liver and in hepatocytes cellular carcinoma) by using the albumin enhancer/promoter (Miyatake *et al*, 1997).

Several systems have been already constructed in which the transgene could be regulated by administration of the GAL:VP16 transactivation system (Oligino *et al.*, 1996), or of tetracycline (Fotaki *et al.*, 1997, Ho *et al.*, 1997). These systems which allow a temporal control, showed activation or repression of the transgene. It seems probable that a combination of these systems will need to be used, in order to have for example both a spatial and temporal control in transgene expression.

The different approaches with HSV-1 vectors show that they may be good tools for cancer therapy. In the future, the vectors will combine several advantages, as for example an immune-stimulatory transgene placed within an oncolytic vector, in order to increase their efficiency (Chase *et al.*, 1998). The earliest clinical target to benefit from the use of HSV-1 vectors will be the treatment of brain tumours as illustrated by ongoing phase I clinical trials. However, many problems remain to be overcome. Also, successes reported herein must not detract from vector induced cytotoxicity or the immunological side effects of viral vector administration which remain to be confronted. Also, little is known of the risk of recombination between viral vector and endogenous latent virus, and associated risks leading to the dissemination of transgenic sequences within human populations. These problems will need to be addressed in the future.

8.4 Adenoviruses as vectors for gene therapy

8.4.1 Adenoviral biology and vectors

8.4.1.1 First generation vectors. Recombinant adenoviruses (RAds) possess several characteristics which make them attractive candidate vehicles for gene therapy: RAds are easily propagated, the non-enveloped virus particles are stable and therefore readily concentrated to high titres and they are potentially safe, as wild-type adenoviruses are only mildly pathogenic in humans and are generally non-integrating. A major advantage over other viral vectors however, is the ability of RAds to mediate efficacious gene transfer to a wide variety of cell types, including post-mitotic cells.

The majority of adenoviral vectors developed to date are derived from human serotypes 2 and 5, since the biology and genetics of these viruses have been intensively investigated. Briefly, the approximately 36 kb linear genome encodes five early transcription units (E1A, E1B, E2, E3 and E4), the products of which regulate viral DNA replication, expression of late genes and viral assembly, and two intermediate transcription units plus a late transcription unit, which mainly encode structural proteins (for a review, see Horwitz, 1994). Adenovirus vectors are rendered replication defective by the deletion of the E1 region (which activates the expression of all the other viral early genes) and are propagated in the human embryonic kidney-derived 293 cell line which provides the E1A and E1B functions in *trans* (Graham *et al.*, 1977). Recombinant adenoviruses encoding a transgene of interest are generated by homologous recombination within 293 cells between the 'backbone' of the viral genome and a plasmid containing the expression cassette flanked by regions of viral DNA (Hitt *et al.*, 1995). The capacity generated by the E1 deletion for the insertion of foreign DNA is limited (approximately 3.2 kb) and as such, the majority of first-generation RAds are also deleted for E3, which is not required for virus replication in cell culture. Double deletions of E1 and E3 allow the insertion of up to 8 kb of foreign DNA.

First generation E1/E3 deleted adenoviruses have been used to successfully deliver genes to a wide variety of tissues and organs in animal models of human disease and also in human phase I clinical trials for cystic fibrosis (Crystal *et al.*, 1994; Knowles *et al.*, 1995) and cancer (Gahery-Segard *et al.*, 1997). These studies have shown that although gene transfer is efficient, the utility of the $E1^-/E3^-$ vectors is limited by their immunogenicity; administration of vector is generally associated with the development of local inflammation at the site of transgene expression, expression of the transgene is consistently transient (typically 3–4 weeks), and furthermore, gene transfer becomes less efficient and the inflammation more severe upon repeated administration of vector. The vector-induced immune response is biphasic and consists of an initial innate response, which rapidly eliminates many transduced cells (Worgall *et al.*, 1997), followed by adaptive cellular and humoral mechanisms (Yang *et al.*, 1994a, 1995a). The precise delineation of the components of the anti-vector immune response has been confounded by differences between host backgrounds, routes of vector administration and particularly by strong immune responses against certain transgenes (Barr *et al.*, 1995; Kaplan *et al.*, 1997; Michou *et al.*, 1997; Morral *et al.*, 1997; Tripathy *et al.*, 1996) however, it has been demonstrated that adenovirus vector antigens are principal factors in triggering destructive immune mechanisms. Although E1-deleted vectors are theoretically incapable of replication or viral gene expression, when cells are infected with high multiplicities of virus, cellular factors can compensate for E1 and activate the expression of early and late genes (Spergel *et al.*, 1992). Viral antigens are subsequently presented on the cell surface by the class I MHC, which activates cytotoxic T lymphocytes to destroy the transduced cells (Yang *et al.*, 1994a). This CTL response is thought to be one of the principal causes of the decline in transgene expression observed after RAd-mediated gene transfer *in vivo*. In addition, the virus capsid proteins stimulate the formation of neutralizing antibodies which preclude repeat administration of the vector (Gahery-Segard *et al.*, 1998; Yang *et al.*, 1995b) (*Figure 8.4*).

8.4.1.2 Second generation adenoviral vectors. The rationale for the design of less-immunogenic, second-generation RAds has been developed either to downregulate the anti-virus immune response by reintroducing viral genes with immunomodulatory functions into the vector backbone, or to reduce 'leaky' viral gene expression by creating further deletions/mutations within additional regions of the genome.

Wild-type adenoviruses have evolved the innate ability to evade host immune surveillance by downregulating anti-virus immune responses (Hayder and Mullbacher, 1996). Region E3 encodes several proteins involved in this downregulation, including gp19K which prevents the translocation of MHC class I molecules to the cell surface,

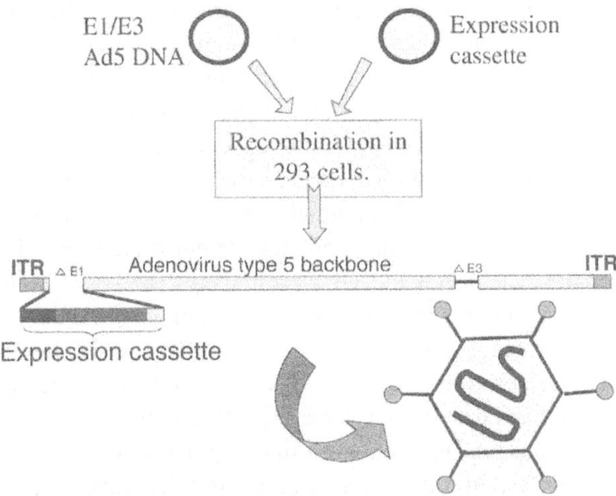

Figure 8.4. Schematic representation of the engineering of first generation adenoviral vectors. The shuttle vector, containing the expression cassette flanked by adenoviral sequences, is co-transfected with the adenoviral genome plasmid into 293 cells. The adenoviral genome plasmid pJM17, containing the full viral genome with a deletion in E3 and the bacterial plasmid pBRX inserted in E1, and the shuttle vector undergo homologous recombination. The resultant adenoviral genome is deficient in E1 and E3 and has the transcription unit inserted in E1. This genome can then be packaged into virions, which are produced by the adenoviral genome in 293 cells, which express the E1A adenoviral transactivator in *trans*.

and 10.4/14.5K and 14.7K which prevent TNF-α-mediated cytolysis of virally infected cells (Wold and Gooding, 1991). E1A gene products modulate the expression of E3 genes from their natural promoter, hence E3 gene expression is defective even in first-generation E1-deleted vectors which retain an intact E3 region. Reintroduction into the vector backbone of the entire E3 region under control of the strong cytomegalovirus (CMV) promoter has demonstrated the importance of E3 in downregulating both humoral and cellular immune responses and promoting sustained transgene expression *in vivo* (Ilan *et al.*, 1997). Constitutive over-expression of gp19K alone has however failed to prolong *in vivo* transgene expression (Schowalter *et al.*, 1997) and an analysis of the role of E3 products in a murine pneumonia model has indicated that expression of the E3 proteins 10.4/14.5K or 14.7K may be more important than expression of gp19K in avoiding immune recognition during acute adenoviral infection (Sparer et al., 1996). In support of this hypothesis, a recent investigation has demonstrated that TNF-α plays a major role in the elimination of adenoviral vectors (Elkon *et al.*, 1997).

Other second-generation RAds have incorporated deletions or mutations within the E2 or E4 regions of the vector genome in addition to the E1/E3 deletions, in order to further cripple the ability of the virus to replicate and express viral proteins. Region E2 can be subdivided into regions E2a and E2b; E2a encodes a 72 kDa DNA binding protein (DBP) which is essential for the initiation of viral DNA synthesis and for the activation of the major late promoter (regulating late gene expression), and E2b encodes a 140 kDa DNA polymerase and an 80 kDa precursor terminal protein, both of which are required for viral DNA replication. E4 gene products are also essential for viral DNA replication and are additionally involved in modulating late gene expression, mRNA splicing and inhibition of host cell protein synthesis (Horwitz, 1994).

An early approach to vector improvement was the further disabling of E1-deleted viruses by creating a temperature-sensitive mutation in E2a (ts125), such that the DNA binding protein is functional at the permissive temperature (allowing the vector to be propagated in cell culture) but non-functional at body temperature. Initial experiments in mice, cotton rats and non-human primates demonstrated that E2a-ts125 virus-mediated gene transfer was correlated with prolonged gene expression and a reduced inflammatory response (Englehardt *et al.*, 1994a,b; Yang *et al*, 1994b; Goldman *et al.*, 1995). Contrary to these results however, a more recent study showed that use of ts125 mutant RAds to deliver genes to mice and dogs failed to prolong transgene expression beyond the duration achieved with first-generation RAds (Fang *et al.*, 1996), indicating that the temperature-sensitive DBP mutation may not assure complete blockage of late gene expression *in vivo*. Additionally, vectors rendered defective by mutation carry the risk of reversion back to the wild-type genotype. More extensive modifications, such as those introduced by deletions within E2 and E4 were therefore considered necessary to significantly improve vector performance and safety. The introduction of additional deletions within the vector backbone also decreases the probability of replication-competent adenovirus (RCA) emerging during propagation in 293 cells, since multiple deletions necessitate multiple recombinogenic events between vector DNA and integrated viral sequences to generate RCA. The recent observation that E4orf6 inhibits the tumour suppressor gene p53 (Moore *et al.*, 1996) provides further justification for the development of E4-deleted vectors for gene therapy.

In order to generate vectors containing lethal mutations within E2 or E4, new packaging cell lines are required which are able to provide the missing functions in *trans*. Generation of such cell lines is not straightforward, since many adenovirus proteins are toxic to the cell when constitutively overexpressed. To date, packaging cell lines have been created which can successfully complement for lethal deletions within E2a (Zhou *et al.*, 1996), E2b (Amalfitano and Chamberlain, 1997) and E4 (Brough *et al.*, 1996; Wang *et al.*, 1995). It has been reported however, that some of the more defective vectors grow to significantly reduced titres even in the appropriate complementing cell line (Brough *et al.*, 1996).

Many different studies have investigated the performance of the E2- or E4-deleted second-generation RAds compared with first-generation vectors. Data from *in vitro* experiments have been encouraging, clearly demonstrating reduced viral DNA replication and significant shut off of late gene expression in the context of E2- or E4-deleted viruses (Dedieu *et al.*, 1997; Lusky *et al.*, 1998). *In vivo* data from different studies have led to interesting and somewhat contradictory conclusions. A study investigating $E1^-/E4^-$ vector-mediated delivery to mouse lung demonstrated prolonged transgene expression over first generation RAds, correlated with a reduction in virus-specific immune responses (Wang *et al.*, 1997). By contrast, other studies have found that although $E1^-/E4^-$ vector DNA persists within transduced tissue longer than $E1^-$ vector DNA, the kinetics of decline in transgene expression are similar from both vectors (Armentano *et al.*, 1997; Dedieu *et al.*, 1997). This suggests that non-immunological mechanisms (e.g. promoter inactivation) may strongly influence the longevity of transgene expression. Moreover, it has been reported that different vector backbones differentially affect the activity of certain promoters (Armentano *et al.*, 1997). Future investigations will likely clarify the utility of the different vector models with different transgenes and promoters. These studies do however indicate that the initial paradigm of transient transgene expression caused by immune responses against vector antigens should be readjusted.

8.4.1.3 High capacity adenoviral vectors. Most recently a new generation of 'gutless' RAds have been developed which are deleted of all viral genes (Kochanek *et al.*, 1996; Parks *et al.*, 1996). Such vectors have a minimum requirement for the extreme termini of the linear adenovirus genome, containing simply the inverted terminal repeats and the packaging signal and, as such they have a potential cloning capacity close to the size of the native genome (36 kb). Gutless vectors are co-propagated in 293 cells with an E1-deleted helper virus, which *trans*-complements for all the missing viral functions. The helper virus is normally subjected to negative selection to allow the amplification of the gutless virus and in the most sophisticated system, the lox-P-flanked packaging signal within the helper virus is removed during reiterative amplification of the gutless virus in 293 cells which stably express the lox-P specific Cre recombinase from bacteriophage P1. Purification by caesium chloride centrifugation is necessary to reduce the titre of helper virus to negligible levels (less than 0.1% of the gutless titre) and this is facilitated in most systems by the differential buoyant densities of the helper and gutless viruses due to different genome sizes. The results of initial *in vivo* studies comparing the performance of gutless adenovirus vectors with first-generation vectors have demonstrated reduced immunogenicity, enhanced safety and prolonged transgene expression associated with gutless vector-mediated gene delivery (Chen *et al.*, 1997; Morsy *et al.*, 1998). Further confirmation of these advantages may signal renewed optimism in the viability of adenoviruses as safe and effective gene delivery vehicles (*Figure 8.5*).

8.4.1.4 Conditionally replicative or attenuated adenoviruses. As an alternative to using adenoviral vector systems for the delivery of therapeutic genes in the treatment of cancer it has been proposed that the lytic property of adenoviruses may be harnessed to selectively kill tumor cells. Systems in which an adenovirus may selectively replicate in a desired tumor cell population have been developed and these adenoviruses show attenuated growth properties in normal cell populations.

One such example of a selectively replicating adenovirus is the ONYX-015 (also called dl 1520) adenovirus, which is E1B 55-kDa deficient. Bischoff *et al.* (1996) proposed that the ONYX-015 virus is able to replicate in p53-deficient tumor cells but not in cells with functional p53. It was thought that loss of p53 was conducive to efficient adenovirus replication and that a functional E1B-55 kDa (known to inactivate p53 function) protein was needed for replication in p53 positive cells, thus making ONYX-015 an ideal treatment for p53-negative tumors. It has since been shown that negative p53 status is not essential for ONYX-015 growth and that it can also replicate in p53-positive cells (Harada and Berk, 1999; Heise *et al.*, 1997; Rothman *et al.*, 1998).

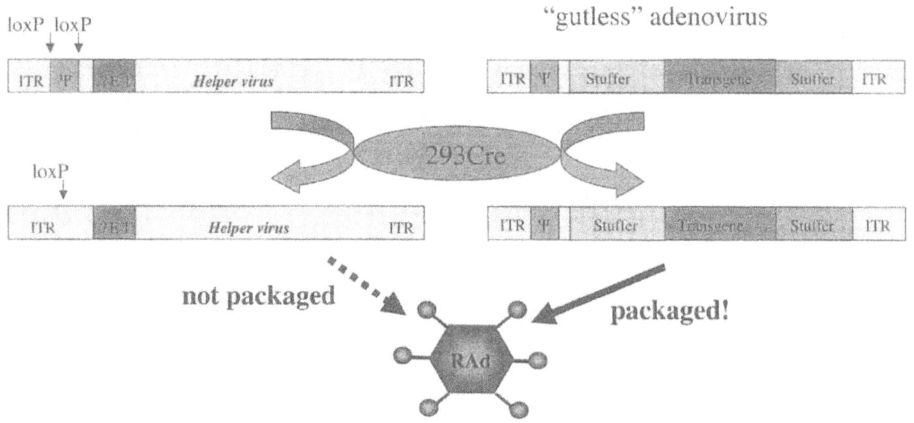

Figure 8.5. Schematic representation of the engineering of high capacity ('gutless') adenoviral vectors. The helper virus provides all necessary functions for vector genome replication and packaging in *trans*. Further, the helper cannot be packaged into virions, because its packaging site (Ψ) is removed through cre-lox recombination during the generation of the high capacity vectors. The high capacity vector genome only contains the left and right viral inverted terminal repeats (ITR), and the adenoviral packaging Ψ site.

Steegenga *et al.* (1999) have subsequently suggested that the efficiency of the ONYX-015 virus to replicate in tumor cells may be determined by the ability to infect cells and to express early adenovirus proteins rather than p53 status. Although ONYX-015 can replicate in p53 positive cells it would appear to have attenuated growth characteristics and its selectivity for tumor cells in clinical trials is encouraging (Kirn *et al.*, 1998, 2000).

A group at Calydon Inc, California are working on attenuated replication-competent adenoviruses for prostate cancer therapy in which the promoters of the E1A and/or E1B genes are replaced with tissue-specific promoters. The (PSA) prostate-specific antigen (a marker for prostate cancer) enhancer/promoter (Rodriguez *et al.*, 1997), prostate-specific rat probasin promoter (Yu *et al.*, 1999a) and (hK2) human glandular kallikrein enhancer/promoter (Yu *et al.*, 1999b) have all been used. These adenoviruses are able to replicate in PSA-positive tumor cells but show attenuated replication in PSA-negative tumor cells.

A further strategy is the use of conditionally replicative adenoviruses in conjunction with suicide gene therapy. The use of E1B 55-kDa deficient adenoviruses which expresses either HSV-1 thymidine kinase (HSV-1/TK) (Wildner *et al.*, 1999) or an HSV-1/TK cytosine deaminase fusion gene (Freytag *et al.*, 1998), has been documented. These adenoviruses show an enhanced tumor cell kill when compared to ONYX-015 (Freytag *et al.*, 1998) or RAd.TK (Wildner *et al.*, 1999) alone.

8.4.1.5 Modification of adenoviral tropism. The natural tropism of the type 2 and 5 human adenoviruses is dependent on the primary binding and secondary internalization steps with target cells. The adenovirus fibre/knob mediates binding to the cellular coxsackie and adenovirus receptor (CAR) (Bergelson *et al.*, 1997). Virus internalization is then mediated by interaction between cell surface integrins $\alpha_v\beta_3/\alpha_v\beta_5$ and the viral penton base (Wickham *et al.*, 1993). This necessity for both CAR and $\alpha_v\beta_3/\alpha_v\beta_5$ integrins on a cell pre-determines adenoviral tropism. By pseudotyping the knob domain (Krasnykh *et al.*, 1996), directly altering the knob (Dmitriev *et al.*, 1998; Krasnykh *et al.*, 1998) or attaching targeting motifs to the C-terminus of the knob (Michael *et al.*, 1995; Wickham *et al.*, 1996a, 1997a) it is possible to alter adenoviral tropism. Alternatively the penton base can be modified (Wickham *et al.*, 1995, 1996b, 1997b) or bispecific ligands between the knob and a target receptor can be used (Douglas *et al.*, 1996; Rogers *et al.*, 1997; Watkins *et al.*, 1997).

In contrast to altering the infectivity range of an adenoviral vector it is possible to achieve targeted gene expression using transcriptional targeting. Cell type specific promoters can be used to directly drive expression of a desired transgene within designated cell populations. These cell type specific promoters can also be used to generate conditionally replicative adenoviruses as discussed previously. To further define or control expression from an adenoviral vector combined cell type specific and inducible systems have also been developed in which an inducer molecule, such as tetracycline (Gossen and Bujard, 1992), is needed to switch on gene expression (Harding *et al.*, 1998; Smith-Arica *et al.*, 2000). Such a system could be used to tightly regulate the levels of gene expression within a target cell population for treatment of a specified disorder.

8.5. Adeno-associated virus (AAV) vectors

8.5.1 AAV biology

Various properties of AAV are attractive for use in gene therapy, i.e. it is non-pathogenic in humans, it is replication-deficient, and its genome is integrated in a site-specific manner, thereby facilitating stable transgene expression without the risk of mutation caused by random integration. In this section, the main features of AAV are illustrated, and various studies using this vector within the context of gene therapy are described. AAV, a small single-stranded linear DNA non-enveloped parvovirus, has been discovered in tissues infected by adenovirus or in adenovirus-purification preparations (Blacklow *et al.*, 1971). This virus is not associated with human disease. So far, five serotypes have been identified and the best characterized is AAV-2.

AAV is an icosahedral virus with a diameter of 20 nm. Its genome is very small (4.7 kb), simple and well characterized (Srivastava *et al.*, 1983). The DNA sequence is composed of two inverted terminal repeats (ITRs) and two open reading frames (orfs); *rep* and *cap*, encoding proteins for the replication and encapsidation of the virion, respectively. The ITR (145 bp) is essentially composed of a palindromic sequence (first 125 bp), forming a hairpin double-stranded structure priming the replication of cDNA. ITRs are also essential for viral encapsidation (Samulski *et al.*, 1989). The AAV genome includes 3 promoters, p5, p19 and p40, which generate 7 mRNAs by alternative splicing, thereby providing 4 regulatory proteins (Rep 78, Rep 68, Rep 52 and Rep 40) and 3 structure proteins (VP1, VP2, VP3). The functions of Rep 78 and Rep 68 have been well documented; these proteins bind DNA and are involved in integration/excision, replication and encapsidation of the virus genome (Chiorini *et al.*, 1994; Labow and Berns, 1988; Ni *et al.*, 1994). In addition, they regulate the level of gene expression (Beaton *et al.*, 1989; Kyostio *et al.*, 1994; Labow *et al.*, 1986; Trempe and Carter, 1988). An interesting feature of AAV is that encapsidation events are completely independent from replication; the majority of the viral DNA is not destined for encapsidation, as ITRs alone are sufficient to package viral particles (Samulski *et al.*, 1989).

Infection of host cells with wild-type AAV alone does not result in virus production (*Figure 8.6*). AAV attachment to and infection of host cell is mediated by membrane heparin sulphate proteoglycans (Summerford and Samulski, 1998). $\alpha_v\beta_5$ integrins have also been described as co-receptors for AAV infection (Summerford *et al.*, 1999). Thereafter AAV remains within the cell as a provirus in a latent phase and is therefore considered to be a defective virus (Atchison and Casto, 1965). Most of the time viral DNA integrates into the host genome, however it can also persist as an episomal form (McLaughlin *et al.*, 1988). In human cells, the AAV DNA sequence integrates into the host genome in a site-specific manner, since it targets a locus (*AAVS1*) in the 19q human chromosome (Kotin and Berns, 1989; Linden *et al.*, 1996). Interestingly, only the ITRs are essential for integration into the host genome (Samulski et al., 1989). Successful propagation of AAV within host cells requires co-infection with a helper virus, usually adenovirus (Ad) or herpes virus (HSV). These helper viruses provide factors in *trans*, which facilitate the excision and replication of AAV DNA. Ad early region gene products (E1a, E1b, E2a and E4) contribute to the transactivation of AAV promoters (Richardson and Westphal, 1981, 1984) and the stabilization of AAV mRNAs, and they also enhance the efficiency of translation of some AAV RNAs (Janik *et al.*, 1989; Jay *et al.*, 1979, 1981).

8.5.2 AAV vectors

AAV is an attractive vector for gene delivery due to the following properties: (i) lack of association with human disease (Blacklow *et al.*, 1968); (ii) infection of a wide range of cells including non-dividing cells (Podsakoff *et al.*, 1994, Russell *et al.*, 1994); (iii) latent phase (Cheung *et al.*, 1980) induced by a stable integration into the host genome in a specific site (Linden *et al.*, 1996); (iv) no viral DNA required for virion encapsidation, giving both advantages of more room for the insert and of minimal viral antigen exposure (Samulski *et al.*, 1989); (v) small genome easily cloned in a plasmid (*Figure 8.6*).

Recombinant AAVs (rAAVs) are generally derived from AAV-2, and more recently from AAV-5. The use of AAV as a vector to transduce mammalian cells was described for the first time in 1984 simultaneously by two groups (Hermonat and Muzyczka 1984; Tratschin *et al.*, 1984). They used a technique of co-infection with Ad and wild-type or partially deleted AAV, however this method generated viral preparations contaminated with wild-type AAV. Subsequently, Samulski *et al.* (1989) developed the technique now commonly used to generate rAAVs. 293 cells are cotransfected with a packaging-competent plasmid containing the expression cassette flanked by the AAV ITRs and a non-packagable plasmid containing the viral genome without the ITRs. Infection with a helper virus is an absolute requirement for activation of *rep* and consequently

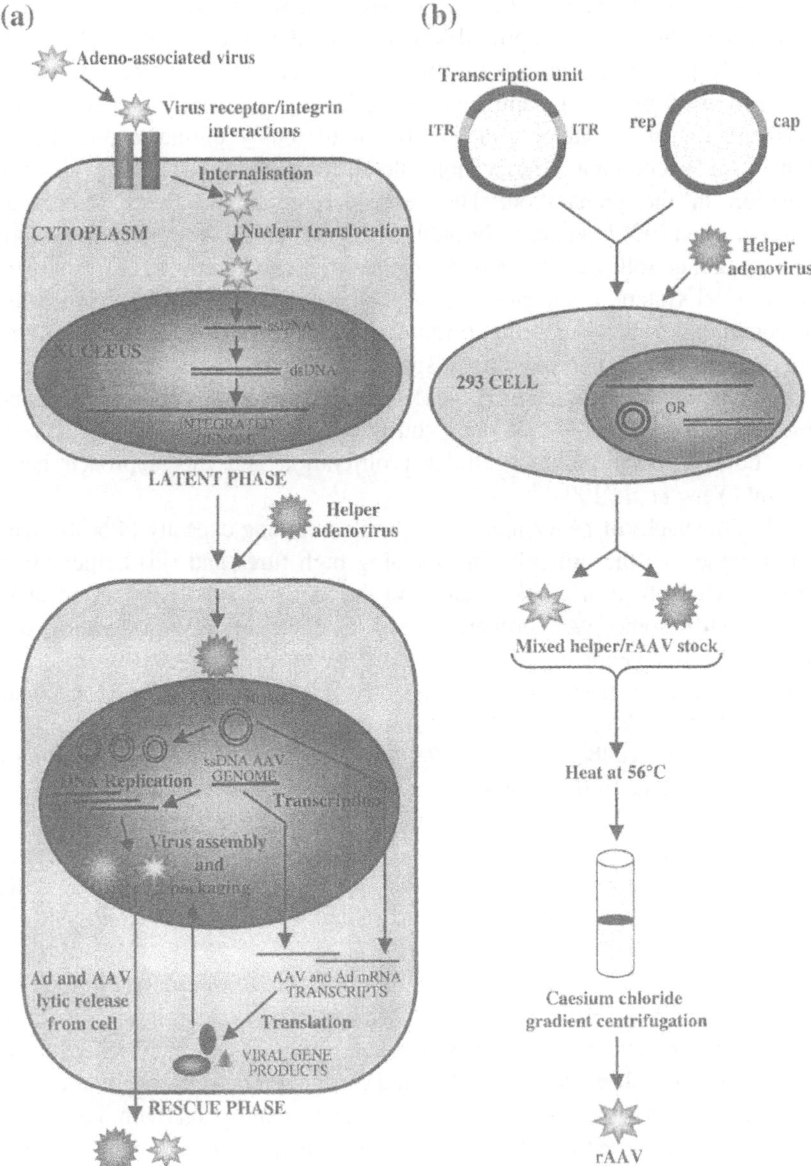

Figure 8.6. AAV life cycle (a). Latency is established following AAV infection of the target cell. AAV virions are rescued following subsequent infection with a helper virus, which provides the sequences needed for viral replication in *trans*. Schematic representation of rAAV production (b). Permissive cells are co-transfected with a plasmid containing the transcription unit flanked by the AAV ITRs and a plasmid containing the AAV genome without the ITRs. Subsequent infection of these cells with a helper adenovirus enables rescue of rAAV virions, following packaging of the ITR containing plasmid, along with helper virus. Helper and rAAV virions can then be separated after heating at 56°C by centrifuging on a caesium chloride gradient (modified from Stone *et al.*, 2000).

replication; the most frequent helper viruses are Ads, which bring E1a, E1b, E2a, E4 and VA in *trans*. This technique provides mixed stocks of rAAV and Ad. Stable rAAV stock is usually purified by inactivating the contaminating Ads by heating (56°C, 45 minutes), followed by CsCl centrifugation, allowing titers between 10^7 and 10^{10} infectious particles ml^{-1}. Helper Ad is difficult to totally eliminate and the use of recombinant Ads encoding a reporter molecule allows evaluation of the levels of helper contamination in the preparation. The rescue technique has been improved and Zolotukhin *et al.* (1999) have recently published novel purification strategies involving iodixanol gradients followed by ion exchange of heparin affinity chromatography. Recently a novel system, completely free of helper virus, has been developed using a plasmid containing a mini-Ad genome (Xiao *et al.*, 1998). This technique promises a better understanding of the immune response induced after *in vivo* gene transfer (Manning *et al.*, 1980, 1998). Since these techniques require cotransfection and co-infection generation of stable cell lines constitutively expressing *rep* and *cap* seems attractive, however Rep inhibits cellular proliferation and this approach has been unsuccessful (Yang *et al.*, 1994).

The main drawbacks of rAAV are: (i) the limited cloning capacity (4.5 kb) available for the transgene; (ii) the difficulty in obtaining high titres and (iii) helper virus-free purified stock (Salvetti *et al.*, 1998), and (iv) the fact that total elimination of helper virus may modify some AAV properties such as infection of non-dividing cells or integration in the host genome.

In vitro, rAAV have been used to efficiently transduce cell lines (Hermonat and Muzycka 1984; Podsakoff *et al.*, 1994; Tratschin *et al.*, 1985) and less efficiently, primary culture cells (Halbert *et al.*, 1997). Moreover, infection of non-dividing cells, such as haematopoietic progenitor cells, has been reported (Goodman *et al.*, 1994, Miller *et al.*, 1994; Zhou *et al.*, 1994). In addition, human cancer cell lines infected with rAAVs expressing foreign proteins; AAV-p53 decreased the growth of neoplastic cells and mediated cytotoxicity (Qazilbash *et al.*, 1997), and AAV-B7 inhibited human multiple myeloma cell proliferation, enhanced the antitumor T cell response and elicited a cytolytic T cell response (Wendtmer *et al.*, 1997).

AAV vectors have been injected *in vivo*. They are able to transduce different tissues in rodent including lung (Flotte *et al.*, 1993; Halbert *et al.*, 1997), brain (Kaplitt *et al.*, 1994b), muscle (Kessler *et al.*, 1996; Xiao *et al.*, 1996), liver (Koeberl *et al.*, 1997), retina (Ali *et al.*, 1996; Jomary *et al.*, 1997) and carotid arteries (Rolling *et al.*, 1997).

Interestingly, rAAVs can achieve long-term gene expression with persistence of transgene expression more than 18 months after percutaneous injection in muscle (Xiao *et al.*, 1996). Conversely to adenovirus, no inflammatory response has been described (Jooss *et al.*, 1998; Xiao *et al.*, 1996), however AAV-specific antibodies have been detected compromising re-administration of the vector. They have been described initially as non-immunogenic vectors (Halbert *et al.*, 1997; Manning et al., 1998) however, other studies have shown that immune responses to AAV-encoded transgenes can be elicited in mice *in vivo*.

Some experiments using rAAV have been performed in large mammals (an important step before human gene therapy trials can be considered). AAV infused in porcine coronary arteries yielded gene expression for more than 6 months without apparent inflammation (Kaplitt *et al.*, 1996). Similarly, vascular gene delivery has been performed in monkeys by intraluminal or intra-adventitial injection of rAAV, and yielded efficient transduction of carotid arteries (Lynch *et al.*, 1997). In addition, instillation of a rAAV with the CFTR gene in non-human primates has shown long-term CFTR expression in lung without inflammation (Conrad *et al.*, 1996). Recently, rAAV encoding the human factor IX (hF-IX) has been tested for safety in a canine model of human haemophilia B; intramuscular injection of this vector resulted in hF-IX expression for up to 10 weeks, and yielded reduction of whole blood clotting time with limited areas of focal lymphocyte infiltration (Monahan *et al.*, 1998). Finally, a clinical protocol with an AAV-CFTR vector has been developed, and the results of the phase I study have been published (Flotte *et al.*, 1996). Recombinant AAV, which is a newer vector for gene therapy, seems to be a promising tool for the future, as it is able to transduce a large variety of cells (including non-dividing cells) or organs, and it yields very low immune responses against the vector.

References

Ali RR, Reichel MB, Thrasher AJ *et al.* (1996) Gene transfer into the mouse retina mediated by an adeno-associated viral vector. *Hum. Mol. Genet.* **5**: 591–594.

Amalfitano A, Chamberlain JS. (1997) Isolation and characterisation of packaging cell lines that co-express the adenovirus E1, DNA polymerase and pre-terminal proteins: implications for gene therapy. *Gene Ther.* **4**: 258–263.

Andreansky S, He B, van Cott J *et al.* (1998) Treatment of intracranial gliomas in immunocompetent mice using herpes simplex viruses that express murine interleukins. *Gene Ther.* **5**: 121–130.

Armentano D, Zabner J, Sacks C *et al.* (1997) Effect of the E4 region on the persistence of transgene expression from adenovirus vectors. *J. Virol.* **71**: 2408–2416.

Arya SK, Zamani M, Kundra P. (1998) Human immunodeficiency virus type 2 vectors for gene transfer: expression and potential for helper virus-free packaging. *Hum. Gene Ther.* **9**: 1371–1380.

Atchison RW, Casto BC. (1965) Adenovirus-associated defective virus particles. *Science* **149**: 754–756.

Barr D, Tubb J, Ferguson D *et al.* (1995) Strain related variations in adenovirally mediated transgene expression from mouse hepatocytes in vivo: comparisons between immunocompetent and immunodeficient inbred strains. *Gene Ther.* **2**: 151–155.

Batchelor AH, O'Hare P. (1990) Regulation and cell-type-specific activity of a promoter located upstream of the latency-associated transcript of herpes simplex virus type 1. *J. Virol.* **64**: 3269–3279.

Beaton A, Palumbo P, Berns KI. (1989) Expression from the adeno-associated virus p5 and p19 promoters is negatively regulated in trans by the rep protein. *J. Virol.* **63**: 4450–4454.

Bergelson JM, Cunningham JA, Droguett G *et al.* (1997) Isolation of a common receptor for Coxsackie B viruses and Adenoviruses 2 and 5. *Science* **275**: 1320–1323.

Bischoff JR, Kirn DH, Williams A *et al.* (1996) An adenovirus mutant that replicates selectively in p53-deficient human tumor cells. *Science* **274**: 373–376.

Blacklow NR, Hoggan MD, Kapikian AZ, Austin JB, Rowe WP. (1968) Epidemiology of adenovirus-associated virus infection in a nursery population. *Am. J. Epidemiol.* **88**: 368–378.

Blacklow NR, Hoggan MD, Sereno MS, Brandt CD, Kim HW, Parrott RH, Chancock RM. (1971) A seroepidemiologic study of adenovirus-associated virus infections in infants and children. *Am. J. Epidemiol.* **94**: 359–366.

Blaese RM, Culver KM, Miller AD *et al.* (1995) T-lymphocyte-directed gene therapy for ADA-SCID: Initial trial results after 4 years. *Science* **270**: 475–480.

Bolovan CA, Sawtell NM, Thompson RL. (1994) ICP34.5 mutants of herpes simplex virus type 1 strain 17syn+ are attenuated for neurovirulence in mice and for replication in confluent primary mouse embryo cell cultures. *J. Virol.* **68**: 48–55.

Bordignon C, Notarangelo LD, Nobili N *et al.* (1995) Gene therapy in peripheral blood lymphocytes and bone marrow for ADA- immunodeficient patients. *Science* **270**: 470–475.

Boviatsis E, Scharf J, Chase M, Harrington K, Kowall NW, Breakefield XO, Chiocca EA. (1994) Antitumor activity and reporter gene transfer into rat brain neoplasms inoculated with herpes simplex virus vectors defective in thymidine kinase or ribonucleotide reductase. *Gene Ther.* **1**: 323–331.

Brough DE, Lizonova A, Hsu C, Kulesa VA, Kovesdi I. (1996) A gene transfer vector-cell line system for complete functional complementation of adenovirus early regions E1 and E4. *J. Virol.* **70**: 6497–6501.

Burns JC, Friedmann T, Driever W, Burrascano M, Yee JK. (1993) Vesicular stomatitis virus G glycoprotein pseudotyped retroviral vectors: concentration to very high titer and efficient gene transfer into mammalian and nonmammalian cells. *Proc. Natl Acad. Sci. USA* **90**: 8033–8037.

Carpenter DE, Stevens JG. (1996) Long term expression of a foreign gene from a unique position in the latent herpes simplex virus genome. *Hum. Gene Ther.* **7**: 1447–1454.

Chambers RJ, Gillepsie GY, Soroceanu L *et al.* (1995) Comparison of genetically engineered herpes simplex viruses for the treatment of brain tumors in a SCID mouse model of human malignant glioma. *Proc. Natl Acad. Sci. USA* **92**: 1411–1415.

Chase M, Chung RY, Chiocca EA. (1998) An oncolytic viral mutant that delivers the CYP2B1 transgene and augments cyclophosphamide chemotherapy. *Nature Biotechnol.* **16**: 444–448.

Chen H-H, Mack LM, Kelly R, Ontell M, Kochanek S, Clemens PR. (1997) Persistence in muscle of an adenoviral vector that lacks all viral genes. *Proc. Natl Acad. Sci. USA* **94**: 1645–1650.

Cheung AK, Hoggan MD, Hauswirth WW, Berns KI. (1980) Integration of adenovirus-associated virus genome into cellular DNA in latency infected human Detroit 6 cells. *J. Virol.* **33**: 739–748.

Chiorini JA, Weitzman MD, Owens RA, Urcelay E, Safer B, Kotin RM. (1994) Biologically active Rep proteins of adeno-associated virus type 2 produced as fusion proteins in Escherichia coli. *J. Virol.* **68**: 797–804.

Chong H, Vile RG. (1996) Replication-competent retrovirus produced by a 'split-function' third generation amphotropic packaging cell line. *Gene Ther.* **3**: 624–629.

Chong H, Starkey W, Vile RG. (1998) A replication-competent retrovirus arising from a split-function packaging cell line was generated by recombination events between the vector, one of the packaging constructs, and endogenous retroviral sequences. *J. Virol.* **72**: 2663–2670.

Chou J, Kern ER, Whitley R, Roizman B. (1990) Mapping of herpes simplex virus neurovirulence to gamma 1 34.5, a gene nonessential for growth in culture. *Science* **250**: 1262–1266.

Coen DM, Kosz-Unenckak M, Jacoson JG *et al.* (1989) Thymidine kinase-negative herpes simplex virus mutants establish latency in mouse trigeminal ganglia but do not reactivate. *Proc. Natl. Acad. Sci. USA* **86**: 4736–4740.

Coffin JM. (1996) Retroviridae: The viruses and their replication. In: *Fundamental Virology*, third edn (eds. BN Fields, DM Knipe, PM Howley). Lippincott-Raven Publishers.

Cone RD, Mulligan RC. (1984) High-efficiency gene transfer into mammalian cells: generation of helper free recombinant retrovirus with broad mammalian host range. *Proc. Natl Acad. Sci. USA* **81**: 6349–6353.

Conrad CK, Allen SS, Afione SA *et al.* (1996) Safety of a single-dose administration of an adeno-associated virus (AAV)-CFTR vector in the primate lung. *Gene Ther.* **3**: 658–668.

Cosset FL, Russell SJ. (1996) Targeting retrovirus entry. *Gene Ther.* **3**: 946–956.

Cosset FL, Takeuchi Y, Battini JL, Weiss RA, Collins MK. (1995) High-titer packaging cells producing recombinant retroviruses resistant to human serum. *J. Virol.* **96**: 7430–7436.

Crystal RG, McElvany NG, Rosenfeld MA *et al.* (1994) Administration of an adenovirus containing the human CFTR cDNA to the respiratory tract of individuals with cystic fibrosis. *Nature Genet.* **8**: 42–51.

Cunningham C, Davison AJ. (1993) A cosmid-based system for constructing mutants of herpes simplex virus type 1. *Virology* **197**: 116–124.

Deatly AM, Spivack JG, Lavi E, Frazer NW. (1987) RNA from an immediate early region of the type 1 herpes simplex virus genome is present in the trigeminal ganglia of latently infected mice. *Proc. Natl Acad. Sci. USA* **84**: 3204–3208.

Dedieu J-F, Vigne E, Torrent C *et al.* (1997) Long-term gene delivery into the livers of immunocompetent mice with E1/E4-defective adenoviruses. *J. Virol.* **71**: 4626–4637.

DeLuca NA, McCarthy AM, Schaffer PA. (1985) Isolation and characterization of deletion mutants of herpes simplex virus type 1 in the gene encoding immediate-early regulatory protein ICP4. *J. Virol.* **59**: 558–570.

Dmitriev I, Krasnykh V, Miller CR *et al.* (1998) An adenovirus vector with genetically modified fibers demonstrates expanded tropism via utilisation of a coxsackievirus and adenovirus receptor-independent cell entry mechanism. *J. Virol.* **72**: 9706–9713.

Dobson AT, Sedarati F, Devi-Rao G *et al.* (1989) Identification of the latency-associated transcript promoter by expression of rabbit beta-globin mRNA in mouse sensory nerve ganglia latently infected with a recombinant herpes simplex virus. *J. Virol.* **63**: 3844–3851.

Donahue RE, Kessler SW, Bodine D *et al.* (1992) Helper virus induced T cell lymphoma in nonhuman primates after retroviral mediated gene transfer. *J. Exp. Med.* **176**: 1125–1135.

Douglas JT, Rogers BE, Rosenfeld ME, Michael SI, Feng M, Curiel DT. (1996) Targeted gene delivery by tropism-modified adenoviral vectors. *Nature Biotechnol.* **14**: 1574–1578.

Dull T, Zufferey R, Kelly M, Mandel RJ, Nguyen M, Trono D, Naldini L. (1998) A third-generation lentivirus vector with a conditional packaging system. *J. Virol.* **72**: 8463–8471.

During MJ, Naegele JR, O'Malley KL, Geller AI. (1994) Long-term behavioral recovery in parkinsonian rats by an HSV vector expressing tyrosine hydroxylase. *Science* **266**: 1399–1403.

Elkon KB, Liu CC, Gall JG *et al.* (1997) Tumor necrosis factor plays a central role in immune-mediated clearance of adenoviral vectors. *Proc. Natl Acad. Sci. USA* **94**: 9814–9819.

Englehardt JF, Ye X, Doranz B, Wilson JM. (1994a) Ablation of E2a in recombinant adenoviruses improves transgene persistence and decreases inflammatory response in mouse liver. *Proc. Natl Acad. Sci. USA* **91**: 6196–6200.

Englehardt JF, Litzky L, Wilson JM. (1994b) Prolonged transgene expression in cotton rat lung with recombinant adenoviruses defective in E2a. *Hum. Gene Ther.* **5**: 1217–1229.

Fang B, Wang H, Gordon G *et al.* (1996) Lack of persistence of E1- recombinant adenoviral vectors containing a temperature-sensitive E2a mutation in immunocompetent mice and hemophilia B dogs. *Gene Ther.* **3**: 217–222.

Federico M. (1999) Lentiviruses as gene delivery vectors. *Curr. Opin. Biotechnol.* **10**: 448–453.

Federoff HJ, Geschwind MD, Geller AI, Kessler JA. (1992) Expression of nerve growth factor in vivo from a defective herpes simplex virus 1 vector prevents effects of axotomy on sympathetic ganglia. *Proc. Natl Acad. Sci. USA* **89**: 1636–1640.

Flotte TR, Afione SA, Conrad C *et al.* (1993) Stable in vivo expression of the cystic fibrosis transmembrane conductance regulator with an adeno-associated virus vector. *Proc. Natl Acad. Sci. USA* **90**: 10613–10617.

Flotte T, Carter B, Conrad C *et al.* (1996) A phase I study of an adeno-associated virus-CFTR vector in adult CF patients with mild lung disease. *Hum. Gene Ther.* **7**: 1145–1159.

Fotaki ME, Pink JR, Mous J. (1997) Tetracycline-responsive gene expression in mouse brain after amplicon-mediated gene transfer. *Gene Ther.* **4**: 901–908.

Fraefel C, Song S, Lim F *et al.* (1996) Helper virus-free transfer of herpes simplex virus type 1 plasmid vectors into neural cells. *J. Virol.* **70**: 7190–7197.

Fraefel C, Jacoby DR, Lage C *et al.* (1997) Gene transfer into hepatocytes mediated by helper-free HSV/AVV hybrid vectors. *Mol. Med.* **3**: 813–825.

Freytag SO, Rogulski KR, Paielli DL, Gilbert JD, Kim JH. (1998) A novel three-pronged approach to kill cancer cells selectively: concomitant viral, double suicide gene, and radiotherapy. *Hum. Gene Ther.* **10**: 1323–1333.

Gahery-Segard H, Moliner-Frenkel V, Le Boulaire *et al.* (1997) Phase I trial of recombinant adenovirus gene transfer in lung cancer. Longitudinal study of the immune responses to trasngene and viral products. *J. Clin. Invest.* **100**: 2218–2226.

Gahery-Segard H, Farace F, Godfrin D *et al.* (1998) Immune responses to recombinant capsid proteins of adenovirus in humans: Anti-fibre and anti-penton base antibodies have a synergistic effect on neutralising activity. *J. Virol.* **72**: 2388–2397.

Geller AI, Breakefield XO. (1988) A defective HSV-1 vector expresses *Escherichia coli* beta-galactosidase in cultured peripheral neurons. *Science* **241**: 1667–1669.

Geller AI, During MJ, Oh YJ, Freese A, O'Malley K. (1995) HSV-1 vector expressing tyrosine hydroxylase causes production and release of L-dopa from cultured rat striatal cells. *J. Neurochem.* **64**: 487–496.

Geschwind MD, Kessler JA, Geller AI, Federoff HJ. (1994) Transfer of the nerve growth factor gene into cell lines and cultured neurons using a defective herpes simplex virus vector. Transfer of the NGF into cells by a HSV-1 vector. *Mol. Brain Res.* **24**: 327–335.

Goldman MJ, Litzky LA, Englehardt JF, Wilson JM. (1995) Transfer of the CFTR gene to the lung of non human primates with E1-deleted E2a-defective recombinant adenoviruses: a preclinical toxicology study. *Hum. Gene Ther.* **6**: 839–851.

Goldman MJ, Lee PS, Yang JS, Wilson JM. (1997) Lentiviral vectors for gene therapy of cystic fibrosis. *Hum. Gene Ther.* **8**: 2261–2268.

Goldstein DJ, Weller S. (1988) Herpes simplex virus type 1-induced ribonucleotide reductase activity is dispensable for virus growth and DNA synthesis: isolation and characterization of an ICP6 lacZ insertion mutant. *J. Virol.* **62**: 196–205.

Goodman S, Xiao X, Donahue RE *et al.* (1994) Recombinant adeno-associated virus-mediated gene transfer into hematopoietic progenitor cells. *Blood* **84**: 1492–1500.

Gossen M, Bujard H (1992) Tight control of gene expression in mammalian cells by tetracycline-responsive promoters. *Proc. Natl Acad. Sci. USA* **89**: 5547–5551.

Graham FL, Smiley J, Russell WC, Nairn R. (1977) Characteristics of a human cell line transformed by DNA from adenovirus type 5. *J. Gen. Virol.* **36**: 59–74.

Halbert CL, Standeart TA, Aitken ML, Alexander IE, Russell DW, Miller AD. (1997) Transduction by adeno-associated virus vectors in the rabbit airway: efficiency, persistence, and readministration. *J. Virol.* **71**: 5932–5941.

Harada JN, Berk AJ. (1999) p53-Independent and dependent requirements for E1B-55K in adenovirus type 5 replication. *J. Virol.* **73**: 5333–5344.

Harding TC, Geddes BJ, Murphy D, Knight D, Uney JB. (1998) Switching expression in the brain using an adenoviral tetracycline-regulatable system. *Nature Biotechnol.* **16**: 553–555.

Hayder H, Mullbacher A. (1996) Molecular basis of immune evasion strategies by adenoviruses. *Immunol. Cell Biol.* **74**: 504–512.

Heise C, Sampson-Johannes A, Williams A, McCormick F, Von Hoff DD, Kirn DH. (1997) ONYX-015, an E1B gene-attenuated adenovirus, causes tumor-specific cytolysis and antitumoral efficacy that can be augmented by standard chemotherapeutic agents. *Nature Med.* **3**: 639–645.

Hermonat PL, Muzyczka N. (1984) Use of adeno-associated virus as a mammalian DNA cloning vector: transduction of neomycin resistance into mammalian tissue culture cells. *Proc. Natl Acad. Sci. USA* **81**: 6466–6470.

Hitt M, Bett AJ, Addison CL, Prevec L, Graham FL. (1995) Techniques for human adenovirus vector construction and characterisation. *Methods Mol. Gen.* **7**: 13–30.

Ho DY, Mocarski ES, Sapolsky RM. (1993) Altering central nervous system physiology with a defective herpes simplex vector expressing the glucose transporter gene. *Proc. Natl Acad. Sci. USA* **90**: 3655–3659.

Ho DY, Saydam TC, Fink SL, Lawrence MS, Sapolsky R. (1995) Defective herpes simplex virus vectors expressing the rat brain glucose transporter protect cultured neurons from necrotic insults. *J. Neurochem.* **65**: 842–850.

Ho DY, McLaughlin JR, Sapolsky RM. (1997) Inducible gene expression from defective herpes simplex virus vectors using the tetracycline-responsive promoter system. *Brain Res. Mol.* **41**: 200–220.

Horwitz MS. (1994) Adenoviridae and their replication. In *Virology*, 2nd edition (ed. Fields BN *et al.*) pp. 1679–1721, Raven Press, New York.

Ilan Y, Droguett G, Chowdhury NR *et al.* (1997) Insertion of adenoviral E3 region into a recombinant viral vector prevents antiviral humoral and cellular immune responses and permits long-term gene expression. *Proc. Natl Acad. Sci. USA* **94**: 2587–2592.

Janik JE, Huston MM, Cho K, Rose JA. (1989) Efficient synthesis of adeno-associated virus structural proteins requires both adenovirus DNA binding protein and VA I RNA. *Virology* **168**: 320–329.

Javier RT, Stevens JG, Dissette VB, Wagner EK. (1988) A herpes simplex virus transcript abundant in latently infected neurons is dipensable for establishment of the latent state. *Virology* **166**: 254–257.

Jay FT, de la Maza LM, Carter BJ. (1979) Parvovirus RNA transcripts containing sequences not present in mature mRNA: a method for isolation of putative mRNA precursor sequences. *Proc. Natl Acad. Sci. USA* **76**: 625–629.

Jay FT, Laughin CA, Carter BJ. (1981) Eukaryotic translation control: adeno-associated virus protein synthesis is affected by a mutation in the adenovirus DNA-binding protein. *Proc. Natl Acad. Sci. USA* **78**: 2927–2931.

Jia WW, McDermott M, Goldie J, Cynader M, Tan J, Tufaro F. (1994). Selective destruction of gliomas in immunocompetent rats by thymidine kinase-defective herpes simplex virus type 1. *J. Natl Cancer Inst.* **86**: 1209–1215.

Johnson PA, Miyanohara A, Levine F, Cahill T, Friedmann T. (1992) Cytoxicity of a replication-defective mutant of herpes simplex virus type 1. *J. Virol.* **66**: 2952–2965.

Johnson KM, Jacoby D, Pechan PA ***et al.*** (1997) HSV/AAV hybrid amplicon vectors extend transgene expression in human glioma cells. *Hum. Gene Ther.* **8**: 359–370.

Johnston JC, Gasmi M, Lim LE ***et al.*** (1999) Minimum requirements for efficient transduction of dividing and nondividing cells by feline immunodeficiency virus vectors. *J. Virol.* **73**: 4991–5000.

Jomary C, Vincent KA, Grist J, Neal MJ, Jones SE. (1997) Rescue of photoreceptor function by AAV-mediated gene transfer in a mouse model of inherited retinal degeneration. *Gene Ther.* **4**: 683–690.

Jooss K, Yang Y, Fisher KJ, Wilson JM. (1998) Transduction of dendritic cells by DNA viral vectors directs the immune response to transgene products in muscle fibers. *J. Virol.* **72**: 4212–4223.

Ju Q, Edelstein D, Brendel MD, Brandhorst D, Brandhorst H, Bretzel RG, Brownlee M. (1998) Transduction of non-dividing adult human pancreatic beta cells by an integrating lentiviral vector. *Diabetologia* **41**: 736–739.

Kafri T, Blomer U, Peterson DA, Gage FH, Verma IM. (1997) Sustained expression of genes delivered directly into liver and muscle by lentiviral vectors. *Nature Genetics* **17**: 314–317.

Kaplan JM, Armentano D, Sparer TE ***et al.*** (1997) Characterisation of factors involved in modulating persistence of transgene expression from recombinant adenovirus in the mouse lung. *Hum. Gene Ther.* **8**: 45–56.

Kaplitt MG, Kwong D, Kleopoulos SP, Mobbs CV, Rabkin SD, Pfaff DW. (1994a) Preproenkephalin promoter yields region-specific and long-term expression in adult brain after direct in vivo gene transfer via a defective herpes simplex viral vector. *Proc. Natl. Acad. Sci. USA* **91**: 8979–8983.

Kaplitt MG, Leone P, Samulski RJ, Xiao X, Pfaff DW, O'Malley KL, During MJ. (1994b) Long-term gene expression and phenotypic correction using adeno-associated virus vectors in the mammalian brain. *Nature Genet.* **8**: 148–154.

Kaplitt MG, Xiao X, Samulski RJ ***et al.*** (1996) Long-term gene transfer in porcine myocardium after coronary infusion of an adeno-associated virus vector. *Ann. Thor. Surg.* **62**: 1669–1676.

Kersey JH. (1997) Fifty years of studies of the biology and therapy of childhood leukemia. *Blood* **90**: 4243–4251.

Kessler PD, Podsakoff GM, Chen X ***et al.*** (1996) Gene delivery to skeletal muscle results in sustained expression and systemic delivery of a therapeutic protein. *Proc. Natl. Acad. Sci. USA* **93**: 14082–14087.

Kim VN, Mitrophanous K, Kingsmann SM, Kingsmann AJ. (1998) Minimal requirement for a lentivirus vector based on human immunodeficiency virus type 1. *J. Virol.* **72**: 811–816.

Kirn D, Hermiston T, McCormick F. (1998a) ONYX-015: clinical data are encouraging. *Nature Med.* **4**: 1341–1342.

Kirn FR, Nemunaitis J, Ganly I. ***et al.*** (2000) A controlled trial of intratumoral ONYX-015, a selectively-replicating adenvirus, in combination with cisplatin and 5-fluorouracil in patients with recurrent head and neck cancer. *Nat. Med.* **6**: 879–885.

Klimatcheva E, Rosenblatt JD, Planelles V. (1999) Lentiviral vectors and Gene Therapy. *Front. Biosci.* **4**: D481–496.

Knowles MR, Hohneker KW, Zhou Z ***et al.*** (1995) A controlled study of adenoviral-vector-mediated gene transfer in the nasal epithelium of patients with cystic fibrosis. *New Engl. J. Med.*

333: 823–831.

Kochanek S, Clemens PR, Mitani K, Chen H-H, Chan S, Caskey CT. (1996) A new adenoviral vector: Replacement of all viral coding sequences with 28kB of DNA independently expressing both full-length dystrophin and β-galactosidase. *Proc. Natl Acad. Sci. USA* **93**: 5731–5736.

Koeberl DD, Alexander IE, Halbert CL, Russell DW, Miller AD. (1997) Persistent expression of human clotting factor IX from mouse liver after intravenous injection of adeno-associated virus vectors. *Proc. Natl Acad. Sci. USA* **94**: 1426–1431.

Kotin RM, Berns KI. (1989) Organization of adeno-associated virus DNA in latency infected Detroit 6 cells. *Virology* **170**: 460–467.

Kramm CM, Chase H, Herrlinger U *et al.* (1997) Therapeutic efficiency and safety of a second-generation replication-conditional HSV-1 vector for brain tumor Gene Therapy. *Hum. Gene Ther.* **8**: 2057–2068.

Krasnykh VN, Mikheeva GV, Douglas JT, Curiel DT. (1996) Generation of recombinant adenovirus vectors with modified fibers for altering viral tropism. *J. Virol.* **70**: 6839–6846.

Krasnykh V, Dmitriev I, Mikheeva G, Miller CR, Belousova N, Curiel DT. (1998) Characterization of an adenovirus vector containing a heterologous peptide epitope in the HI loop of the fiber knob. *J. Virol.* **72**: 1844–1852.

Kyostio SR, Owens RA, Weistman MD, Antoni BA, Chejanovsky N, Carter BJ. (1994) Analysis of adeno-associated virus (AAV) wild-type and mutant Rep proteins for their abilities to negatively regulate AAV p5 and p9 mRNA levels. *J. Virol.* **68**: 2947–2957.

Labow MA, Berns KI. (1988) The adeno-associated virus rep gene inhibits replication of an adeno-associated virus/simian virus 40 hybrid genome in cos-7 cells. *J. Virol.* **62**: 1705–1712.

Labow MA, Hermonat PL, Berns KI. (1986) Positive and negative autoregulation of the adeno-associated virus type 2 genome. *J. Virol.* **60**: 251–258.

Lachman RH, Efstathiou S. (1997) Utilization of the herpes simplex virus type 1 latency-associated regulatory region to drive stable reporter gene expression in the nervous system. *J. Virol.* **71**: 3197–3207.

Lachman RH, Brown C, Efstathiou S. (1996) A murine RNA polymerase I promoter inserted into the herpes simplex virus type 1 genome is functional during lytic, but not latent, infection. *J. Gen. Virol.* **77**: 2575–2582.

Lasner TM, Tal-Singer R, Kesari S, Lee VM, Trojanowski JQ, Fraser NW. (1998) Toxicity and neuronal infection of a HSV-1 ICP34.5 mutant in nude mice. *J. Neurovirol.* **3**: 100–105.

Lawrence MS, Ho DY, Dash R, Sapolsky RM. (1995) Herpes simplex virus vectors overexpressing the glucose transporter gene protect against seizure-induced neuron loss. *Proc. Natl Acad. Sci. USA* **92**: 7247–7251.

Leib DA, Olivo PD. (1993) Gene delivery to neurons: is herpes simplex virus the right tool for the job? *Bioessays* **15**: 547–554.

Lewis PF, Emerman M. (1994) Passage through mitosis is required for oncoretroviruses but not for human immunodeficiency virus. *J. Virol.* **68**: 510–516.

Lewis P, Hensel M, Emerman M. (1992) Human immunodeficiency virus infection of cells arrested in the cell cycle. *EMBO J.* **11**: 3053–3058.

Lim F, Hartley D, Starr P *et al.* (1996) Generation of high titer defective HSV-1 vectors using an IE2 deletion mutant and quantitative study of expression in cultured cortical cells. *Biotechniques* **20**: 460–469.

Linden RM, Ward P, Giraud C, Winocour E, Berns K. (1996) Site-specific integration by adeno-associated virus. *Proc. Natl Acad. Sci. USA* **93**: 11288–11294.

Lokensgard JR, Bloom DC, Dobson AT, Feldman LT. (1994) Long-term promoter activity during herpes simplex latency. *J. Virol.* **68**: 7148–7158.

Lusky M, Christ M, Rittner K *et al.* (1998) In vitro and in vivo biology of recombinant adenovirus vectors with E1, E1/E2a, or E4 deleted. *J. Virol.* **72**: 2022–2032.

Lynch CM, Hara PS, Leonard JC, Williams JK, Dean RH, Geary RL. (1997) Adeno-associated virus vectors for vascular gene delivery. *Circulation Res.* **80**: 497–505.

Mann R, Mulligan RC, Baltimore D. (1983) Construction of a retrovirus packaging mutant and its use to produce helper-free defective retrovirus. *Cell* **33**: 153–159.

Manning WC, Paliard X, Zhou S *et al.* (1980) Adenovirus helper function for growth of adeno-associated virus: effect of temperature-sensitive mutations in adenovirus early gene region 2. *J. Virol.* **35**: 65–75.

Manning WC, Zhou S, Pat Bland M, Escobedo JA, Dwarki V. (1998) Transient immunosuppression allows transgene expression following readministration of adeno-associated viral vectors. *Hum. Gene Ther.* **9**: 477–485.

Marbert JM, Malick A, Coen DM, Martuza RL. (1993) Reduction and elimination of encephalitis in an experimental glioma therapy model with attenuated herpes simplex mutants that retain susceptibility to acyclovir. *Neurosurgery* **35**: 597–603.

Markowitz D, Goff S, Bank A. (1988a) A safe packaging line for gene transfer: separating viral genes on two different plasmids. *J. Virol.* **62**: 1120–1124.

Markowitz D, Goff S, Bank A. (1988b) Construction and use of a safe and efficient packaging cell line. *Virology* **167**: 400–406.

Martuza R, Malick A, Markert JM *et al.* (1991) Experimental therapy of human gliomas by means of a genetically engineered virus mutant. *Science* **252**: 854–855.

McKie EA, Graham DI, Brown SM. (1998) Selective astrocyte transgene expression *in vitro* and *in vivo* from the GFAP promoter in a HSV RL1 null mutant vector – potential glioblastoma targeting. *Gene Ther.* **5**: 440–450.

McLaughlin SK, Collis P, Hermonat PL, Muzyczka N. (1988) Adeno-associated virus general transduction vectors: analysis of proviral structures. *J. Virol.* **62**: 1963–1973.

McLean AR, Ul-Fareed M, Roberston L, Harland J, Brown SM. (1991) Herpes simplex virus type 1 variants 1714 and 1716 pinpoint neurovirulence-related sequences in Glascow strain 17+ between immediate early gene 1 and the "a" sequence. *J. Gen. Virol.* **72**: 631–639.

Michael SI, Hong JS, Curiel DT, Engler JA. (1995) Addition of a short peptide ligand to the adenovirus fiber protein. *Gene Ther.* **2**: 660–668.

Michou AI, Santoro L, Cgrist M, Juillard V, Pavirani A, Mehtali M. (1997) Adenovirus-mediated gene transfer: influence of transgene, mouse strain and type of immune response on persistence of transgene expression. *Gene Ther.* **4**: 473–482.

Miller AD, Buttimore C. (1986) Resign of retrovirus packaging cell lines to avoid recombination leading to helper virus production. *Mol. Cell. Biol.* **6**: 2895–2902.

Miller JL, Donahue RE, Sellers SE, Samulski RJ, Young NS, Nienhuis AW. (1994) Recombinant adeno-associated virus (rAAV)-mediated expression of a human β-globin gene in human progenitor-derived erythroid cells. *Proc. Natl Acad. Sci. USA* **91**: 10183–10187.

Mineta T, Rabkin SD, Martuza RL. (1994) Treatment of malignant gliomas using gancyclovir hypersensitive ribonucleotide reductase deficient herpes simplex virus mutant. *Cancer Res.* **54**: 3963–3966.

Mineta T, Rabkin S, Yazaki T, Hunter WD, Martuza RL. (1995) Attenuated multi-mutated herpes simplex virus type1 for the treatment of malignant gliomas. *Nature Med.* **1**: 938–944.

Mitrophanous KA, Yoon S, Rohll JB *et al.* (1999) Stable gene transfer to the nervous system using a non-primate lentiviral vector. *Gene Ther.* **6**: 1808–1818.

Miyatake SI, Martuza RL, Rabkin SD. (1997a) Defective herpes simplex virus vectors expressing thymidine kinase for the treatment of malignant glioma. *Cancer Gene Ther.* **4**: 222–228.

Miyatake SI, Iyer A, Martuza RL, Rabkin SD. (1997b) Transcriptional targeting of herpes simplex virus for cell-specific replication. *J. Virol.* **71**: 5124–5132.

Miyoshi H, Takahashi M, Gage FH, Verma IM. (1997) Stable and efficient gene transfer into the retina using an HIV-based lentiviral vector. *Proc. Natl Acad. Sci. USA* **94**: 10319–10323.

Miyoshi H, Blomer U, Takahashi M, Gage FH, Verma IM. (1998) Development of a self-inactivating lentivirus vector. *J. Virol.* **72**: 8150–8157.

Monahan PE, Samulski RJ, Taxelaar J *et al.* (1998) Direct intramuscular injection with recombinant AAV vectors results in sustained expression in a dog model of hemophilia. *Gene Ther.* **5**: 40–49.

Moore M, Horikoshi N, Shenk T. (1996) Oncogeneic potential of the adenovirus E4orf6 protein. *Proc. Natl Acad. Sci. USA* **93**: 11295–11301.

Morral N, O'Neal W, Zhou H, Langston C, Beaudet A. (1997) Immune responses to reporter proteins and high viral dose limit duration of expression with adenoviral vectors: comparison of E2a wild-type and E2a deleted vectors. *Hum. Gene Ther.* **8**: 1275–1286.

Morsy MA, Gu MC, Motzel S *et al.* (1998) An adenoviral vector deleted for all viral coding seqences results in enhanced safety and extended expression of a leptin transgene. *Proc. Natl Acad. Sci. USA* **95**: 7866–7871.

Mselli-Lakhal L, Favier C, Da Silva Teixeira MF *et al.* (1998) Defective RNA packaging is responsible for low transduction efficiency of CAEV-based vectors. *Arch. Virol.* **143**: 681–695.

Muenchau DD, Freeman SM, Cornetta K, Zweibel JA, Anderson WF. (1990) Analysis of retroviral packaging lines for generation of replication-competent virus. *Virology* **176**: 262–265.

Naldini L. (1998) Lentiviruses as gene transfer agents for delivery to non-dividing cells. *Curr. Opin. Biotechnol.* **9**: 457–463.

Naldini L, Blomer U, Gallay P *et al.* (1996a) In vivo gene delivery and stable transduction of nondividing cells by a lentiviral vector. *Science* **272**: 263–267.

Naldini L, Blomer U, Gage FH, Trono D, Verma IM. (1996b) Efficient transfer, integration, and sustained long-term expression of the transgene in adult rat brains injected with a lentiviral vector. *Proc. Natl Acad. Sci. USA* **93**: 11382–11388.

Ni TH, Zhou X, McCarty DM, Zolotukhin I, Muzyczka N. (1994) In vitro replication of adeno-associated virus DNA. *J. Virol.* **68**: 1128–1138.

Oligino T, Poliani PL, Marconi P, Bender MA, Schmidt MC, Fink DJ, Glorioso JC. (1996) In vivo transgene activation from an HSV-based gene therapy vector by GAL4:VP16. *Gene Ther.* **3**: 892–899.

Olsen JC. (1998) Gene transfer vectors derived from equine infectious anemia virus. *Gene Ther.* **5**: 1481–1487.

Pakzaban P, Geller AI, Isacson O. (1994) Effect of exogenous nerve growth factor on neurotoxicity of and neuronal gene delivery by a herpes simplex amplicon vector in the rat brain. *Hum. Gene Ther.* **5**: 987–995.

Parks RJ, Chen L, Anton M, Sankar U, Rudnicki MA, Graham FL. (1996) A helper-dependent adenovirus vector system: Removal of helper virus by Cre-mediated excision of the viral packaging signal. *Proc. Natl Acad. Sci. USA* **93**: 13565–13570.

Podsakoff, G, Wong KKJr, Chatterjee S. (1994) Efficient gene transfer into nondividing cells by adeno-associated virus-based vectors. *J. Virol.* **68**: 5656–5666.

Poeschla E, Gilbert J, Li X, Huang S, Ho A, Wong-Stall F. (1998a) Identification of a human immunodeficiency virus type 2 (HIV-2) encapsidation determinant and transduction of nondividing human cells by HIV-2 based lentivirus vectors. *J. Virol.* **72**: 6527–6536.

Poeschla E, Wong-Staal F, Looney DJ. (1998b) Efficient transduction of non-dividing human cells by feline immunodeficiency virus lentiviral vectors. *Nature Med.* **4**: 354–357.

Post LE, Roizman B. (1981) A generalized technique for deletion of specific genes in large genomes: alpha gene 22 of herpes simplex virus is not essential for growth. *Cell* **25**: 227–232.

Qazilbash MH, Xiao X, Seth P, Cowan KH, Walsh CE. (1997) Cancer gene therapy using a novel adeno-associated virus vector expressing human wild-type p53. *Gene Ther.* **4**: 675–682.

Rebolledo MA, Krogstad, P, Chen F, Shannon KM, Klitzner TS. (1998) Infection of human fetal cardiac myocytes by a human immunodeficiency virus-1-derived vector. *Circulation Res.* **83**: 738–742.

Richardson WD, Westphal H. (1981) A cascade of adenovirus early functions is required for expression of adeno-associated virus. *Cell* **27**: 133–141.

Richardson WD, Westphal H. (1984) Requirement for either early region 1a or early region 1b adenovirus gene products in the helper effect for adeno-associated virus. *J. Virol.* **51**: 404–410.

Rock DL, Frazer NW. (1983) Detection of HSV-1 genome in central nervous system of latently infected mice. *Nature* **302**: 523–525.

Rodriguez R, Schuur ER, Lim HY, Henderson GA, Simons JW, Henderson DR. (1997) Prostate attenuated replication competent adenovirus (ARCA) CN706: a selective cytotoxic of prostate-specific antigen-positive prostate cancer cells. *Cancer Res.* **57**: 2559–2563.

Roe TY, Reynolds TC, Yu G, Brown PO. (1993) Integration of murine leukaemia virus DNA depends on mitosis. *EMBO J.* **12**: 2099–2108.

Roemer K, Johnson PA, Friedmann T. (1991) Activity of the simian virus 40 early promoter-enhancer in herpes simplex type 1 vectors is dependent on its position, the infected cell type, and the presence of Vmw175. *J. Virol.* **65**: 6900–6912.

Rogers BE, Douglas JT, Ahlem C, Buchsbaum DJ, Frincke J, Curiel DT. (1997) Use of a novel cross-linking method to modify adenovirus tropism. *Gene Ther.* **4**: 1387–1392.

Roizman B. (1996) The function of herpes simplex virus genes: a primer for genetic engineering of novel vectors. *Proc. Natl Acad. Sci. USA* **93**: 11307–11312.

Roizman B, Sears A. (1993) Herpes Simplex Virus and their replication. In: *The Human Herpesvirus* (eds B Roizman, R Whitley, C Lopez). Raven Press Ltd, New York.

Rolling F, Nong Z, Pisvin S, Collen D. (1997) Adeno-associated virus-mediated gene transfer into rat carotid arteries. *Gene Ther.* **4**: 757–761.

Rother RP, Squinto SP, Mason JM, Rollins SA. (1995) Protection of retroviral vector particles in human blood through complement inhibition. *Hum. Gene Ther.* **6**: 429–435.

Rothmann T, Hengstermann A, Whitaker NJ, Scheffner M, zur Hausen H. (1998) Replication of ONYX-015, a potential anticancer adenovirus, is independent of p53 status in tumor cells. *J. Virol.* **72**: 9470–9478.

Russell DW, Miller AD, Alexander IE. (1994) Adeno-associated virus vectors preferentially transduce cells in S phase. *Proc. Natl Acad. Sci. USA* **91**: 8915–8919.

Salvetti A, Orève S, Chadeuf G *et al.* (1998) Factors influencing recombinant adeno-associated virus production. *Hum. Gene Ther.* **9**: 695–706.

Samulski RJ, Chang LS, Shenk T. (1989) Helper-free stocks of recombinant adeno-associated viruses: normal integration does not require viral gene expression. *J. Virol.* **63**: 3822–3828.

Savard N, Cosset FL, Epstein AL. (1997) Defective herpes simplex virus type 1 vectors harboring gag, pol, env genes can be used to rescue defective retrovirus vectors. *J. Virol.* **7**: 4111–4117.

Scarpa M, Cournoyer D, Muzny DM, Moore KA, Belmont JW, Caskey CT. (1991) Characterisation of recombinant helper retroviruses from moloney-based vectors in ecotropic and amphotropic packaging cell lines. *Virology* **180**: 849–852.

Schowalter DB, Tubb JC, Liu M, Wilson CB, Kay MA. (1997) Heterologous expression of adenovirus E3-gp19K in an E1a-deleted adenovirus vector inhibits MHC I expression in vitro, but does not prolong transgene expression in vivo. *Gene Ther.* **4**: 351–360.

Sheldrick P, Berthelot N. (1975). Inverted repetitions in the chromosome of herpes simplex virus. *Cold Spring Harbor Symp. Quant. Biol.* **39**: 667–6788.

Smith-Arica J, Morelli AE, Larregina AT, Smith J, Lowenstein PR, Castro MG. (2000) Cell-type specific and regulatable transgenesis in the adult brain: adenovirus-encoded combined transcriptional targeting and inducible transgene expression. *Mol. Therapy* (in press).

Sommerfelt MA. (1999) Retrovirus receptors. *J. Gen. Virol.* **80**: 3049–3064.

Song S, Wang Y, Bak SY *et al.* (1997) An HSV-1 vector containing the rat hydroxylase promoter enhances both long-term and cell-specific expression in the midbrain. *J. Neurochem.* **68**: 1792–1803.

Spaete RR, Frenkel N. (1982) The herpes simplex virus amplicon: a new eukaryotic defective-virus cloning-amplifying vector. *Cell* **30**: 295–304.

Spaete RR, Frenkel N. (1985) The herpes simplex virus amplicon: analyses of cis-acting replication functions. *Proc. Natl Acad. Sci. USA* **82**: 692–698.

Sparer TE, Tripp RA, Dillehay DI, Hermiston TW, Wols WSM, Gooding LR. (1996) The role of human adenovirus early region 3 proteins (gp19K, 10.4K, 14.5K and 14.7K) in a murine pneumonia model. *J. Virol.* **70**: 2431–2439.

Spergel JM, Hsu W, Akira S, Thimmappaya B, Kishimoto T, Chen-Kiang S. (1992) NF-IL6, a member of the C/EBP family, regulates E1A-responsive promoters in the absence of E1A. *J. Virol.* **66**: 1021–1030.

Srivastava A, Lusby EW, Berns KI. (1983) Nucleotide sequence and organization of the adeno-associated virus 2 genome. *J. Virol.* **45**: 555–564.

Steegenga WT, Riteco N, Bos JL. (1999) Infectivity of the early adenovirus proteins are important regulators of wild-type and DeltaE1B adenovirus replication in human cells. *Oncogene* **18**: 5032–5043.

Steiner I, Spivack JG, Lirette RP *et al.* (1989) Herpes simplex virus type 1 latency-associated transcripts are evidently not essential for latent infection. *EMBO J.* **8**: 505–511.

Stevens JG, Wagner EK, Devi-Rao GB, Cook ML, Feldman LT. (1987) RNA complementary to a herpes virus alpha gene mRNA is prominent in latently infected neurons. *Science* **235**: 1056–1059.

Stone D, David A, Bolognani F, Lowenstein PR, Castro MG. (2000) Viral vectors for gene delivery and gene therapy within the endocrine system. *J. Endocrinol.* **164**: 103–118.

Stow N, McMonagle E. (1983) Characterization of the TRs/IRs origin of DNA replication. *Virology* **130**: 427–438.

Stravropoulos TA, Strathdee CA. (1998) An enhanced packaging system for helper-dependent herpes simplex virus vectors. *J. Virol.* **72**: 7137–7143.

Summerford C, Samulski RJ. (1998) Membrane-associated heparan sulfate proteoglycan is a receptor for adeno-associated virus type 2 virions. *J. Virol.* **72**: 1438–1445.

Summerford C, Bartlett JS, Samulski RJ. (1999) AlphaVbeta5 integrin: a co-receptor for adeno-associated virus type 2 infection. *Nature Med.* **5**: 78–82.

Takeuchi Y, Cosset FL, Lachmann PJ, Okada H, Weiss RA, Collins MK. (1994) Type C retrovirus inactivation by human complement is determined by both the viral genome and the producer cells. *J. Virol.* **68**: 8001–8007.

Takeuchi Y, Porter CD, Strachan KM. (1996) Sensitisation of cells and retroviruses to human serum by (α1–3) galactosyltransferase. *Nature* **379**: 85–88.

Toda M, Martuza RL, Kojima H, Rabkin SD. (1998) In situ cancer vaccination: an IL-12 defective vector/replication-competent herpes simplex virus combination induces local and systemic antitumor activity. *J. Immunol.* **160**: 4457–4464.

Tratschin JD, West MH, Sandbank T, Carte BJ. (1984) A human parvovirus, adeno-associated virus, as a eucaryotic vector: transient expression and encapsidation of the procaryotic gene for chloramphenicol acetyltransferase. *Mol. Cell. Biol.* **4**: 2072–2081.

Tratschin JD, Miller IL, Smith MG, Carter BJ. (1985) Adeno-associated virus for high-frequency integration, expression, and rescue of genes in mammalian cells. *Mol. Cell. Biol.* **5**: 3251–3260.

Trempe JP, Carter BJ. (1988) Regulation of adeno-associated virus gene expression in 293 cells: control of mRNA abundance and translation. *J. Virol.* **62**: 68–74.

Tripathy SK, Black HN, Goldwasser E, Leiden JM. (1996) Immune responses to transgene-encoded proteins limit the stability of gene expression after injection of replication-defective adenovirus vectors. *Nature Med.* **2**: 545–550.

Uchida N, Sutton RE, Friera AM *et al* (1998) HIV, but not murine leukemia virus, vectors mediate high efficiency gene transfer into freshly isolated G0/G1 human hematopoietic stem cells. *Proc. Natl Acad. Sci. USA* **95**: 11939–11944.

Wang Q, Jia X-C, Finer MH. (1995) A packaging cell line for propagation of recombinant adenovirus vectors containing 2 lethal gene-region deletions. *Gene Ther.* **2**: 775–783.

Wang Q, Greenburg G, Bunch D, Farson D, Finer MH. (1997) Persistent transgene expression in mouse liver following in vivo gene transfer with a $E1^-/E4^-$ adenovirus vector. *Gene Ther.* **4**: 393–400.

Wang S, Vos JM. (1996) A hybrid herpesvirus infectious vector based on Epstein-Barr virus and herpes simplex virus type 1 for gene transfer into human cells in vitro and in vivo. *J. Virol.* **70**: 8422–8430.

Watkins SJ, Mesyanzhinov VV, Kurochkina LP, Hawkins RE. (1997) The 'adenobody' approach to viral targeting: specific and enhanced adenoviral gene delivery. *Gene Ther.* **4**: 1004–1012.

Weinberg JB, Matthews TJ, Cullen BR, Malim MH. (1991) Productive human immunodeficiency virus type 1 (HIV-1) infection of nonproliferating human monocytes. *J. Exp. Med.* **174**: 1477–1482.

Weiss RA. (1993) Cellular receptors and viral glycoproteins involved in retrovirus entry. *Retroviridae* **2**: 1–108.

Weiss RA, Tailor CS. (1995) Retrovirus receptors. *Cell* **82**: 531–533.

Weller S, Sparado A, Schaffer J, Murray A, Maxam A, Schaffer P. (1985) Cloning, sequencing and functional analysis of oriL, a herpes simplex virus type 1 origin of DNA replication. *Mol. Cell. Biol.* **5**: 930–942.

Welsh RM, Cooper NR, Jensen FC, Oldstone MBA. (1975) Human serum lyses RNA tumor viruses. *Nature* **257**: 612–614.

Wendtmer C-M, Nolte A, Mangold E *et al.* (1997) Gene transfer of costimulatory molecules B7-1 and B7-2 into multiple myeloma cells by recombinant adeno-associated virus enhances the cytolytic T cell response. *Gene Ther.* **4**: 726–735.

White SM, Renda M, Nam NY, *et al.* (1999) Lentivirus vectors using human and simian immunodeficiency virus elements. *J. Virol.* **73**: 2832–2840.

Wickham TJ, Mathias P, Cheresh DA, Nemerow GR. (1993) Integrins alpha v beta 3 and alpha v beta 5 promote adenovirus internalization but not virus attachment. *Cell* **73**: 309–319.

Wickham TJ, Carrion ME, Kovesdi I. (1995) Targeting of adenovirus penton base to new receptors through replacement of its RGD motif with other receptor-specific peptide motifs. *Gene Ther.* **2**: 750–756.

Wickham TJ, Roelvink PW, Brough DE, Kovesdi I. (1996a) Adenovirus targeted to heparan-containing receptors increases its gene delivery efficiency to multiple cell types. *Nature Biotechnol.* **14**: 1570–1573.

Wickham TJ, Segal DM, Roelvink PW, Carrion ME, Lizonova A, Lee GM, Kovesdi I. (1996b) Targeted adenovirus gene transfer to endothelial and smooth muscle cells using bispecific antibodies. *J. Virol.* **70**: 6831–6838.

Wickham TJ, Tzeng E, Shears LLN *et al.* (1997a) Increased in vitro and in vivo gene transfer by adenovirus vectors containing chimeric fiber proteins. *J. Virol.* **71**: 8221–8229.

Wickham TJ, Lee GM, Titus JA, Sconocchia G, Bakacs T, Kovesdi I, Segal DM. (1997b) Targeted adenovirus-mediated gene delivery to T cells via CD3 cells. *J. Virol.* **71**: 7663–7669.

Wildner O, Morris JC, Vahanian NN, Ford Jr H, Ramsey WJ, Blaese RM. (1999) Adenoviral vectors capable of HSVtk/GCV suicide gene therapy of cancer. *Gene Ther.* **6**: 57–62.

Wilkinson GWG, Darley RL, Lowenstein PR. (1994) Viral vectors for gene therapy. In: *From Genetics to Gene Transfer*, (ed. DS.Latchman) pp. 161–193. Oxford: Bios Scientific Publishers Ltd.

Wold WSM, Gooding LR. (1991) Region E3 of adenovirus: A cassette of genes involved in host immunosurveillance and virus-cell interactions. *Virology* **184**: 1–8.

Worgall S, Wolff G, Falck-Pedersen E, Crystal RG. (1997) Innate immune mechanisms dominate elimination of adenoviral vectors following in vivo administration. *Hum. Gene Ther.* **8**: 37–44.

Wu N, Watkins SC, Schaffer PA, De Luca NA. (1996) Prolonged gene expression and cell survival after infection by a herpes simplex virus mutant defective in the immediate-early genes encoding ICP4, ICP27, and ICP22. *J. Virol.* **70**: 6358–6369.

Xiao X, Li J, Samulski RJ. (1996) Efficient long-term gene transfer into muscle tissue of immunocompetent mice by adeno-associated virus vector. *J. Virol.* **70**: 8098–8108.

Xiao X, Li J, Samulski RJ. (1998) Production of high-titer recombinant adeno-associated virus vectors in the absence of helper adenovirus. *J. Virol.* **72**: 2224–2232.

Yang Q, Chen F, Trempe JP. (1994) Characterization of cell lines that inducibly express the adeno-associated virus Rep proteins. *J. Virol.* **68**: 4847–4856.

Yang Y, Ertl HCJ, Wilson JM. (1994a) MHC class I-restricted cytoxic T lymphocytes to viral antigens destroy hepatocytes in mice infected with E1-deleted recombinant adenoviruses. *Immunity* **1**: 433–442.

Yang Y, Nunes FA, Berencsi K, Gonczol E, Englehadt JF, Wilson JM. (1994b) Inactivation of E2a in recombinant adenoviruses imroves the prospect for Gene Therapy in cystic fibrosis. *Nature Gen.* **7**: 362–369.

Yang Y, Li Q, Ertl HCJ, Wilson JM. (1995a) Cellular and humoral immune responses to viral antigens create barriers to lung-directed gene therapy with recombinant adenoviruses. *J. Virol.* **69**: 2004–2015.

Yang Y, Trinchieri G, Wilson JM. (1995b) Recombinant IL-12 prevents formation of blocking IgA antibodies to recombinant adenoviruses and allows repeated gene therapy to mouse lung. *Nature Med.* **1**: 890–893.

Yu DC, Chen Y, Seng M, Dilley J, Henderson DR. (1999a) The addition of adenovirus type 5 region E3 enables calydon virus 787 to eliminate distant prostate tumor xenografts. *Cancer Res.* **59**: 4200–4203.

Yu DC, Sakamoto GT, Henderson DR. (1999b) Identification of the transcriptional regulatory sequences of human kallikrein 2 and their use in the construction of calydon virus 764, an attenuated replication competent adenovirus for prostate cancer therapy. *Cancer Res.* **59**: 1498–1504.

Zhou H, O'Neal W, Morral N, Beaudet AL. (1996) Development of a complementation cell line and a system for construction of adenoviruses with E1 and E2a deleted. *J. Virol.* **70**: 7030–7038.

Zhou SZ, Cooper S, Kang LY, Ruggieri L, Heimfeld S, Srivastava A, Broxmeyer HE. (1994)

Adeno-associated virus 2-mediated high efficiency gene transfer into immature and mature subsets of hematopoietic progenitor cells in human umbilical cord blood. *J. Exp. Med.* **179**: 1867–1875.

Zolotukhin S, Byrne BJ, Mason E *et al.* (1999) Recombinant adeno-associated virus purification using novel methods improves infectious titer and yield. *Gene Ther.* **6**: 973–985.

Zuffrey R, Nagy D, Mandel RJ, Naldini L, Trono D. (1997) Multiply attenuated lentiviral vector achieves efficient gene delivery in vivo. *Nature Biotechnol.* **15**: 871–875.

Zuffrey R, Dull T, Mandel RJ, Bukovsky A, Quiroz D, Naldini L, Trono D. (1998) Self-inactivating lentivirus vector for safe and efficient in vivo gene delivery. *J. Virol.* **72**: 9873–9880.

Zwaagstra JC, Ghiasi H, Slanina SM *et al.* (1990) Activity of herpes simplex virus type 1 latency associated transcript (LAT) promoter in neuron-derived cells: evidence for neuron specificity and for a large LAT transcript. *J. Virol.* **64**: 5019–5028.

Chapter 9

The safety of work with genetically modified viruses

Simon R. Warne

9.1 Introduction

When considering the safety of work with genetically modified viruses there tends to be a polarization of opinion around one or other of the following viewpoints. Some people argue that it is extremely unlikely that a novel pathogen might be created as a consequence of genetic modification and that current controls are, if anything, too stringent. This viewpoint is based on the assertion that in most cases the genetically modified viruses used in the laboratory bear little resemblance to the naturally occurring pathogenic strains that are the scourge of mankind. This is because the vast majority of the virus projects involve the use of strains whose replicative capacity has been disabled. Moreover, even in cases where a non-disabled virus is used there is evidence to show that the presence of foreign DNA represents a replicative burden, which makes the modified virus less fit than the wild-type. As a consequence foreign inserts are often rapidly deleted by selective pressures. Proponents of such arguments can also point to the fact that in the 25 years that genetic modification techniques have been applied there are no documented cases of serious ill health or environmental harm. Taken together these arguments form a reasonable case but they should not obscure the existence of an alternative view. This is that we should proceed with great caution and that our current approach to safety may not always make enough allowance for uncertainties in our knowledge. This viewpoint is supported by the fact that there are already examples where naturally occurring viruses with drastically altered pathogenic properties have arisen due to the operation of the processes by which genetic material from related viruses is recombined or reassorted. Indeed, epidemic strains have emerged as a consequence of relatively minor changes at the genetic level (Ryan *et al.*, 1990; Webster *et al.*, 1986). Therefore, it can be argued that it is not beyond the realms of possibility that strains with novel pathogenic properties might be generated, either as a direct consequence of genetic modification or as a consequence of a genetically modified virus being accidentally released from containment and recombining with a related non-modified virus. It is noteworthy that we are already starting to see examples where artificially introduced modifications are enhancing particular pathogenic traits, if not the overall pathogenicity and fitness of a modified virus (Alcami and Smith, 1992; Baulcombe *et al.*, 1993; Ding *et al.*, 1996; Perng *et al.*, 1999).

Genetically Engineered Viruses: Development and Applications, edited by C.J.A. Ring and E.D. Blair

As someone who is closely involved in regulatory issues, my work involves an attempt to straddle the ground between these two polarized viewpoints. It is, therefore, from this perspective that I will outline my personal viewpoint as to the key safety issues relating to virus work. Full implementation of the measures outlined should go a long way to ensuring that the existing exemplary safety record is maintained and that serious incidents are avoided. The importance of this cannot be overstated since if an accident were to occur, that resulted in even a limited spread of a genetically engineered pathogen from the laboratory into the community, this would have major implications for the perception of the technology.

In this chapter I will draw on experience gained during my work with the UK Health and Safety Executive as a specialist in the safety of genetic modification work. During my work I regularly inspect laboratories undertaking work with genetically modified viruses and I am also involved in the technical review of risk assessments. I will subdivide the safety issues relating to virus work into three broad areas: physical containment, experimental design and risk assessment, and safety management issues. I will then discuss the regulatory framework covering safety and related matters, including a discussion of the extent to which ethical issues are considered, and then finally bring the chapter to a close with some concluding points aimed at stimulating further discussion.

9.2 Physical containment

9.2.1 The four containment levels

Although there are variations in the legislation around the world covering the safety of work with viruses, there is broad agreement over the notion that it is helpful to rank micro-organisms that are hazardous to humans into four hazard groups. In general, micro-organisms in hazard group 1 can be regarded as safe, those in hazard group 2 cause diseases which can either be treated or are not normally fatal, those in hazard group 3 cause diseases that are often fatal and those in hazard group 4 pose a risk of causing widespread deaths in the general population. In parallel with the assignment of micro-organisms to hazard groups, four levels of containment or biosafety levels have been designated. These range from level 1, where there are minimal measures to prevent the micro-organism infecting those who handle it, to level 4, where there is complete physical separation of the workers from the micro-organisms, either by the use of enclosed equipment accessed by glove ports or by equipping the workers with protective suits fed by air lines. Thus once a micro-organism has been assigned to a hazard group this forms the basis for the assignment of appropriate containment measures, with hazard group 2 micro-organisms being handled at containment level 2 and hazard group 3 micro-organisms at containment level 3, etc.

As stated above, whilst there is broad agreement around the world over the requirements for the four containment levels, there are minor variations at a national level. In view of this and for reasons of brevity it is not possible to go into detail over the exact requirements and standards for the four containment levels. Therefore, having cited the UK requirements as an example (ACGM, 2000; HSE, 2000), this section will be restricted to some more general thoughts.

9.2.2 Environmental protection

It should be noted that the containment levels described above were drawn up with a view to worker protection and not with environmental considerations specifically in mind. Therefore, if work is carried out on viruses that represent a hazard to either fauna or flora assignment to one of the four containment levels should only be seen as the start of the risk assessment process. Detailed consideration should be given to whether, once the appropriate measures to protect human health have been put in place, there remains a risk to the environment. A key aspect of this process will be to consider whether the arrangements for waste disposal are adequate. Thus it is possible to envisage that a laboratory used for work with an environmental pathogen might correspond in many aspects of physical containment to containment level 1 but have in place additional controls to protect the environment, such as the restriction of access by a card-key system and the requirement to inactivate all waste in an autoclave located within the facility.

One key part of any environmental assessment will be to determine whether the virus represents a hazard to animals or plants used in agriculture. All of the main developed countries have legislation controlling work with such viruses and so it is advisable to contact the relevant Government Department for advice. There will generally be national lists of pathogens of plants and animals used in agriculture and they may be divided into hazard groups with corresponding containment levels, as described above for human pathogens.

9.2.3 Examples of containment issues that are often overlooked

The minimum requirements set down for the four containment levels described in Section 9.2.1 should only be seen as a framework on which to base local codes of practice for containment. Although some of the requirements are self-explanatory (e.g. the requirements for benches to be impervious to water and resistant to chemicals and for an observation window to be present, so that laboratory workers can be seen), others need to be supplemented by detailed local guidance. In order to illustrate the kinds of matters which need to be covered in local codes of practice this section will outline a few examples of containment issues that are often overlooked.

9.2.3.1 Use of disinfectants. At first sight the requirement that a containment facility should have specified disinfection procedures would seem straightforward. However, in reality the issue of disinfection is beset with complications. When choosing a disinfectant there are a number of factors which need to be considered. These include the efficacy of the disinfectant against the biological agents in use, the extent to which the disinfectant is inactivated by the presence of organic matter and the extent to which the activity of the disinfectant falls off with time when a solution has been made up from stock. Overall there would not appear to be any one disinfectant that should be regarded as suitable for all purposes with all types of viruses.

A prime example of the limitations on the efficacy of some disinfectants is that phenolic disinfectants are not fully active against non-enveloped viruses. Thus, even though phenolic disinfectants are highly efficacious against some important biological agents (e.g. *Mycobacterium tuberculosis*) and have the advantage that they function reasonably well in the presence of organic matter, their use in virology laboratories requires careful consideration. Hypochlorite disinfectants on the other hand exhibit a high level of efficacy against viruses but suffer from the disadvantage of being considerably inactivated by organic matter.

In addition to the well-established disinfectants, such as hypochlorites and phenolics, increasingly disinfectants that are a complex formulation of chemicals are also being used. For example, one disinfectant in common usage is a stabilized blend of peroxygen compounds, surfactant, organic acids and an inorganic buffer system. Whilst this disinfectant has some advantages it is important that users read the manufacturers instructions on matters such as the treatment of spillages and the decontamination of cell culture media. For example, it should be noted that the directions from the manufacturer state that spillages of virus in cell culture media should be treated with the powdered stock of disinfectant. In spite of this it is not uncommon to see local rules for disinfection stating that a solution of the disinfectant at the normal working concentration would be used to treat a spillage. Given that spillage treatment is the most likely emergency situation to arise in a laboratory, the inappropriate application of a disinfectant for this purpose represents a serious shortcoming.

9.2.3.2 Siting and use of safety cabinets. At containment levels 2 and 3 one of the most important safety features of a laboratory is the use of microbiological safety cabinets. The vast majority of safety cabinets in use within virology laboratories are of class II design. Such cabinets are popular because they provide not only operator protection but also protect any biological reagents that are being handled within them from airborne contamination. Class II cabinets are open-fronted and are designed so that there is an inflow of air at the work aperture which discourages the emission of any harmful aerosols.

In view of the widespread use of class II cabinets, it is surprising that many laboratory managers are unaware of the importance of siting them in an appropriate location. The airflows within these cabinets are complex and can be readily disturbed by outside perturbations. For this reason, considerable care needs to be taken in choosing a suitable location for a class II cabinet within a laboratory. It is important that there should be no air currents across the front of the cabinet to interfere with protective airflows that are generated within it. Such interference can be generated by changes in air pressure when a door is opened, by movement of staff across the front of the cabinet or by air currents associated with a general room ventilation system. Even the cooling effect of a closed window immediately adjacent to a cabinet can generate a side draught which interferes with the performance of a cabinet.

Given the above, it is important that an operator protection test is carried out as part of the commissioning process when a cabinet is first installed. One relatively straightforward way of undertaking these tests is to use the potassium iodide method

(Clark and Goff, 1981). This involves using special test equipment to generate a fine mist of potassium iodide droplets within the cabinet as a challenge aerosol. The extent to which potassium iodide particles are able to move against the inward airflow generated by the cabinet and pass out of the work aperture is then measured using a bank of air samplers mounted outside the cabinet.

It is also important that staff are trained in the correct use of safety cabinets (Collins, 1993). For example, since the movement of hands into and out of the work aperture disrupts the protective airflows it is important that the number of such movements is restricted as far as possible. This means that it is certainly not good practice to place a table immediately outside the work aperture and repeatedly take experimental materials into and out of the cabinet during the course of work. Instead, all the experimental materials and equipment that are required for a particular laboratory procedure should be collected together and placed within the cabinet before commencing work. The operator should also work well within the cabinet and not near to the front edge. When using a class II cabinet, it is considered to be best practice to keep the hands and arms just above the bottom grill because this is the part of the work aperture where the protective airflows are strongest.

9.2.3.3 Room fumigation. Many projects undertaken at containment level 3 and all projects undertaken at containment level 4 require a laboratory that is sealable for fumigation. Room fumigations are generally carried out using formaldehyde, although there has been recent interest in the use of alternative fumigants such as hydrogen peroxide. Fumigation is a complex procedure and the methodology used needs to be well documented. If room fumigation is to be used as part of the procedure for decontaminating a laboratory following a spillage, it is important that the whole process has been thought through and preferably practised in advance. For a start it should be clear to everyone involved what kind of spillage, in terms of size and location, would necessitate room fumigation. Then the procedure for fumigation should be well defined, since it is imperative that the steps to be taken should be known to all concerned and not improvised at the time of the emergency.

It should be recognized that fumigation is primarily a way of inactivating small deposits of infective agent formed as a consequence of the splashes and aerosols generated during a spillage. It should not be seen as a way of inactivating the bulk of liquid within the spill itself. Thus fumigation should be undertaken alongside measures to treat the main spill with an appropriate disinfectant. For example a logical sequence of events would be to evacuate the laboratory, then fumigate, then disinfect the bulk spillage and remove all debris and then possibly fumigate again, if this is felt to be necessary. In cases where the infective agent is transmissible by the airborne route, such a procedure would be likely to necessitate the use of respiratory protective equipment during the process of disinfecting the spillage. If so staff would need to be trained in the use of such equipment and it would have to be maintained on a regular basis. Common problems that can cause fumigation to be ineffective are a failure to generate a sufficient concentration of fumigant when fumigating a large room and a failure to generate a high enough humidity when fumigating with formaldehyde. These points should therefore be addressed at the outset.

9.3 Experimental design and risk assessment

9.3.1 Use of disabled vectors

In the vast majority of cases a high level of safety can be built into a project by the choice of an appropriate disabled vector at the outset. As can be seen from the preceding chapters, disabled vectors have now been produced from a wide range of human viruses. Moreover, as genetic technology has developed the safety features that have been incorporated into the design of viral vectors have become more sophisticated. The strategies by which safety features have been incorporated into vector design include the deletion of essential replication or structural genes and the construction of corresponding packaging lines (Fallaux *et al.*, 1998; Miller and Buttimore, 1986; Speck *et al.*, 1996), the formation of vectors which are only activated by an *in vitro* enzymatic cleavage step (Berglund *et al.*, 1993), the incorporation of mutations which make the vector acutely sensitive to UV light or the host immune system, and the incorporation of expression signals or other genetic elements which would be non-functional following an accidental infection (Bilbao *et al.*, 1999).

The progress that has been made in the development of safe vectors is perhaps best illustrated by improvements that have been made, over the years, to retroviral gene transfer systems. Vectors based on mouse leukaemia viruses have been developed by forming replication-defective derivatives of the full-length virus from which the viral *gag, pol* and *env* genes have been deleted. These vectors require the viral genes to be provided in *trans* by a packaging cell line in order for the vector to be packaged into particles capable of gene transfer, by a single round of abortive infection. The first packaging cell lines that were constructed contained a provirus that had a mutation in its packaging sequence thereby preventing the packaging of its own transcript into virus particles (Mann *et al.*, 1983). Although useful for some experimental purposes, these 'first generation' packaging lines were unsatisfactory because viral stocks were consistently contaminated by replication-competent retroviruses (RCR), formed by recombination between the helper viral sequences in the packaging cells and the vector. With these 'first generation' packaging lines a single homologous recombination event was sufficient to generate RCR and as a consequence wild-type virus could even be found in harvests made following transient transfections. There was thus a need to develop 'second generation' packaging lines, such as PA317, in which the helper provirus not only lacked the packaging signal but also had the 3′ LTR deleted (Miller and Buttimore, 1986). Thus with these packaging lines two separate recombination events were required to generate RCR. However, whilst these 'second generation' lines were much improved, they still generated significant amounts of RCR. This shortcoming led to the development of 'third generation' packaging lines in which, rather than have the helper sequences on a single segment of DNA, the helper genes *gag*, *pol* and *env* were introduced as separate transcriptional units. In such 'split-function' packaging lines, such as GP+envAM12 (Markowitz *et al.*, 1988), at least three recombination events are required to generate RCR. Whilst there is some evidence that these third generation packaging lines can generate RCR on extremely rare occasions (Chong *et al.*, 1998) they have generally been found to provide a satisfactory level of safety. This is particularly the case when they are used in conjunction with vectors that have been manipulated to minimize, or even eliminate, sequence overlap with the helper segments.

The development of vectors based on other viruses has proceeded along similar lines and we are now at the stage where a range of safe vector systems are available. The use of such systems is an important aspect of ensuring the safety of virus work. Moreover, the selection of a safe vector system at the outset of a project can save time, effort and expense in the long term. For example, it has been known for a viral construct to be made in a particular way, because of short-term expediencies, only for it to be realized at a later stage that if a different construct had been made the containment requirements for future experiments would be much less stringent.

9.3.2 Choice of cloning site

In most cases disabled vector systems are designed in such a way that foreign genes are inserted at a position that in the unmodified parental virus would have been occupied by an essential replication or structural gene. Thus as a matter of experimental convenience foreign genes tend to be inserted within disabled vectors at the site of the disabling mutation. What is not sometimes realized is that the cloning of foreign genes in this way provides an important safety factor.

The safety implications of the choice of cloning site become apparent when the possibility of a genetically modified virus recombining with the unmodified parental virus is considered. Such a possibility cannot always be easily dismissed as there are a number of possible scenarios whereby recombination could occur. Thus, it could occur as the result of cross contamination with a wild-type virus being used in the same laboratory, or by recombination with viral sequences in a packaging cell line or following infection of a laboratory worker already infected with the wild-type virus (particularly in the case of a virus that is present in the general population in a latent or persistent form).

Once the problem of recombination is recognized the possibility that needs to be guarded against is the inadvertent formation of a replication-competent virus carrying the foreign gene. In view of this it is clearly preferable if the foreign gene is inserted at the main site of disablement. Construction of a recombinant virus in this way means that, in the event of a disabled virus becoming replication competent following a homologous recombination event involving the flanking sequences on either side of the disabling mutation, the end product of the recombination would be a wild-type virus lacking the foreign gene. Thus whilst the recombinant would be replication competent at least the foreign sequences in the disabled vector would be biologically contained.

From a safety viewpoint the site chosen to insert a foreign gene into a vector becomes particularly important where there is a known mechanism whereby the gene could give rise to a harmful phenotype, if it were to become incorporated into a replication-competent virus. In cases where two foreign sequences are being inserted at two separate locations within a vector care should be taken to ensure that, at least the sequence which would be most harmful in the context of a fully replication-competent virus is inserted at a site of disablement, even if the other sequence is not.

9.3.3 Awareness of the key ways in which novel hazards may arise

When genetic modification technology was first developed initial safety concerns in relation to viruses were focused on the cloning of DNA from oncogenic animal viruses into bacterial plasmids (Berg *et al.*, 1974). Although there is now much less concern about this specific type of experiment, it is still common for people's consideration of the risks of new viral projects to be focused on similar types of scenario for hazards arising. In these scenarios the focus of concern is novel hazards that might arise due to the direct effects of an expressed gene product, with the vector itself being seen as little more than a gene delivery vehicle. Examples of such scenarios are situations where the sequence inserted into a virus vector is a toxin, a cytokine or an allergen. Although hazards of this kind need careful consideration there are other issues of at least equal importance that are sometimes not fully considered. These involve recognizing that many viral vectors retain some of the harmful properties of the wild-type viruses from which they were derived and that it is not acceptable to assume that this inherent pathogenicity will necessarily be unaltered following genetic modification. Thus it is important to consider whether the sequences that are inserted into a viral vector could act alongside the existing pathogenic traits of a viral vector and so contribute to an increase in its overall pathogenicity. Examples of ways in which a foreign sequence may act alongside the existing pathogenic traits of a viral vector, and so exacerbate its inherent pathogenicity, include situations where there may be an alteration of either tissue tropism or the ability of the virus to evade the host defence mechanisms.

The safety implications of alterations in tissue tropism are topical at the moment and it is, therefore, worthwhile discussing this issue in more detail. The tissue tropism of a virus, i.e. the tendency of a virus to colonize a particular cell type and thereby become confined to particular tissues in the host species, is one important factor in determining the type and severity of its pathological symptoms. Therefore, in terms of safety the concern over an alteration in tissue tropism relates to the possibility that a modified virus might be more readily transmitted or be capable of spreading more widely within the body of an infected individual. Given this possibility, it is of considerable relevance that there is now a good deal of work underway which is aimed at deliberately altering components of viruses which determine tissue tropism. The main impetus for this work is the design of improved vectors for use in gene therapy but there is also some work which is aimed at studying the fundamental biology of the infective process. Most of the work on developing vectors for gene therapy involves the further modification of disabled vectors and so, as long as the issue of recombination with wild-type virus is satisfactorily addressed, it does not tend to raise major safety implications. Of more concern is work on the fundamental biology of viruses where an alteration in tissue tropism may be an unforeseen consequence of the work.

To illustrate the points in the above paragraph a project that was recently undertaken will now be outlined as an example. The discussion of this example should not be taken as an indication that, as it turned out, the project was necessarily hazardous. Rather the example is raised to illustrate the kind of project where an alteration of tissue tropism could have arisen as a consequence of the work, even though it was undertaken for other reasons. The project involved the formation of a hybrid virus in which the envelope glycoproteins of measles virus (MV) were replaced by a complete or altered G protein from vesicular stomatitis virus (VSV) (Spielhofer *et al.*, 1998). Thus the hybrid viruses were essentially derivatives of MV in which the normal outer envelope had been replaced with that of VSV. This work was undertaken at containment level 2. The aim was primarily to demonstrate the feasibility of constructing MV variants bearing envelopes from different viruses for use as vaccines but it is possible that the modified viruses could have been endowed with a significantly altered tissue tropism. This is because MV and VSV produce different symptoms and affect different tissues within an infected individual. Moreover, an additional factor which might be regarded as militating against the use of the VSV envelope gene for this kind of work is the fact that VSV has an extremely broad host range, extending from nearly all mammals to insects, and is noted for its ability to grow rapidly in culture to high titres. Now that the work has progressed there may well be good evidence to show that the containment was adequate but this does not detract from the general concern relating to this kind of project. This is that, given the uncertainties that exist when a project of this type is started, it is difficult to ensure that a large enough safety margin is built in to allow for a worst case scenario in which an unexpected alteration in tropism leads to increased pathogenicity.

Before concluding this section it is worthwhile noting that, whilst the main focus of a risk assessment tends to be on making a judgement as to the likelihood that the modified virus might cause harm to a laboratory worker who became accidentally infected, another factor which is also relevant is the fitness of the modified virus (i.e. its ability to establish infections *in vivo* and spread and replicate outside of the controlled laboratory environment without deleting the foreign sequences). Indeed, it can be argued that when considering the worst case scenario, of the potential of a modified virus to cause an outbreak of infection in the general community, the fitness of the virus will be at least as important as the severity of the symptoms suffered by any laboratory worker who might initially become infected. However, whilst fitness is clearly a relevant issue to be considered in a risk assessment, there is currently little data on which to predict the fitness of a particular construct. Therefore, whilst it is common for risk assessments to refer to the broad generalization that modified viruses will tend to have a greatly reduced fitness, as compared to their naturally occurring counterparts, it is debatable how much weight should be given to such blanket assertions. Clearly once a particular construct has been shown to be unstable then viral fitness can be used as a major argument in any risk assessment of future work with that construct. However, until a reduced viral fitness has actually been demonstrated, any arguments to justify safety on the basis of predictions about viral fitness should be regarded as tenuous.

A number of projects in a similar vein to the MV/VSV hybrid virus described above have already been undertaken or are in progress. It may turn out that all of these current projects generate viruses that, for reasons including the issue of viral fitness, are no more hazardous than the parental virus. There can, however, be no guarantee that this will always be the case, particularly when it is realized that as our knowledge increases we will become more astute in designing functional hybrid viruses from parental viruses that are not closely related.

9.3.4 Risk assessment forms

In order to avoid project leaders being faced with a blank piece of paper when they sit down to undertake the risk assessment of new work, it is extremely useful if some prior thought has been given to the design of a risk assessment form. The aim of designing such a form should be to provide a structure that will assist the proposers of new work in structuring their thought processes as they go through the assessment. To this end, the form should contain specific questions that address properties of the vector and final construct which are important to the evaluation of safety. Some of the key issues which need to be addressed in the risk assessment of viral projects are listed in the following bullet points:

- Is the viral vector disabled?
- Where has the inserted gene been cloned? Has it been inserted at the site of vector disablement?
- What are the chances of the disabled vector reverting to a replication competent derivative either by recombination with helper sequences in a packaging cell line or by recombination with wild-type virus?
- Does the inserted sequence encode a gene that either predisposes cells to tumour formation or whose protein product is inherently harmful e.g. an oncogene, a toxin, an allergen, a modified prion or an inflammatory cytokine?
- Does the insert encode sequences that might alter the existing pathogenic properties of the viral vector e.g. does the insert encode a gene that may assist the virus in evading the host immune system or does the insert encode the envelope gene of a related virus which could confer on the modified virus an altered host range or tissue tropism?
- Are there potentially harmful sequences in the final construct that could be transferred to the wild-type virus by homologous recombination?
- Will the work involve any non-standard operations that could increase the chances of operator exposure e.g. growth of high titre virus, use of needles or other laboratory sharps, inoculation of animals likely to excrete virus?

Risk assessment forms should be tailored to meet the needs of individual research establishments. Thus whilst the set of questions presented above are aimed specifically at the risk assessment of research on human viruses, it would be possible to adapt the questions so that they covered work where the main concern was the possibility of environmental harm. Similarly if a centre is involved in gene therapy the safety issues that are relevant to this type of work should be specifically raised on the risk assessment form.

Another factor which needs to be borne in mind when designing a risk assessment form is that if care is not taken, proposers of new projects may be required to undertake considerable amounts of paperwork regardless of the type of work being proposed. One way in which this issue can be addressed is to have a form that comes in two parts. The first part could then require a project proposer to prepare one or two paragraphs outlining the proposed work, and the scientific goals which it is seeking to address, whilst the second part of the form could cover the more detailed risk assessment issues of the type listed above. It is then possible that, as experience in the process of drawing up and reviewing risk assessments develops within a particular research institute, a stage may be reached where it is acceptable for some projects to be risk assessed on the basis of the completion of a project outline alone. In such a case the proposers would only have to complete the first part of the form and would thus be saved unnecessary paperwork. For example, there would be justification for handling the risk assessment of projects involving the expression of the thymidine kinase gene from a range of disabled murine retroviral vectors or the insertion of marker genes into an E1-deleted adenovirus vector using such a streamlined procedure.

9.4 Competency of staff and safety management issues

9.4.1 Training

In discussing the training of staff in safety issues it is helpful to subdivide the subject into two distinct areas. Perhaps the most obvious aspect of training is the need to ensure that staff are fully aware of the safety aspects of the work procedures and local rules they follow during their laboratory work. However, no less important is the process of ensuring that staff are fully aware of the hazards associated with their work and of the basic biology of the viral vectors they use. The former aspect of training will be covered in the remainder of this section and the latter issue will be covered in Section 9.4.2.

Whilst formal group training courses are of value when training workers in safety procedures, it is likely that the most important training will take place 'on the job'. This will involve a number of points. It is important that workers familiarize themselves with written protocols and local safety rules before starting laboratory work. The training process is also greatly assisted if new workers can first shadow an experienced worker and then go on to an intermediate stage where they undertake the procedure whilst being supervised. Like all aspects of training, the extent to which such familiarization processes should be formalized will depend on the level of hazard and the previous experience of an individual worker. For example, whilst at containment level 1 it is quite acceptable for new workers to be working independently within a matter of days, training at containment level 3 should be much more formalized. For example it would be appropriate when training a new worker at containment level 3 for them to spend several months training before they are allowed to work unsupervised. As part of this training they might perhaps first undertake work at containment level 2 and then go on to a stage where they are closely supervised for several weeks at containment level 3.

A useful way of keeping training issues to the fore is to formalize the process by which the supervisor of a new worker makes the final judgement as to when they are competent to work unsupervised. Whilst some formal checks of competence are appropriate in all laboratories, this issue needs careful consideration at the highest levels of containment. The process of deciding that a worker is competent to work unsupervised can be facilitated by preparing a form for the supervisor to sign off at the appropriate time. In order to make the process fully transparent it would be helpful if this form contained a series of tick boxes for the supervisor to indicate that the new worker had demonstrated a satisfactory knowledge of key safety issues, such as emergency procedures and the handling of waste.

There are particular difficulties in providing a suitable training regime for a worker who is starting work in a new laboratory having had previous experience in a quite separate research institution. In such cases the signing off process outlined above can provide a useful mechanism for aiding decisions.

9.4.2 Staff awareness of hazards

Staff should be fully aware of the hazards associated with their work and any biological containment that has been built into the experimental design, for example by the use of disabled vectors. Whilst this may appear to some to be self-evident it is surprising how often shortcomings relating to this issue come to light. It is a particular problem in situations where viral vectors are being used in research establishments that have no background in classical virology. In such situations there is the possibility that a viral vector will be seen merely as another laboratory reagent rather than as a derivative of a pathogen which could in some circumstances regain some of its pathogenic characteristics. It can thus be argued that all laboratory workers should have at least a rudimentary knowledge of the basic biology of the viral vectors they are using. This point is illustrated in the following example which, whilst it is to some extent hypothetical, is nevertheless based on experience gained during the inspection of laboratories undertaking genetic modification work.

In laboratories where a range of retroviral vectors are in use it is not unknown to find that some individual workers lack a basic knowledge of the nature of the disabling mutations which are present in the particular vectors they are using. It is thus possible to envisage a situation where the local safety committee may have decided that in order to express a particular harmful gene it was important to utilize a ‘split-function’ packaging line, such as GP+envAM12 (see Section 9.3.1), but where the laboratory worker actually undertaking the work had no knowledge of the significance in safety terms of the choice of vector. In such a situation it does not take a tremendous leap of imagination to suggest that the laboratory worker might encounter problems undertaking a particular transfection and turn to a colleague in another laboratory for advice. This colleague might have experience of using the packaging line PA317 (see Section 9.3.1) and suggest that this be used as an alternative, based on the fact that it generates virus at a 40-fold higher titre (Miller, 1997). Thus, acting only in the interests of progressing the science, a laboratory worker might unknowingly remove an important safety factor in the experimental design.

9.4.3 Safety management

Whilst the above two sections have focused on specific issues in which workers should receive training, it needs to be recognized that there are other more subtle factors which determine the mind-set with which safety issues are actually approached. It is one thing for a worker to be aware of the requirement to undertake a procedure a particular way but it is another to be able to move on from this and be confident that the worker will invariably undertake the procedure in the prescribed way. In order to address the additional factors which determine the state of mind with which workers are likely to address safety issues it is necessary to consider the management of safety.

Clearly one way in which safety can be managed is by very rigid management controls in a hierarchical structure. Whilst such a management system may be achieved in some industrial applications of biotechnology, it is clearly far from the norm in most research environments, particularly in academia. Instead the heads of research establishment tend, where possible, to free staff from excessive managerial controls and give them considerable freedom to pursue their scientific goals. In doing this they tend to rely on the experience and expertise of their senior staff as a key safeguard. There are of course problems in managing safety in such an informal environment and in order for it to work successfully additional measures must be taken to replace the conventional managerial controls. Where there is less by way of formal structured control through job descriptions, management by objectives, rigid accountability etc. there must be an even greater emphasis on effective monitoring by locally organized safety inspections, competence assurance and the establishment of a positive safety culture.

Some readers may not have previously come across the concept of safety culture and so it may be helpful to explain this term more fully. In essence it is the prevailing attitude of the staff of an organization to safety issues and represents the sum total of the beliefs and habitual responses of the individuals within that organization. In other words it can be regarded as a measure of what would be regarded as acceptable behaviour in a particular organization when dealing with a safety issue. Organizations with a positive safety culture are characterized by good communications between all levels of staff on safety issues, with information being exchanged without undue emphasis on the assignment of blame, and by a shared perception of the importance of safety.

As has already been said, it would be unduly onerous to expect a full and regimented implementation of formal management controls in every laboratory undertaking genetic modification work, regardless of the hazards involved. Nevertheless, some more formal measures for safety management do need to be applied when undertaking high-risk work at containment levels 3 and 4. Indeed it is fair to say that one of the most important aspects of high containment work is the mind-set of the workers involved. Thus at high containment the safety aspects of experimental protocols should be well documented and agreed procedures rigorously adhered to. This should ensure that safety procedures are thought through in advance, when there is time to consider various options in a careful and systematic way, rather than improvised by a particular worker when, for example, an accident occurs. Another important aspect of safety at high containment should be the establishment of clear lines of accountability in which the responsibilities of the senior scientific staff are identified alongside the limits of the roles of safety advisers and safety officers.

9.5 The regulatory framework and ethical issues

9.5.1 Legislation covering genetically modified viruses

All of the main developed countries have some form of legislative control over the safety of work with genetically modified organisms. For example, in the European Union there is Directive 98/81/EC covering the contained use of genetically modified micro-organisms while in the USA the NIH Guidelines provide the framework for control. Whilst such regulations covering the general use of genetic modification techniques tend to be well publicized it should be noted that, depending on the nature of the work, there may also be other legislation covering safety-related issues which must be complied with. Thus depending on the structure of the legislation in an individual country there may be additional legislation covering the following: the infection of laboratory animals with genetically modified viruses, the growth of genetically modified viruses for the purpose of manufacturing vaccines or medicines, the inoculation of genetically modified viruses into patients during gene therapy, the construction of genetically modified derivatives of plant pests or pathogens of domesticated animals and the treatment of waste from the laboratory facilities. As an example of the way in which work can be covered by a number of different pieces of legislation it is instructive to consider the legislative controls that would apply in the UK to work on the experimental infection of transgenic pigs with genetically modified derivatives of swine vesicular disease virus. In addition to general health and safety legislation such work would be specifically covered by the Genetically Modified Organisms (Contained Use) Regulations (2000), the Genetically Modified Organisms (Risk Assessment) (Records and Exemptions) Regulations (1996), the Environmental Protection Act (1990), the Animals (Scientific Procedures) Act (1986) and the Specified Animal Pathogens Order (1998).

Taking the UK as an example it is interesting to note that, of the several pieces of legislation and guidance specifically covering work with genetically modified viruses, ethical issues are only covered in two cases. These are under the Animals (Scientific Procedures) Act (1986) and during the review of human gene therapy proposals by the Gene Therapy Advisory Committee (GTAC). In the former case, the Act recognizes that scientific procedures should only be permitted when the benefits that the work is likely to bring outweigh any pain or stress that the animals may experience. In the latter case, GTAC are asked to consider the acceptability of proposals, on ethical grounds, taking into account the scientific merits of the proposals and the potential benefits and risks.

9.5.2 The case for risk–benefit analysis to be incorporated into the legislation covering safety

As can be seen from the above, the safety legislation covering *in vitro* laboratory research with genetically modified viruses does not, in general, make any provision for consideration of ethical issues, such as a risk–benefit analysis of proposed work. Thus the general spirit of the legislation tends to be that the only situation in which the regulatory authorities can withhold their approval for a proposed project is when they judge that the proposed containment is inadequate. This means that during the safety evaluation there is currently no question of asking whether the potential benefits of the work and the quality of the science outweigh the residual risks at the proposed level of containment. Indeed many would argue strongly that, since the quality and applicability of the science will have been closely evaluated as part of the process of seeking funding, it is only right that these issues are not addressed again during the safety evaluation. However, as will be outlined below there are arguments to back an alternative view.

The arguments to support the idea of risk–benefit analysis forming part of safety legislation are based on the fact that, even if the very best advice on the procedures for risk assessment is followed and the most knowledgeable scientists are called upon for peer review, there is no way in which the assignment of containment to work with genetically modified viruses can be regarded to be an exact science. Although judgements about containment generally err on the side of caution, the possibility that on rare occasions a modified virus will turn out to be significantly more pathogenic than would be predicted from the available evidence can never be completely ruled out. Experience suggests that concerns of this nature tend to be more acute when considering work on replication-competent viruses that are already to some extent pathogenic prior to modification. Therefore, it can be argued that the more projects of this type that are undertaken, the greater are the chances of safety being compromised due to a misjudgement in the risk assessment process. There is thus a case for keeping close control over such projects where there are significant hazards. Therefore, those who advocate a cautious approach to safety issues are starting to suggest that in such cases the safety evaluation should include some form of risk–benefit analysis, with only those projects offering major benefits being approved.

Any proposal to include risk–benefit analysis in safety evaluation would, as has been said above, be regarded by many in the field as unacceptable. It is, therefore, worth noting that it can be argued that the listing of pathogens in hazard groups already involves an element of pragmatism. For example when HIV first came to attention there was no treatment or prophylaxis available and so it would seem that its designation to hazard group 3, rather than 4, owed as much to the practicalities of getting a major research programme underway, as to a strict interpretation of the criteria. Although there is currently no indication that the position on the applicability of risk–benefit will change in the near future, it would seem only to be a matter of time before a debate on this issue comes into the public arena.

9.6 Discussion and conclusions

In the above sections, a range of issues that are relevant to safety in laboratories undertaking work on genetically modified viruses have been considered. Deciding on priorities amongst these various aspects of safety is to some extent a matter of opinion. Nevertheless, there is a case to say that, whilst it is clearly a key issue to ensure that standards of physical containment are satisfactory, this aspect should not be given overriding prominence. If anything, more emphasis should be placed on maintaining the quality of risk assessments and ensuring that staff view issues relating to safety in a positive way. In other words, whilst it is important to be satisfied that laboratories fully comply with the specifications of the various containment levels and that safety equipment is adequately maintained, perhaps the most crucial issue in ensuring high standards of biosafety is the competence and commitment of staff with regard to safety issues.

In order to reinforce this conclusion, it may be useful at this stage to discuss a specific example which illustrates the problems that can arise if there is an over-reliance on physical containment with insufficient attention being paid to issues relating to the management of safety. This example relates to the Australian Animal Health Laboratory (AAHL). This facility was set up with the objective of providing Australia with the capability to diagnose exotic animal diseases. The decision to build this facility met opposition right from the planning stage because its remit was, not only to study disease agents that already existed in Australia, but also to import, for the purposes of research, disease agents that were non-indigenous to the country. In order to counteract this adverse publicity, it was made known that the facility would be the most advanced microbiological containment facility of its type in the world and that this would make a breach of containment virtually impossible. On the basis of such arguments approval for the laboratory was eventually given despite considerable protests. The laboratory was eventually completed in 1985 with state-of-the-art physical containment. Facilities for handling harmful pathogens were in a sealed building with a gradation of increasingly negative air pressures moving from the entrance to the innermost laboratories where the most hazardous work was done. Moreover staff were required to use a change of protective clothing and showering procedures were put in place to maintain microbiological security. However, despite these high levels of physical containment an accident that could have had serious ramifications did occur within 2 years of the laboratory opening (Della-Porta and Murray, 1999; Morgan, 1987).

The accident at AAHL involved work on Newcastle disease virus (NDV), which is the causative agent of a major disease of poultry. Virulent forms of NDV cause major economic losses to poultry farmers. NDV can also occasionally cause human infections with symptoms ranging from mild influenza-like illness to conjunctivitis. The seriousness of the accident related to the fact that it involved a virulent strain that does not occur naturally in Australia (Australia only has low virulence and avirulent forms of the virus). The accident occurred when a laboratory technician was concentrating a preparation of virus using ultra-filtration. The filter membrane was not properly in place within the filtration unit and so when pressure was applied a significant quantity of virus was squirted into her eye. Following the accident the technician cleared up the spill and showered out of the laboratory suite. An accident report was made and the technician received counselling that she might develop conjunctivitis. She was told to let the laboratory know if anything happened but the incident was generally treated as a minor one. As a consequence the environmental risks of the incident were not fully addressed. The technician went home and the following week reported that she had developed conjunctivitis. Because the environmental risks had not been adequately addressed during her counselling the tissues she used to rub her eyes were thrown into her rubbish bin at home. There was thus a chance that the virus could have been picked up by birds that forage on rubbish tips. Senior members of management were absent from the site at the time of the accident and there were problems with internal communications. As a consequence it was only when the technician developed conjunctivitis that the appropriate senior staff became aware of the accident. Thus there was a delay in making a full response to the incident.

Following the incident there were extensive investigations which identified a number of underlying problems. It was found that there had been significant communication problems within the laboratory and that there had been no written or agreed procedures for the use of the filtration equipment. It was also recognized that increased vigilance was necessary when dealing with agents that could infect humans as well as animals. One outcome of the investigations was therefore an increased awareness that in such cases more stringent guidelines should be applied over and above the measures required to protect the environment. This was backed-up by the implementation of standard operating procedures throughout AAHL. In addition procedures for the training of staff and the handling of accidents were thoroughly overhauled. Another outcome of the incident was the establishment within the AAHL site of a motel-like unit to be used for accommodating any staff member involved in an accident that represented a threat to microbiological security.

Before leaving this example it should be pointed out that, even with the shortcomings that were identified, the inquiry that followed the above incident expressed confidence that there was not at any time a significant risk of Newcastle disease being transmitted to the Australian poultry industry. Thus, whilst the incident is an undoubted blemish on the otherwise excellent safety record of the biotechnology industry, it can also be taken as evidence that a precautionary approach is generally taken and that there are many layers of safety in operation. Thus, even when some of the safety measures break down the situation reached may still be some way from the worst case scenario.

In discussing the safety of work involving the genetic modification of viruses another issue which cannot be over-stressed is the need for high-quality risk assessments. There needs to be continuing vigilance to ensure that risk assessments are undertaken in a rigorous way and that they are not merely seen as a bureaucratic exercise. There is a danger that there may be an air of false confidence arising from the fact that, to date, there have been no documented cases of serious ill health or environmental harm as a consequence of genetic modification work. In this context it is worth noting that whilst genetic modification technology has been applied to viruses for nearly 25 years, many of the early years were spent undertaking inherently safe procedures such as deletion mutagenesis and the cloning of viral genes in bacteria. It is really only in the last four or five years that we have had the knowledge to start to understand in detail the ways in which viruses function at a molecular level. This has now enabled us to start dissecting out from viral genomes the determinants of pathogenicity with a view to splicing together functional hybrid viruses with altered pathogenic properties. Therefore, although most viral work will continue to be relatively safe due to the use of disabled vectors, if anything, increasing vigilance will be necessary to ensure that work on modified viruses that are replication competent is closely controlled.

All of the technical information in this chapter is in line with the current guidance provided by the Health and Safety Executive (HSE) on the safety of work with genetically modified organisms. The specific GM safety legislation that is enforced by HSE does not cover ethical issues and so any views on such matters that are expressed in this chapter are merely those of the author.

References

ACGM. (2000) Advisory Committee on Genetic Modification: Compendium of Guidance (ISBN 0 7176 1763 7). HSE Books, Sudbury, UK.

Alcami A, Smith GL. (1992) A soluble receptor for interleukin-1B encoded by vaccinia virus: A novel mechanism of virus modulation of the host response to infection. *Cell* **71**: 153–167.

Baulcombe DC, Lloyd J, Manoussopoulos IN, Roberts IM, Harrison BD. (1993) Signal for potyvirus-dependent aphid transmission of potato aucuba mosaic virus and the effect of its transfer to potato virus X. *J. Gen. Virol.* **74**: 1245–1253.

Berg P, Baltimore D, Boyer HW *et al.* (1974) Potential hazards of recombinant DNA molecules. *Science* **185**: 303.

Berglund P, Sjoberg M, Garoff H, Atkins GJ, Sheahan BJ, Liljestrom P. (1993) Semliki Forest Virus Expression System: Production of conditionally infectious recombinant particles. *Biotechnology* **11**: 916–920.

Bilbao G, Zhang H, Contreras JL *et al.* (1999) Construction of a recombinant adenovirus vector encoding Fas ligand with a CRE/Loxp inducible system. *Transplant. Proc.* **31**: 792–793.

Chong H, Starkey W, Vile RG. (1998) A replication-competent retrovirus arising from a split function packaging cell line was generated by recombination events between the vector, one of the packaging constructs and endogenous retroviral sequences. *J. Virol.* **72**: 2663–2670.

Clark RP, Goff MR. (1981) The potassium iodide method for determining protection factors in open-fronted microbiological safety cabinets. *J. Appl. Bacteriol.* **51**: 439–460.

Collins CH (1993) *Laboratory-acquired infections*. 3rd edn. Butterworth-Heinemann, Oxford.

Della-Porta AJ, Murray PK. (1999) Management of Biosafety. In: *Anthology of Biosafety I. Perspectives on laboratory design* (ed. JY Richmond), pp. 1–23, American Biological Safety Association, Mundelein.

Ding SW, Shi BJ, Li WX, Symons RH. (1996) An interspecies hybrid RNA virus is significantly more virulent than either parental virus. *Proc. Natl Acad. Sci. USA* **93**: 7470–7474.

Fallaux FJ, Bout A van der Velde I *et al.* (1998) New helper cells and matched early region 1-deleted adenovirus vectors prevent generation of replication-competent adenoviruses. *Human Gene Ther.* **9**: 1909–1917.

HSE. (2000) Health and Safety Executive: A guide to the Genetically Modified Organisms (Contained Use) Regulations, 2000 (ISBN 0 7176 1758 0). HSE Books, Sudbury, UK.

Mann R, Mulligan RC, Baltimore D. (1983) Construction of a retrovirus packaging mutant and its use to produce helper-free defective retrovirus. *Cell* **33**: 153–159.

Markowitz D, Goff S, Bank A. (1988) Construction and use of a safe and efficient amphotropic packaging cell line. *Virology* **167**: 400–406.

Miller AD. (1997) Development and application of retroviral vectors. In: *Retroviruses* (eds JM Coffin, SH Hughes, HE Vermus), pp. 437–473, Cold Spring Harbor Laboratory Press, New York.

Miller AD, Buttimore C. (1986) Redesign of retrovirus packaging cell lines to avoid recombination leading to helper virus production. *Mol. Cell. Biol.* **6**: 2895–2902.

Morgan C. (1987) Import of animal viruses opposed after accident at laboratory. *Nature* **328**: 8.

Perng G-C Slanina SM, Yukht A *et al.* (1999) A herpes simplex virus type 1 latency-associated transcript mutant with increased virulence and reduced spontaneous reactivation. *J. Virol.* **73**: 920–929.

Ryan MD, Jenkins O, Hughes PJ *et al.* (1990) The complete nucleotide sequence of enterovirus type 70: relationships with other members of the Picornaviridae. *J. Gen. Virol.* **71**: 2291–2299.

Speck PG, Efstathiou S, Minson AC. (1996) In vivo complementation studies of a glycoprotein H-deleted herpes simplex virus-based vector. *J. Gen. Virol.* **77**: 2563–2568.

Spielhofer P, Bachi T, Fehr T *et al.* (1998) Chimeric measles viruses with a foreign envelope. *J. Virol.* **72**: 2150-2159.

Webster RG, Kawaoka Y, Bean WJ. (1986) Molecular changes in A/chicken/Pennsylvania/83 (H5N2) influenza virus associated with acquisition of virulence. *Virology* **149**: 165–173.

Index

For Product Safety Concerns and Information please contact our EU representative GPSR@taylorandfrancis.com
Taylor & Francis Verlag GmbH, Kaufingerstraße 24, 80331 München, Germany

www.ingramcontent.com/pod-product-compliance
Ingram Content Group UK Ltd.
Pitfield, Milton Keynes, MK11 3LW, UK
UKHW010752250425
457613UK00046B/904

* 9 7 8 1 8 5 9 9 6 1 0 3 2 *